P9-DCV-438

EXPLORING
THE OCEAN WORLD

EXPLORING THE

A HISTORY OF
OCEANOGRAPHY

Revised

OCEAN WORLD

C. P. Idyll, *EDITOR*

Thomas Y. Crowell Company / New York / Established 1834

76 3662

PARKWAY SOUTH JUNIOR HIGH SCHOOL
LIBRARY

BY *C. P. Idyll*

ABYSS: The Deep Sea and the Creatures That Live in It

THE SEA AGAINST HUNGER

EXPLORING THE OCEAN WORLD: A History of Oceanography

Copyright © 1972, 1969 by Thomas Y. Crowell Company, Inc.

All rights reserved. Except for use in a review,
the reproduction or utilization of this work in
any form or by any electronic, mechanical, or
other means, now known or hereafter invented,
including xerography, photocopying, and recording, and in any information storage and retrieval
system is forbidden without the written permission
of the publisher. Published simultaneously in
Canada by Fitzhenry & Whiteside Limited, Toronto.

Manufactured in the United States of America

L.C. Card 72–78266

ISBN 0–690–28610–4

1 2 3 4 5 6 7 8 9 10

Preface

This is the "Age of Oceanography." With diminishing land resources of food, minerals, and energy, men are looking with sharper attention to what many believe to be inexhaustible stores in the ocean. Never before has public interest in the sea been as high, nor scientific activity in the field as great, as in our time. The glittering promise of "sea farming" dazzles practically all nations, large and small, old and new. Countries that have exploited the sea for decades or centuries are striving for even greater wealth and advantage while other nations (including some newly created) that have not had the ability to exploit the sea, or in many cases have even lacked knowledge of what lay off their shores, are insisting on part of the harvest.

The press of more people in every land results in more boats crowding ocean waters, especially close to shore, and this leads to many an abrasive conflict, and frequently to fouling of the waters. Indeed, all the new promise of the sea, all its potential importance to the welfare of mankind, may be cancelled unless we cease to treat the ocean as a dump of infinite capacity. Pollution of the sea poses a threat newer than nuclear weapons and perhaps more menacing.

With such spurs as these urging attention to the sea, it comes as an unpleasant surprise that no well-organized, codified set of laws exists to regulate man's activities there. Several attempts have been made since World War II to erect a universally acceptable law of the sea, but with only partial success. Now the nations of the world are preparing to try again. It is to be fervently hoped that the new Law of the Sea Conference, requested by the United Nations for 1973, will form the basis for equitable, orderly sharing and conservation of the ocean's bounty, and for the protection of that vast environment against destruction by pollution.

The increased attention now focused on the sea makes it appropriate that we understand how we have arrived at this dynamic era in our history, and that is the purpose of this book. As always, the story of the past will illuminate the probable paths of the future.

Oceanography is not a science in the same sense as chemistry, biology, or physics. Instead, it is a composite of all the basic sciences, using their techniques and accumulated knowledge to solve problems of the oceans—the character of the watery environment and of its plant and animal inhabitants.

Oceanography came of scientific age when it began to integrate accumulated stores of knowledge and to plan future observations that would fill gaps and test proposed principles. The movement of waters and the chemicals they contain are now being studied less for themselves alone and more as interacting systems and as environments for living creatures. Plants and animals are being studied less as individual types and more as populations whose distributions, abundance, and well-being are affected by changes in the ocean itself. As capricious as the sea and its inhabitants may appear at times, it is increasingly clear that all physical and biological changes are in fact "programmed," and that the patterns of cause and effect can be determined and reduced to mathematical descriptions if man has the required skill and patience.

Fundamentally, every oceanographer is first of all a physicist, a biologist, a chemist, or a geologist, but to achieve the necessary integration marine scientists work in teams, and individuals must become proficient in basic sciences other than their own.

However, in spite of the tendency for marine scientists to be drawn ever closer together in concepts, aims, and even procedures, there are still separate branches of oceanography corresponding to the classical basic sciences. These branches are physical, chemical, biological, and geological oceanography.

Physical oceanography is concentrated on the circulation, density, and temperature of water, and on its penetration by light. All of these shift and vary over the breadth and depth of the seas, and each influences the others. Further, the variations that occur in these conditions affect, and are affected by, conditions that come under the scrutiny of the other branches. *Chemical oceanography* is concerned with the nature and distribution of the dissolved substances in the sea. The total amount of dissolved salts affects the density of various parts of the sea, and this affects ocean movements. *Biological oceanography* deals with the kinds, abundance, distribution, and interrelationships of the bewildering assemblage of plants and animals in the sea. All these complicated relationships are affected by current, temperature, saltiness, density; in turn, the sea's living inhabitants change its chemical content, color, and other characteristics. *Geological oceanography* deals with the shape of the ocean basins and the character of the rocks and sediments that underlie them. These facts, too, are affected by, and affect, all the others—physical, chemical, and biological.

Oceanography has gradually shifted from the necessary early preoccupation with description to a quantitative approach to the nature of observed phenomena. The shift from *what* to *how much*, and to *why*, opens up enormous and exciting vistas of prediction. Once we understand causes and effects of oceanic phenomena, the occurrence of a particular set of happenings can lead to increasingly confident prediction of the next set that will come along. This mastery will result some day in our vastly improved ability to control the sea as a mode of transportation and as a source of food, minerals, and power. It will give the best-informed nations a strong, and probably crucial, power in war, whether for offense or defense. It will help to improve weather prediction and may even lead to weather control, since the sea and the air are two interconnected and intimately interdependent mechanisms.

Every branch of oceanography has its own history, intertwined as each may be with the others. Following an overview in Chapter One of the history of ocean science as a whole, the stories of these branches are told in individual chapters. In addition, the applied disciplines—fishery science, mineral and energy extraction, archaeology, aquaculture—are treated separately. A new chapter has been added for this second edition on marine ecology and pollution, outlining the kinds of detrimental changes taking place in the ocean and the questions that must be answered if we are to guard the seas from destruction.

C. P. Idyll
Rosenstiel School of Marine and
   Atmospheric Science
Miami, Florida

Contents

Color Essays

1 THE SCIENCE OF

by C. P. Idyll

The birth of oceanography as a science can be pinpointed with remarkable accuracy. It occurred on January 3, 1873, when scientists aboard HMS *Challenger* began collecting data, making five soundings and three dredge hauls west of Lisbon. Three and a half years and 68,890 miles later, *Challenger* returned to England bearing notebooks crammed with data, museum bottles aswim with astonishing creatures, and lockers full of rocks and mud from the deep-sea floor. The implications of all this so excited the scientific world that the sea became a subject of eager scrutiny.

Of course, mankind already had been accumulating knowledge about the sea for millennia. The very first man to stumble out of the jungle and look upon the great water probably noted that it was salty, that it moved in and out, and that it contained strange animals, some of which made tasty meals. But neither his observations nor most of those made in the ages that followed were scientific: they did not form part of an integrated whole, leading to principles and the testing of those principles with further planned observations. Instead, knowledge of the sea was gathered in bits and pieces by curious individuals, on the seashore or in the shallows, or incidentally during voyages of exploration, and passed on, bit by bit, through word of mouth.

No doubt it was the food of the sea that persuaded early men to live on the shore. Paleolithic men made bone harpoons for spearing fish. Neolithic men added hooks, nets, and boats. The boats, probably for fishing alone originally, provided more scope and opportunity for observing currents, tides, and animals of the sea. As their owners grew bolder, they began to make larger craft and to go voyaging. This, in turn, created the need for more knowledge about the sea and its movements.

The systematic accumulation of knowledge of the sea started in the Mediterranean. The wonderfully clear and logical minds of the early Greeks, and the accuracy of their observations, are revealed in yet another way by a chronicle of some of their accomplishments in this area. Miletus, in the sixth century B.C., and Herodotus, in the fifth, regarded the earth as a sphere—an unusual insight in the face of casual appearances. In the fourth century B.C. Aristotle focused what may have been the most remarkable intellect of all time on the creatures of the sea. He carefully described 180 species of marine animals. He noted certain patterns of structure and habit among them, and he showed how they could be grouped by body form. In the first Christian century, Pliny the Elder could account for only 176 species of marine animals—four fewer than Aristotle had catalogued 400 years before. Pliny ingenuously believed that he had listed all the animals occurring in the sea. "Surely, then," said he, "everyone must allow that it is quite impossible to comprise every species of terrestrial animal in one general view for the information of mankind. And yet, by Hercules! in the sea and in the Ocean, vast as it is, there exists nothing that is unknown to us, and, a truly marvellous fact, it is with those things that Nature has concealed in the deep that we are best acquainted!"

Naturally, the Greeks knew the Mediterranean best. Their name for this nearly landlocked sea was Thalassa, and much later, in the nineteenth century, the word "thalassography" was coined to denote "the science of the sea." But this name would be much too restrictive; even the ancient Greeks had early realized that there was more to the oceans than Thalassa. Homer, about 1000 B.C., had regarded the world as a flat disk, with the land portion encircling the Mediterranean, but at about the same time Hesiod referred to lands beyond the Sea—the Isles of the Blessed and the Hesperides. The Babylonians, too, conceived of lands surrounded by Ocean, outside of which was the Dawn.

The Greeks of a later date believed that the earth was bisected east to west by the Mediterranean Sea

THE SEA

into two continents, while a mysterious and foreboding River Oceanus circled it ceaselessly beyond the Pillars of Hercules. Exotic lands were said to exist in this River. For example, in the third century B.C. Plato, apparently repeating an old Egyptian tale, described the lost land of Atlantis, destroyed by a sudden cataclysm.

By now no man could doubt the existence of the great Ocean, and as time went on, and as more and more voyages of exploration were launched, it became increasingly clear that Mediterranean lands were not the only part of the habitable world.

One of the earliest voyages of exploration was undertaken by the Egyptian Queen Hatshepsut. Carvings tell that in about 1500 B.C. she went by sea to the "Land of Punt," probably the Somali Republic.

The Minoans of Crete, whose sailors wore kilts and sandals, were the first real seafarers of the Mediterranean, but the Phoenicians must be given credit for expanding man's horizons most. They voyaged from Tyre and Sidon to ports in Italy, Spain, Greece, and North Africa. Once outside the Pillars of Hercules, they visited other strange lands, even distant Britain. In 600 B.C. the Phoenicians claimed to have made a voyage around Africa from east to west. This was doubted by contemporaries and by many later historians. One statement in particular seemed to dispute the veracity of the Phoenician sailors: They said that at one point the sun was on their right hand instead of on the familiar left. Of course, this is where it should have been if the ship was in the Southern Hemisphere, and instead of casting doubt on the epic voyage, the statement strengthens its likelihood. The journey took a long time, since when the party ran short of food they beached their vessels, planted grain, and waited for it to ripen before sailing on the next leg of their journey. Three years were to pass before the ships came home again.

In the first century B.C. Hippalus noted the behavior of the monsoon winds of the Indian Ocean, blowing first one way during half the year and then reversing their direction. This important oceanographic discovery made possible voyages across the open ocean if the proper season were chosen, greatly reducing the length of journeys.

At about the beginning of the Christian era, Poseidonius, a Greek geographer, measured the depth of the Mediterranean near Sardinia, recording about 1,000 fathoms. In modern terms, this measure is equivalent to 6,000 feet. The fathom is a seaman's measure equal to six feet—roughly the distance between the fingertips of each hand when a man's arms are fully outstretched—and is the maximum convenient length for gathering up the line used to sound the depth. Actually, most people have an arm span of slightly less than six feet, and in days gone by a fathom was commonly equated with five and a half feet. The fathom also roughly equals the height of a man; thus, on the human scale, the oceans are enormously deep.

As in every sphere of knowledge, during the Dark Ages many of the bright truths and daring (and often accurate) theories that the ancients had evolved about the sea were clouded and distorted. The earth became flat again, and knowledge derived from observation was thrust rudely aside by the rule of authority—no matter how false.

Explorations did not stop in this misty period of history, but they were poorly reported and far less use was made of them than they deserved. Some of the bravest voyages of all time were made by the Vikings, who reached Iceland, Greenland, and eventually North America about the year 1000. They must have accumulated and used a great store of oceanographic information to have accomplished these journeys over some of the stormiest seas of the world.

The Great Explorers

The Golden Age of Exploration began in the fifteenth century. Under the influence of Prince

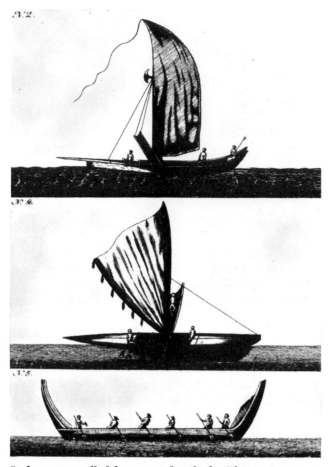

Henry the Navigator, the Portuguese were the leaders, sailing ships under their own flag as well as in the service of other monarchs. In one of the epic voyages of man, Bartholomeu Diaz sailed around the Cape of Good Hope in 1488 and back to Portugal. Christopher Columbus rediscovered America in 1492. Neither his shrunken view of the globe, which caused him to persist in regarding the New World as part of Asia, nor the fact that the Vikings had been ahead of him, in any way diminishes the magnificence of his accomplishment or its influence on exploration and the science of the sea. Spurred by Columbus, Vasco da Gama ventured around the Cape of Good Hope in 1497 and opened up the important trade route to India. Magellan's voyage into the Pacific, as part of the first circumnavigation of the world, is another great landmark in oceanographic history. Magellan was destined not to finish the voyage: He was killed in 1521 by the natives at Cebu, in the Philippines.

Now the emerging maritime nations of Europe scrambled to find an easy and quick way across the top of the world to India: The search for a northwest passage was under way. British sailors were

In boats propelled by oar and sail, the Phoenicians journeyed from Tyre and Sidon to ports of the western Mediterranean and even as far as Britain. In 600 B.C. they claimed to have sailed around Africa from east to west.

The Greek astronomer Eratosthenes correctly calculated the circumference of the earth in the third century B.C. A hundred years later Ptolemy miscalculated it to be 18,000 miles—about 7,000 miles smaller than it actually is. This Ptolemaic map was prepared in 1482 by Nicholaus Germanus. (Map Division, Library of Congress)

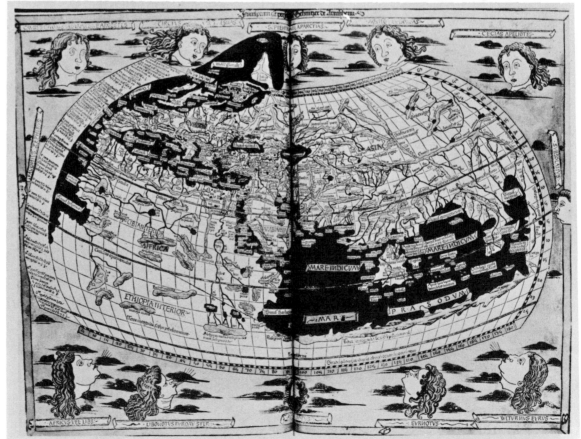

Ptolemy's miscalculation persisted for more than a millennium and a half. In 1492 it may have caused Columbus (above) to believe he had reached the Indies.

Although he did not personally finish the voyage, Portuguese navigator Ferdinand Magellan is credited with being the first to sail around the earth. He was killed en route in the Philippines in 1521.

especially active. Martin Frobisher tried in 1576–78, as did John Davis in 1585–87. Early in the seventeenth century Henry Hudson and William Baffin searched; they were followed in the eighteenth century by James Cook and Vitus Bering (a Dane working for Russia). It was not until 1850–53 that Sir Robert McClure proved the existence of an east-west passage to the north of the American continent, which he traversed from the west to a point that Sir William Parry had already reached from the east; but part of his journey was over ice. The Norwegian Roald Amundsen at last fulfilled the old dream of navigating a maritime route, landing in Nome in 1906. While the U.S. nuclear submarines *Nautilus* and *Skate* negotiated it in 1958 by going under the ice, the route seemed of no importance commercially until 1969 when the icebreaking tanker S.S. *Manhattan* made the trip in a test of the feasibility of transporting oil from Alaska's north slope by ship.

But commerce was served richly by most of these wonderful voyages of exploration, and so was the science of the sea. Of necessity captains and navigators were obliged to look at the ocean with increasing concentration to gain the skill in navigation and survival demanded by their far-ranging voyages. In particular, surface phenomena were recorded and cataloged, and the patterned behavior of tides, currents, salinity, temperature—the basic data of oceanography—became better and better understood.

Maps were improved. The sixteenth-century Flemish geographic genius, Gerhardus Mercator, produced new and far more useful charts of the world. One of his theories was erroneous, since he held that land and sea must occupy about equal areas of the globe, and this led to the supposition that there must be an enormous land mass over the South Pole. Exploration by Abel Janzoon Tasman, a Dutchman, and later by the great English navigator James Cook, showed that the land mass was far smaller than Mercator's concept had suggested.

Among the many facts about the sea that emerged from the voyages of exploration was the realization that the seas, unlike the land masses, are all interconnected. By now the ocean had lost some

Early in the seventeenth century, Henry Hudson searched in vain for a Northwest Passage to India. A longitudinal section of the *Half Moon* is shown here.

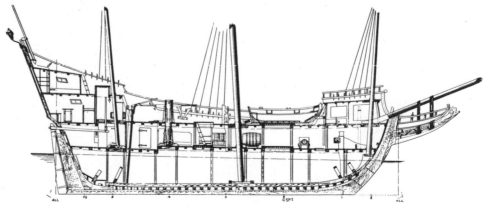

GERARDI MERCATORIS RVPELMVNDANI EFFIGIEM ANNOR.
DVORVM ET SEX — AGINTA, SVI ERGA IPSVM STVDII
CAVSA DEPINGI CVRABAT FRANC. HOG. CIꓛ. Iꓛ. LXXIV.

Flemish geographer Gerhard Mercator revolutionized the process of map making in the sixteenth century. He theorized incorrectly, however, that the land and sea must occupy equal areas of the globe.

of its menace, and instead of being regarded as a barrier it was beginning to be looked upon as a convenient avenue connecting lands and peoples.

Captain James Cook holds a first-rank place among history's great seamen. But he was more: a scientist with insight much ahead of his time. The first of Cook's three voyages, in 1768–71 to the South Pacific, was not primarily oceanographic; he was sent by Britain's Royal Society to observe the transit of Venus across the face of the sun and, secondarily, to investigate the supposition that a vast continent existed around the South Pole. But oceanography was given a powerful thrust forward by Captain Cook, not only from the impressive amount of data he accumulated about the sea but from the ordered and disciplined approach to these observations that characterized his leadership. He made soundings to depths of 200 fathoms, as well as frequent and accurate observations of winds and currents and the temperature of the water. By the time Cook was killed by Sandwich Islanders, in 1779—ironically,

after having treated the natives far more kindly than most European sailors—his three voyages had made the Pacific a much better known part of the world.

The First Scientists

In the early part of the eighteenth century great impetus was given to marine biology by the work of Carolus Linnaeus, a Swedish botanist. He provided the orderliness of a rigid classification of organisms, thus sweeping away a great underbrush of confusion. By giving concise descriptions and double names to living creatures he encouraged armies of biologists to scurry over the world collecting, naming, and describing plants and animals. From there it was a short and logical step to study the biology and interrelationships of many creatures that had been unknown before.

In Italy, Count Luigi Ferdinando Marsigli, a native of Bologna, conducted early but little-recognized oceanographic studies. Marsigli was most famous as a soldier, serving in the Austrian army against the Turks, and fighting in the War of the Spanish Suc-

In the first scientific oceanic expeditions in history, Captain James Cook cataloged and accumulated data about the sea in an orderly fashion. He made soundings to depths of 200 fathoms and accurate observations of winds, currents, and temperatures.

cession. He appears to have been the first to use the naturalists' dredge to collect marine animals from the bottom, and his observations of the system of currents in the Bosporus in the 1680's were amazing in terms of the knowledge and techniques of the time. He showed that a surface current flows from the Black Sea to the Mediterranean, balanced by a deep countercurrent in the opposite direction. To determine this, Marsigli built a current meter with a propeller for subsurface measurements. He proposed the theory that the currents were produced by density differences: The water of the Mediterranean was saltier, and therefore denser, than the water of the Black Sea, said Marsigli, because of the higher rate of evaporation in the Mediterranean region; as a result, the Mediterranean water sank to form the subsurface current flowing through the Bosporus into the Black Sea, and this was replaced by the surface current of lighter water from the Black Sea. A true scientist, Marsigli tested his hypothesis by taking samples at different levels and noting the density and salinity of the different water masses. He even made a model of the system in a large tank, placing valves near the top and the bottom of a partition and observing the pattern of flow of a dense bottom layer and a lighter top layer of water. In Cassis, a fishing village near Marseilles, Marsigli engaged in no less monumental a task than "research on the history of the sea, where I hope to treat the nature of the water of the sea and its diverse movements; of the differences of the bottoms of the sea, which seem to me to be related to the structure of the mountains; of the effect of winds on this water; of the nature of fish developed by means of analysis of the vegetation growing on the bottom of the sea." Physical and chemical oceanographers, marine geologists, meteorologists, and ecologists will all see the essentials of their disciplines set forth in this succinct statement. Marsigli made a noteworthy contribution to at least part of this ambitious program, and he set forth the results in his *The Physical History of the Sea* in 1711.

Under the influence of Captain James Cook some historic early observations were made by Johann Reinhold Forster and his son Georg. During Cook's second voyage they discovered that subsurface temperatures were often different from those at the surface, and they attempted some of the earliest deep-sea soundings, bringing blue mud from the Pacific floor from 683 fathoms. Some years later Georg Forster became a friend of the famous German scientist Baron Alexander von Humboldt. It may well be that some of Cook's wisdom and disciplined approach to observations of natural phenomena rubbed off on

German scientist Alexander von Humboldt showed that the volcanoes of the New World were arranged in a linear pattern. The Humboldt Current, which sweeps northward along the west coast of South America, is named after him.

Humboldt by way of Forster. In any case, Humboldt deserves a place in an account of the history of ocean science for his observations off the west coast of South America.

He was fascinated by the teeming life of the great, cold, ocean stream that sweeps northward along the western coast of South America—the current that now bears his name. He reported to Europe his observations of the efficiency of guano (the droppings of the enormous populations of sea birds) as fertilizer, and this stimulated use of the material in European farms, to the rich benefit of the Peruvian economy.

Through his observations of tropical storms and his mapping of the volcanoes of the New World (in which he showed that these were set in linear patterns), Humboldt contributed to land, air, and sea science. He is given credit for laying the foundation of the science of meteorology.

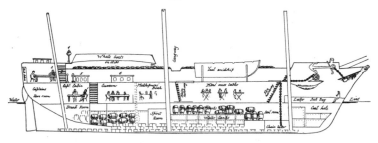

Charles Darwin spent five seasick years in the crowded quarters of the H.M.S. *Beagle*. He made a major contribution to marine biology with his theory of the formation of coral reefs.

Charles Darwin was influenced by Humboldt, having read with interest the German's *Personal Narrative* describing his South American travels. Darwin was not primarily a marine scientist, but during his voyage as an unpaid naturalist on the *Beagle* between 1831 and 1836 he made a major contribution to marine biology with his theory of the formation of coral reefs. This has held up in its major aspects to this day. Darwin used a tow-net for collecting plankton long before the German, Johannes Müller, although Müller is often given the credit for the invention of this technique in 1844.

It is fascinating to learn that Darwin suffered agonies of seasickness, and that his quarters for five long years on the *Beagle* have been described as "a wretched little cabin." The voyage broke Darwin's health permanently; yet from these unpromising conditions came the theories of evolution and natural selection that altered the course of human thought.

An early American contribution to oceanography was the U.S. Exploring Expedition of 1839–42, commonly called the "Wilkes Expedition," after its leader, Captain Charles Wilkes. Copper wire was used in making soundings in the Antarctic Ocean, a technique that did not come into general use until much later.

Oceanography also owes a debt to another early American, Benjamin Franklin. When he was serving as postmaster general for the Colonies Franklin received complaints that mails from England were often delayed two weeks. He discovered that this was because many of the mail ships bucked the east-flowing Gulf Stream. Ships sailing south of it on the voyage from Europe, as the American whalers had long since learned to do, were able to make considerably faster trips. Timothy Folger, a Nantucket whaling captain, explained the situation to Franklin:

. . . the American captains were acquainted with the Gulf Stream, while those of the English packets were not. We are all acquainted with that stream, because of our pursuit of whales, which keep near the sides of it but are not met within it, we run along the side and frequently cross it to change our side, and in crossing it have sometime met and spoke with those packets who were in the middle and stemming it. We have informed them that they were stemming a current that was against them to the value of three miles an hour and advised them to cross it, but they were too wise to be counselled by simple American fishermen.

With the help of Captain Folger, Franklin drew a historic chart of the surface currents of the North Atlantic for the General Post Office. We know now that Franklin's chart was greatly oversimplified, and

As postmaster general for the Colonies, Benjamin Franklin discovered that the westbound mails from England were habitually delayed by the Gulf Stream. With the help of American whaling captains, he completed the first map of this important oceanic current. (Historical Society of Pennsylvania)

that the Gulf Stream is a highly complex and variable current system, but his energy in this matter was of great help to ships and was a stimulation to early physical oceanography.

The Two "Founding Fathers"

Sometimes progress seems to rise to a crescendo of achievement and then to slacken or even to retrogress for a time. There was little notable achievement for seventy years after Cook's contributions. But then came another Britisher, Edward Forbes, and an American, Matthew Maury, whose devoted and energetic work led them each to be called "The Father of Oceanography."

Forbes, who taught at the University of Edinburgh, made his contribution in the field of marine biology. He improved the naturalists' dredge for collecting bottom animals and used it in deep water off the Isle of Man and in the Aegean Sea. In 1840 he announced that the abundance of animals decreased steadily with depth. He established eight

Britisher Edward Forbes dredged animals from the ocean bottom off the Isle of Man and in the Aegean Sea. He theorized incorrectly that life ceases at depths of 300 fathoms.

Twenty years before Forbes announced his theory, Sir John Ross had captured a starfish and other living creatures from a depth of 1,050 fathoms. His work, however, was not widely published.

zones of abundance and theorized incorrectly that life ceases at 300 fathoms.

Much has been made of this error by Forbes in discussions of the history of marine biology. Indeed, it represents one of the curious mental lapses of mankind, since when Forbes announced his theory it had already been vividly shown to be wrong. Over twenty years previously, Sir John Ross, exploring Baffin Bay in search of a northwest passage, had become interested enough in the possibility of animals living on the bottom of the deep sea to instruct the ship's blacksmith to construct him a "deep-sea clamm" to take bites of the bottom mud. He employed this historic piece of oceanographic gear as far down as 1,050 fathoms, capturing several pounds of greenish mud from that depth. In the mud were living worms and other animals, and in the line just above the sampler a starfish was entangled. Perhaps it was because Ross's work was not widely publicized that Forbes and others forgot it; in all likelihood, too, there was general unwillingness to accept the idea that any living creature could withstand the enormous pressures imposed by tons of water. But animal bodies are largely water, with water's incompressibility; and without lungs or other gas-filled cavities deep-sea creatures can live without difficulty in the deepest parts of the sea. Sir James Clark Ross, a nephew of Sir John's, confirmed his uncle's observations in 1839–43, this time in the Antarctic, where he brought

aboard abundant and varied animals. He expressed the opinion that "from however great a depth we may be able to bring the mud and stones of the bed of the ocean, we shall find them teeming with animal life." As early as 1850 the Norwegian Michael Sars and his son, G. O. Sars, dredged animals from depths between one and two miles. Many a later expedition produced abundant evidence that life exists at the greatest depths.

Unfortunate as Forbes's error was, he did pioneer in relating marine animals to their physical environment. A measure of his enthusiasm and his personality can be obtained from his full-blown description of dredging:

Beneath the waves there are many dominions yet to be visited and kingdoms to be discovered, and he who venturously brings up from the abyss enough of their inhabitants to display the physiognomy of the country, will taste that cup of delight, the sweetness of whose draught those only who have made a discovery know. Well do I remember the first day when I saw the dredge hauled up after it had been dragging along the sea bottom, at a depth of more than 100 fathoms . . . anxiously separating every trace of organic matter from the enveloping mud, and gazing with delighted eye on creatures hitherto unknown, or on groups of living shapes, the true habitats of which had never been ascertained before, nor had their aspect, when in the full vigour and beauty of life, ever before delighted the eye of a naturalist. And when, at close of day, our active labours over, we counted the bodies of the slain, or curiously watched the proceedings of those whom we had selected as prisoners, and confined in crystal vases, filled with a limited allowance of their native element, our feelings of exaltation were as vivid, and surely as pardonable, as the triumphant satisfaction of some old Spanish "Conquistador" musing over his siege of a wondrous Astaln [Aztec] city, and reckoning the number of painted Indians he had brought to the ground by the prowess of his stalwart arm.

This kind of enthusiasm is contagious, and Forbes stimulated many an able man to look more closely at the sea. His influence was wide and salutary.

Matthew Fontaine Maury may have an even greater claim than Forbes to being the founder of oceanography, but since the American's field was concentrated on the physical aspects and the Englishman's on the biological, perhaps they can comfortably share this honored place in history. Maury traveled widely over the world ocean as a lieutenant in the United States Navy prior to 1839, when he was forced to take a shore post following a stagecoach accident in which he was lamed for life. The event was a tragedy for Maury, both personally and professionally, but many a sailor and oceanographer has since had occasion to be thankful for it. Maury recognized the necessity for better observations of oceanographic conditions, and his organizing genius in collecting and interpreting them saved enormous sums of money for shipowners and set the pattern for ocean observations for generations of scientists to follow.

Beginning in 1842, as Officer in Charge of the Depot of Charts and Instruments, Lieutenant Maury received observations of winds and currents made by Navy ships. Before this time little use had been made of these data, and sailors ran their ships as best they could from their own experience. Maury saw the opportunity that existed to correlate the great amount of information then being collected, and the even greater vistas that would open up if all ships could be persuaded to record temperatures, current direction and strength, winds, and weather. At an international conference held in Brussels in 1853 at Maury's suggestion, he urged that this be done. "In a little while," he says in his epic book, *The Physical Geography of the Sea*, "there were more than a thousand navigators engaged day and night, and in all parts of the ocean, in making and recording observations according to a uniform plan, and in furthering this attempt to increase our knowledge as to the winds and currents of the sea, and other phenomena that relate to its safe navigation and physical geography." In 1854 the

As a charts and instruments officer of the U.S. Navy, Lieutenant Matthew Fontaine Maury recorded observations of winds and currents and prepared the first depth map of the North Atlantic. His observations shortened the passage from Britain to California by as much as thirty days. (The Mariners Museum, Newport News, Va.)

president of the British Association estimated that Maury's contributions had already saved ocean commerce millions of dollars: "the sailing directions have shortened the passage [from Britain] to California 30 days, to Australia 20, and to Rio de Janeiro 10."

Maury was interested in more than the surface phenomena of the sea. Noting the paucity of deep-sea soundings, he stimulated the invention of an improved sounding apparatus by another U.S. Naval officer, Midshipman John Mercer Brooke. In 1854, with the information obtained by this machine, Maury prepared the first depth map of the North Atlantic Ocean, down to 4,000 fathoms. When the Civil War broke out, Maury resigned his position in Washington to fight for his native Virginia, but he returned to the Depot after the war. His enthusiasm in stimulating the collection of oceanographic information and his genius in making it useful to the ships of many nations constituted an "immense service to commerce," in the words of Sir Alister Hardy, the eminent English oceanographer.

The Cable Layers

In the last quarter of the nineteenth century the problems associated with the laying of the transatlantic telegraph cable made better knowledge of the sea a practical necessity. Here marine biology, physical oceanography, and marine geology came together. Engineers had to know what currents sweeping the

In the late nineteenth century, the problems of laying transatlantic cables necessitated a better knowledge of the sea. Engineers had to know about currents, depths, and bottom irregularities that might affect their work. This watercolor by Robert Dudley shows the Atlantic Telegraph Fleet assembled at Berehaven, Ireland. (Metropolitan Museum of Art, gift of Cyrus W. Field, 1892)

Dr. Wyville Thomson, professor of natural history at Edinburgh, studied the seas near Britain aboard the *Porcupine*. This illustration of the stern deck, showing the dredge and the accumulator, is from Thomson's book *The Depths of the Sea.*

One of the many creatures the *Porcupine's* trawl net rolled in was a living fossil: a sea urchin that "panted." Thomson described it as having a "perfectly flexible leather-like" skeleton. All living sea urchins known previously had rigid shells.

area might affect the laying of the cable or its stability afterward; they had to know the precise depths and the irregularities of the bottom where the cable was laid; they sought knowledge of the animals that might live on and around their cable, perhaps fouling it or chewing on its covering.

The British took the lead in this work, and now the waters north and west of their homeland were the sites of the pioneering deep-sea investigations that culminated in the founding of scientific oceanography. The man of greatest influence was Dr. Wyville Thomson, professor of natural history at the University of Edinburgh. Said Thomson, "The wonderful project of establishing a telegraphic communication between the old world and the new directed the attention of practical men to a region about which there had been a great deal of hazy misconception— the bottom of the deep sea." Under his leadership, and that of Dr. W. B. Carpenter, a series of highly important voyages took place. In 1868 the *Lightning,* and in 1869 the *Porcupine* worked the seas near Britain. Their temperature observations showed the hitherto unguessed existence of two closely adjacent areas where the bottom temperature differed as much as 15° F. They came to the conclusion that this was the boundary of two great masses of water of different origins, maintaining separate existences. Deep-sea dredgings scraped up some creatures that excited the biologists greatly. One was a large scarlet sea urchin that "panted." This seemed contrary to nature, since living sea urchins known to that day had rigid shells that would not permit of such movement. But from ancient rocks of the Cretaceous period, 100 million years ago, fossil urchins had been uncovered with articulated skeletons; so here from the deep sea was a living fossil! This and another lively ghost, a stalked sea lily from the Sars' dredges, led biologists to believe that the deep sea might hide all manner of antique forms.

Following these significant voyages, the Royal Society, led by Carpenter and Thomson, was able without undue difficulty to persuade the British Admiralty to organize the first deep-sea expedition on a large scale for the sole purpose of conducting scientific oceanography. "The proposal to defray the expense of such an Expedition out of public funds received the cordial assent of the House of Commons," according to the official account, and the Admiralty assigned H.M.S. *Challenger,* a 226-foot navy ship.

An Epic Voyage

Challenger's task was to investigate "the conditions of the Deep Sea throughout the Great Oceanic Basins." She was carefully refitted by Thomson, who had the advice of a distinguished committee of the Royal Society, including Professor T. H. Huxley. All but two of her guns were removed, and laboratories, special winches, scientific equipment, and "spirits of wine" were placed aboard.

The genius of Thomson's advisers seemed limited by the tools available to them. The ship itself, a spar-decked sailing vessel with auxiliary steam power, was slow and clumsy. She had the habit of rolling incredibly—46 degrees one way and 52 degrees the other!

On a three-and-a-half-year voyage around the world, H.M.S. *Challenger* collected information for fifty folio volumes of reports. The carefully cataloged studies laid the scientific foundation for every major branch of oceanography.

Yet in three and a half years she went around the world, took over 350 stations in all the oceans of the world except the Arctic, and logged 68,890 nautical miles over an area of 140 million square miles of sea surface. The equipment was the best of the time— 1872—but it was primitive by modern standards, and most of it was experimental and unproven. In spite of the inadequacies of ship and gear, the *Challenger* expedition laid the scientific foundation for every major branch of oceanography. Although *Challenger's* dredges were small and clumsy devices pulled by hempen rope, her scientists mapped the ocean bottoms, labeling vast areas according to newly defined classifications of deep-sea sediments: globigerina, radiolarian, or diatom oozes, or red clay. Wyville Thomson's chief scientific assistant, John Murray, laid the foundation for marine geology with his brilliant study of the bottom samples.

Hempen lines wound on bulky reels pulled simple trawls, but the ship dragged in water as deep as 4,475 fathoms, trailing as much as eight miles of line. A haul sometimes required twelve or fourteen hours to complete, and not infrequently the net came up empty. Yet the ship's biologists captured myriads of new and fascinating forms with immense implications for science. No longer could there be any doubt that life existed in abundance to great depths. Well-documented theories were advanced that deep-sea animals were derived from shallow-water forms instead of being uniquely different.

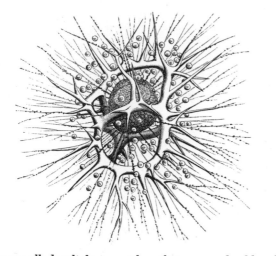

A one-celled radiolarian with a silicious, or glasslike, skeleton was one of the many curious and delicate forms of life described by *Challenger* biologists.

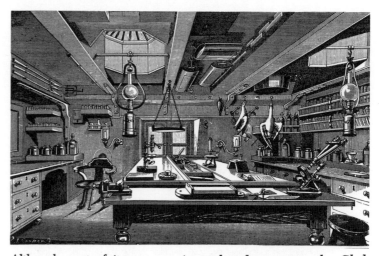

Although most of it was experimental and unproven, the *Challenger's* equipment was the best of its time. The naturalists' workroom, during a calm, was shown in the *Challenger Reports*.

Challenger's plankton nets were simple bags of muslin or silk attached to iron rings one foot in diameter; yet they captured scores of new planktonic forms and permanently shaped the branch of marine biology that deals with such forms.

The ship's thermometers were primitive and easily damaged; yet enough temperature observations were made to show that the deep sea below 2,000 fathoms was icy cold—of a temperature eminently suitable to cool the champagne for the ship's officers.

Current meters were no better than the remainder of the gear, yet *Challenger* scientists were able to construct accurate charts of the principal surface currents and some of the subsurface currents of the world ocean.

From seventy-seven water samples taken throughout the world ocean, J. Y. Buchanan, the expedition's chemist, derived data that formed the foundation of chemical oceanography. Buchanan also exploded the fascinating myth of Bathybius, thought by Wyville Thomson, T. H. Huxley, and other first-rate minds of the time to be an exceedingly primitive amorphous form of life from the deep sea. The British vessel H.M.S. *Cyclops* had discovered a strange, gray, gelatinous material in ooze dredged from the bottom of the Atlantic. The German evolutionist Ernst Haeckel called this material "the mother of protoplasm," and Huxley dubbed it Bathybius. When *Challenger* samples included some of this material Buchanan showed that it was merely a precipitate of calcium sulfate deposited when alcohol was added to the water to preserve the specimens.

Challenger was run by a naval crew under two dedicated captains, George S. Nares and then Frank

T. Thomson, who succeeded Nares when the latter was recalled by the Admiralty to take another post. The crew did a splendid job, but it is not to be expected that they would share all the enthusiasm of the scientists. The latter were willing to suffer hours of boredom and agonies of frustration over gear failures in exchange for the thrills of encountering animals never seen by the eyes of man, or of handling ooze from the bottom of the abyss that had lain in utter darkness for millions of years.

Dredging [reports one of the naval officers] was our *bête noire*. The romance of deep-sea dredging or trawling in the *Challenger,* when repeated several hundred times, was regarded from two points of view; the one was the naval officer's who had to stand for 10 or 12 hours at a stretch, carrying on the work. The other was the naturalist's, to whom some new worm, coral, or echinoderm is a joy forever, who retires to a comfortable cabin to describe with enthusiasm this new animal, which we, without much enthusiasm, and with much weariness of spirit, to the rumbling tune of the donkey engine only, had dragged up for him from the bottom of the sea.

The *Challenger* expedition's results were to fill fifty bulky folio volumes of scientific reports. But it was not so much the impressive volume of material that caused the expedition to be considered the beginning of oceanography as a disciplined science; instead it was the pattern that the scientists established of scrupulously precise observations, of unflagging, enthusiastic devotion, of great efficiency. "Never did an expedition cost so little and produce such momentous results for human knowledge." This opinion of Sir Ray Lancaster has been echoed one way or another by scores of oceanographers in the last century.

Agassiz and the *Albatross*

Biological oceanography was given strong support in America by the activities of Alexander Agassiz, son of the distinguished Swiss-American teacher Louis Agassiz. Alexander Agassiz conducted epochal voyages of the steamer *Blake* in the Caribbean Sea and the Gulf of Mexico and off the Atlantic coast of Florida in 1877–80. He was the first to use steel cables for deep-sea dredging, and he and Captain Charles Dwight Sigsbee of the *Blake* invented a new kind of dredge that would work equally well no matter which way it fell on the bottom. These and other improvements in gear were partly the result of Agassiz's training as a mining engineer. This career also had another important effect on oceanography. Agassiz developed the rich Calumet and Hecla copper mines, becoming wealthy in the process. He used his money freely to finance scientific voyages and to support the Museum of Comparative Zoology at Harvard, where much exceedingly valuable marine work has been done ever since.

Agassiz covered more than 100,000 miles in the course of his oceanographic voyages, first on the *Blake* and later on the U.S. fish commission steamer *Albatross*. The first ship built especially for scientific exploration of the oceans, *Albatross* was ordered by Spencer F. Baird, head of the U.S. Commission of Fish and Fisheries. He had called for "a thoroughly sea-worthy steamer capable of making extensive cruises and working with dredge and trawls in all depths to 3,000 fathoms." *Albatross* amply fulfilled these requirements, and more. Said Alexander Agassiz later, "While I knew of course the great facilities the ship afforded, I did not fully realize the capacity of the equipment until I came to make use of it."

The combination of an able ship and a brilliant scientific direction made the *Albatross* and Agassiz a productive team. Sir John Murray, *Challenger* biologist and Thomson's successor as custodian of the expedition's material, appraised Agassiz's contribution in glowing terms: "If we can say that we now know the physical and biological conditions of the great ocean basins in their broad general outlines— and I believe that we can do so—the present state of our knowledge is due to the combined work and observations of a great many men belonging to many nationalities, but most probably more to the work and inspiration of Alexander Agassiz than to any other single man." After the ship was retired in 1925

The first ship built especially for oceanographic exploration, the steamer *Albatross* was ordered by U.S. fish commissioner Spencer F. Baird. Thomas Edison designed the electrical system. (National Archives)

Alexander Agassiz (center) posed with the ship's officers aboard the *Albatross* in Yokohama harbor, March 1900. With her advanced equipment, the *Albatross* could bring up more deep-sea fish specimens in one haul than the *Challenger* had collected in her entire voyage.

her services were praised by the U.S. Commissioner of Fish and Fisheries, C. H. Townsend: "The *Challenger* was a pioneer ship in oceanographic work and must remain the leader in the literature of the science. The *Albatross* entered the field much later, but, thanks to her more modern equipment and longer service, her collections were naturally much more extensive and the bulk of her published results was perhaps also greater." (As a testimony to this, it might be recorded that in one haul from 1,760 fathoms the *Albatross* brought up more specimens of deep-sea fishes than the *Challenger* had collected in the three and a half years of her expedition.) "If ever the American people received the fullest possible value from a government ship, they received it from this one," said Townsend. "The benefits to science, the fisheries, and commerce springing from her almost continuous investigations . . . are incalculable."

A Heady and Productive Era

The 1880's and 1890's were heady and productive times in oceanography, as ships and men of many nations bustled to answer some of the questions raised by the voyages of the *Challenger*, the *Albatross*, and other ships—and, inevitably, to raise more questions of their own. In 1889 the German "Plankton expedition" sailed under the leadership of Victor Hensen. During this expedition the name "plankton," from the Greek word for "floaters," was first used. The cruise also marked the first time quantitative sampling of plankton had been attempted, and the results of Hensen's work were influential on investigations of the biological economy of the sea.

The Russians entered the field of open-sea oceanography in 1886–89, when the *Vitiaz*, cruising around the world under Stepan O. Makarov, made pioneer observations of temperature and density in the North Pacific. Following this essay into ocean science, the Russians did little until modern times, but now a second *Vitiaz*, one of the biggest of oceanographic ships, along with scores of other large and well-equipped Soviet research vessels, plies the seas of the world. Currently the Soviet Union is among the leaders in oceanography.

Beginning in 1885 a glamorous newcomer, the prince of Monaco, entered the field. Prince Albert was no rich or idle dilettante, but a hard-working and productive oceanographer. He fitted out a series of yachts as research vessels, each better planned and better equipped than the last. The prince worked chiefly in the Mediterranean and the North Atlantic, recording hydrographic data, but in particular making notable collections of deep-sea animals. Prince Albert and his colleagues developed several new and ingenious gears, including bottom and mid-water trawls, and closing nets to fix more precisely the depths of his catches. In particular he employed several new kinds of deep-sea traps, some involving the use of lights to attract animals to the gear. In 1895, when the *Princesse Alice* was working near the Azores, local whalers mortally wounded a whale, which died under the yacht. In its death throes it spewed up parts of a great squid new to science. Excited about the possibility of finding still more specimens, the prince installed whale-catching gear. As a result of the activities that followed he added a considerable body of new knowledge concerning the food of whales. He found that certain species feed mainly on squids, including some with tentacles as thick as a man's arm and studded with great suckers bearing powerful catlike claws. Prince Albert founded and endowed the now famous museum and laboratory at

Prince Albert of Monaco outfitted a series of oceanographic vessels, which he equipped with new and ingenious gears. He founded the marine museum and laboratory at Monte Carlo. (Musée Océanographique de Monaco)

Monaco, whose present director is the colorful Frenchman Jacques-Yves Cousteau.

The Great Laboratories

In the latter part of the nineteenth century three influences stimulated the founding of laboratories for marine biology and fishery research. These were the exciting deep-sea cruises, especially that of the *Challenger;* the desire of biologists to study live animals instead of pickled specimens; and the requirement to learn more about the economically important animals that support commercial fisheries. Anton Dohrn is given credit for founding the first of the important and permanent laboratories in 1872, when he established the Stazione Zoologica in Naples, Italy. The next year Louis Agassiz founded a short-lived but significant station at Penikese Island, near Cape Cod. This laboratory was succeeded by the Marine Biological Laboratory at Woods Hole, Massachusetts, in 1888. The Woods Hole Laboratory of the U.S. Commission of Fish and Fisheries had started in 1875. The second of the commission's laboratories, all of which are dedicated to research on the commercial fisheries, was opened in Beaufort, North Carolina, in 1899.

Before the turn of the century marine stations arose in many countries of Europe and in other parts of the world: in Villefranche, France; Santander, Spain; Port Erin, Millport, and Dover, Great Britain; Mount Desert Island, Maine. The Marine Biological

15

The first marine biological laboratory in the United States was founded by Louis Agassiz at Penikese Island, near Cape Cod, in 1873. It was succeeded by the Marine Bio-logical Laboratory at Woods Hole in 1888. (Houghton Mifflin Company)

The U.S. fish commission opened its Woods Hole Laboratory in 1875. The workroom is shown here as it appeared about ten years later. (Paul S. Galtsoff)

Association Laboratory in Plymouth, England, was founded in 1879, and it has remained one of the most distinguished and productive stations in all the years since.

In a number of cases marine institutions were founded to house the collections of particular scientists. Notable among these were the Musée Océanographique at Monaco, which received the collections of Prince Albert, started in 1906; and the Bingham Oceanographic Laboratory at Yale University (1908), where the material collected from Harry Payne Bingham's yacht, *Pawnee*, was housed.

On the Atlantic coast of the United States several university-associated marine stations were founded in the present century, including the Chesapeake Biological Laboratory (University of Maryland), in 1925; the Narragansett Marine Laboratory (University of Rhode Island), in 1937; the Duke University Marine Laboratory at Beaufort, North Carolina, in 1937; and the Marine Laboratory (now Institute of Marine Sciences) of the University of Miami, in 1943.

On the Pacific coast, David Starr Jordan, an outstanding ichthyologist who was president of Stanford University, stimulated the founding of the Hopkins

Marine Station in 1892. The importance of the fisheries to the state of Washington caused the creation in 1919 of the College of Fisheries at the University of Washington, and of a School of Oceanography in 1951. Scripps Institution of Oceanography began as a biological station in 1905 and was taken over by the University of California in 1912. In 1924 it expanded its activities to physical oceanography and has been active in all branches of ocean science since. Texas A&M University and Oregon State University are other American schools having major laboratories.

The Carnegie Institution of Washington, D.C., operated the Dry Tortugas Laboratory from 1904 to 1939, when logistical problems forced the closing of a highly productive institution. Other private institutes established laboratories, including the California Academy of Sciences in San Francisco in 1853, the Academy of Natural Science of Philadelphia in 1947, and the Allen Hancock Foundation for Scientific Research in Los Angeles in 1940.

In other parts of the world, laboratories were also established in a number of places, including Helgoland, Germany; Viña del Mar, Chile; Recife and São Paulo, Brazil; Nanaimo (British Columbia) and St. Andrews (New Brunswick), Canada; Tokyo, Kochi, and many other places in Japan.

In more recent years, especially in the last decade, marine laboratories and research stations have proliferated at an enormous rate, even in inland states, as oceanography has become "fashionable."

Fishing Becomes a Science

In the 1870's marine biologists began to turn their attention to the commercial fisheries. For the first time it seemed possible that even the teeming fish of the sea could be affected by the activities of man. Thus, about 1890 another branch of oceanog-

Commissioner of Fish and Fisheries Spencer F. Baird pushes off the boat for a collecting expedition at Woods Hole. (Norman T. Allen)

The Scripps Institution of Oceanography began as a biological station in 1905. This aerial shot was taken about 1916. (Scripps Institution of Oceanography, San Diego)

raphy, fishery science, was founded. Alexander Agassiz in the United States, Frank Buckland in England, and W. C. McIntosh in Scotland had been among the early men who had approached ocean science from the point of view of improving the harvest of fish, and their work led in the last decade of the century to organized studies on the life histories, migrations, and growth of fish, and on signs of overfishing.

T. W. Fulton of the Fishery Board for Scotland related changes in the pattern of ocean currents to supplies of commercial species; he also did pioneer work on the fecundity of fish and on their size at first maturity. A Dane, C. G. J. Petersen, holds an important place in the profession for his theory of over-

fishing, in terms of growth, mortality, and fishing pressure. He invented one of the first effective fish tags, two metal or plastic buttons fastened to the fish with a metal pin. The "Petersen tag" is still widely used, and volumes of important information have been obtained with this technique. E. W. L. Holt of England may deserve the title of "Father of Fishery Science," since it was he who established firmly the pattern of combining a fundamental knowledge of the biology of the fish and the pattern of their landings. Holt emphasized another important precept—that fishery work must be done on the boats and out among the fishermen if it is to make practical sense in the end.

The rising concern over the fisheries caused the International Hydrographic Congress to be held in 1899 in Sweden. This resulted in the creation in 1902 of the International Council for the Exploration of the Sea, which became the coordinating body for fishery research in Europe. Emphasis was placed on the study of stocks of fish rather than of individuals; "fishery hydrography," meaning the influence of changes in the ocean environment on fish populations, came into being.

Now began a period when Scandinavian scientists were particularly active. In 1910 the *Michael Sars* expedition was launched under the leadership of a Norwegian, Johan Hjort. This major undertaking was financed personally by Sir John Murray of *Challenger* fame. Impressed with Hjort, Murray proposed to the Norwegian Government that if it would supply the ship and assign Hjort as scientific leader, he would pay the expenses. A highly productive voyage resulted: Greatly improved knowledge of the North Atlantic came from tow-netting, trawling, and dredging in deep water. Hjort noted that the animals of the ocean differ in characteristics depending on the depth they occupy. In the well-lighted top layer of the ocean they are typically blue or transparent; in depths from about 450 to 1,500 feet they are gray or silvery; deeper than this, many animals are black or red.

Swedish, Norwegian, and Danish physical oceanographers were among the first to show that the ocean of water and the ocean of air function as two enormous, interconnected "heat machines" for distributing heat from the tropics to the higher latitudes. Hence, meteorologists like Vilhelm Bjerknes, Björn Helland-Hansen, and Sven Ekman became oceanographers, the better to understand these complex processes.

The Carlsberg Foundation, supported by the Danish brewing company, supported two deep-sea

17

cruises of the *Dana* in 1920–22 and 1928–30. Part of the stimulus for this work was provided by the International Council for the Exploration of the Sea, which was heavily oriented toward fishery research. From the *Dana* work Johannes Schmidt came forward with the astonishing story of the eel.

Prior to this time the life history of the eel had been a mystery. It seemed to have no young, since eggs and larvae were never found in fresh water where the adults lived. Aristotle had suggested that eels were generated out of the mud; others believed that horsehairs became long, thin worms and then eels. Schmidt solved the mystery by filtering enormous amounts of Atlantic Ocean water through his plankton nets and noting that the eel larvae captured showed a definite pattern of distribution. Larvae found closest to the mouths of European rivers were biggest, with the size diminishing progressively to

The main biological laboratory of the *Discovery*, first research ship of Britain's antarctic studies of the whale, is shown in a wide-angle sketch made on the voyage by Sir Alister Hardy. The sketch appeared in Hardy's book, *Great Waters.* (William Collins Sons & Co.)

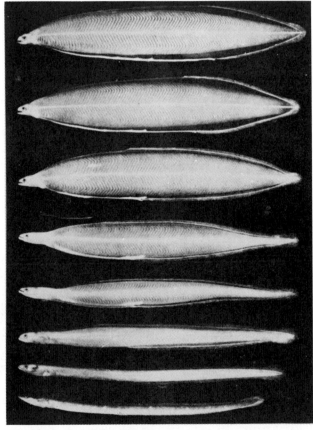

The life cycle of the European eel, whose eggs or young had never been seen, mystified scientists until its spawning place was discovered in the western Atlantic, thousands of miles away from its European rivers. The larvae take three years to drift to these rivers, during which time they change from the leaflike stage shown at the top to the smaller, slender shape shown at the bottom. (Johannes Schmidt)

the westward. Their spawning place was discovered thousands of miles away in the Sargasso Sea, a great eddy of the western Atlantic. Eggs and newly hatched larvae were caught at depths of 350 to 550 feet. Later work showed that larvae take three years to get back to the mouths of their European rivers, transforming during this period from a strange leaflike "leptocephalus" larva to the "silver eel" stage that ascends the streams in prodigious numbers.

Another fishery problem, that of the whale, stimulated an enormously productive series of British cruises. In 1925 Britain assigned Scott's famous antarctic exploratory ship, *Discovery*, to the task of gaining information on the biology of antarctic whales and their environment. The next year the *William Scoresby* joined *Discovery*, and in 1930 *Discovery II*, a ship specially designed for rugged antarctic oceanographic research, replaced her famous namesake. These investigations have made the antarctic waters among the best known of any section of the ocean, although through no fault of the scientists involved they have not been able to stem the calamitous decline of the whale stocks caused by overfishing.

A New Age of Discovery

Among the mineral constituents of seawater, gold had been detected. Early reports put the quantities far too high, and this gave false encouragement that gold could profitably be extracted from the ocean. Even hardheaded scientists were misled. After World War I, when the Allies imposed heavy reparations on the German Government that were payable only in gold, Germany's most famous chemist, Fritz Haber, was assigned the task of getting this gold from the sea. He launched the *Meteor* expedition in 1925; the ship cruised the South Atlantic, vainly trying to extract useful quantities of gold. The attempt failed because it costs far more to recover gold than the product is worth. But if the dream of gold from the sea was blasted by the work of the *Meteor,* her time was not wasted, since a great amount of valuable data was accumulated on the distribution of nutrient salts in relation to plankton and other living materials.

Two important round-the-world oceanographic cruises were taken by the Swedish *Albatross* in 1947–48 and the Danish *Galathea* in 1950–51. The Swedish expedition, headed by Hans Pettersson, emphasized marine geology, and with an ingenious coring device invented by its geologist, Dr. Börje Kullenberg, *Albatross* took some remarkable deep-sea cores. She also made dredge hauls to 25,900 feet in the Atlantic.

Deep dredgings were the specialty of the Royal Danish Research Vessel *Galathea.* She brought up fish in her trawls from 23,200 feet and anemones clinging to rocks from the prodigious depth of 32,800 feet. Clearly, life exists in the deepest part of the abyss.

Man has gone under the sea in one kind of apparatus or another for thousands of years—perhaps starting in 330 B.C. with Alexander the Great, who is said to have had himself lowered gently in a barrel. Through the centuries a fascinating array of underwater ships have culminated in the cruiser-sized Polaris submarines of the U.S. Navy. Meanwhile, the ability to live and work in the sea has been greatly assisted by the invention in 1943 of "scuba" (self-contained underwater breathing apparatus) by Jacques Cousteau and Emile Gagnan. Cousteau's *Conshelf* and the Navy's Sealabs are pushing this frontier rapidly forward. Man's ability to exist and work underwater is already well advanced. Whether he will ever actually live in underwater cities is another matter.

In May 1950 a new *Challenger,* named for the famous old ship whose world cruise founded the science of oceanography, set out on a two-year voyage covering the Atlantic, Pacific, and Indian Oceans and the Mediterranean Sea. Her special mission was to make precise soundings of the ocean depths, employing seismic techniques that would have astonished Wyville Thomson and his companions on their nineteenth-century ship. On the old *Challenger* it had required two and a half hours to make a sounding in deep water—lowering a hempen rope until the rate of descent told watchful scientists that the weight must have struck the bottom, then laboriously winding it up again. In the same two-and-a-half-hour pe-riod, the new *Challenger* could make thousands of depth measurements, each far more accurate than those of the original *Challenger,* thanks to modern echo-sounding devices that bounce ultrasonic impulses—traveling at about 4,900 feet per second—off the sea floor and time the intervals between sending of the impulses and arrival of their reflections at the ship in millionths of a second. Working in the Mariana Trench in the western Pacific in 1951, *Challenger* recorded a depth of 36,200 feet and gave her name (and thus, fittingly, the name of her illustrious namesake) to the deepest part of the ocean.

In 1960 Lieutenant Donald Walsh and Jacques Piccard took the remarkable submersible vessel, the bathyscaphe *Trieste,* to the bottom of the sea in the Challenger Deep. On that bottom, seven miles below the surface, they saw a sole-like fish and a shrimp.

The cruise of the new *Challenger* in 1950–51 resulted in another highly portentous discovery—the rift valley that runs down most of the enormous length of the Mid-Atlantic Ridge. This ridge was first mapped on the basis of soundings made in preparation for the laying of the transatlantic cables in the second half of the nineteenth century. Since soundings were scarce at first, the mountains appeared to

Seven miles below the surface in the Challenger Deep, scientists in the bathyscaphe *Trieste* saw a sole-like fish and a shrimp on the ocean floor. The *Trieste* is shown here on a hoist above the Mediterranean. (official U.S. Navy photograph)

In 1957, a south-drifting current below the Gulf Stream was found by a cooperative effort of the British *Discovery II* and the American *Atlantis*. The British ship is shown here. (John E. Long)

be grouped in isolated ranges. Gradually it dawned on oceanographers that the ridge twists down the whole length of the Atlantic for more than 10,000 miles, from above Jan Mayen Island north of Iceland to beyond Tristan da Cunha and Bouvet Islands near the Antarctic continent. Most of its peaks are drowned beneath thousands of feet of water, but a few are high enough to break the surface and become familiar as islands: the Azores, Ascension Island, St. Helena, the Rocks of St. Peter and St. Paul. As soundings increased in number it became clear that mid-ocean ridges exist in the Pacific and Indian Oceans, too, and that all the ridges together comprise the most immense natural feature of the globe.

The rift valley in the Mid-Atlantic Ridge is a crack twenty to thirty miles across and half a mile to one and a half miles deep. It adds credence to the theory that the continents are slowly drifting apart, a proposition that was put forward in the nineteenth century, rejected out of hand by most geologists, and is now widely accepted once again.

Another interesting discovery of very recent years is the existence of enormous subsurface currents. These had been predicted on theoretical grounds by some physical oceanographers. In 1957 a south-drifting current was found beneath the Gulf Stream by a cooperative investigation of the British *Discovery II* and the American *Atlantis*. The ships centered their observations on that part of the stream that is east of the Blake Plateau, off Charleston, South Carolina. A new tool invented by Dr. John C. Swallow, a British oceanographer, was of great assistance. This is a float whose buoyancy can be delicately adjusted so that it is made to sink to a predetermined depth and remain there, floating in water of the same density as itself. The float has a small ultrasonic transmitter that announces its position to tracking vessels. At a depth of 6,500 feet a current was discovered flowing southward at speeds ranging from two to eight miles a day.

In 1951 an immense subsurface current was discovered stretching at least halfway across the Pacific.

This current, perhaps the largest in the Pacific except for the Kuroshio Current off Japan, carries more than half as much water as the Gulf Stream, at nearly as fast a rate. The current came to light when scientists of the U.S. Fish and Wildlife Service, fishing for tuna at the equator south of the Hawaiian Islands, had set their gear at a depth of about 165 feet in an area where the surface waters flowed westward. They were surprised to see the buoys head unmistakably to the east. Apparently there was an easterly current only a hundred feet or so below the surface. Later study by several oceanographers, including Townsend Cromwell, for whom the current is named, showed that it is one of the major components of the Pacific system; it is at least 3,500 miles long and it may stretch for 8,000 miles. It is shallow—only sixty-five feet in places—and its center is exactly on the equator.

Oceanography Today and Tomorrow

Research ships still make long voyages with multiple objectives, but the pattern of oceanographic exploration is changing. Cruises are now usually planned over relatively limited expanses of the ocean, with particular circumscribed problems to solve. Since the broad picture has been painted by the investigations of preceding decades, scientists are now engaged in filling in the finer details. Further, it has become clear that isolated observations fail to depict the true situation in so vast and so quickly changing an environment as the sea. Of far greater value are data collected simultaneously over a wide area. If repeated sets of observations can be made in the same region at a later time, the information becomes more useful still. This has led to oceanographic expeditions in which many ships, often of several nations, cooperate in achieving a common goal. Large-scale cooperation started during the International Geophysical Year in 1957–58, whose programs included massive oceanographic studies. During the International Indian Ocean Expedition of 1959–65, scientists of twenty-two nations collected data under a plan coordinated by the Scientific Committee on Oceanographic Research (SCOR) of UNESCO. Far more information of much greater use was gained than would have been if each ship had gone its own way. As a result, a massive plan has been developed with the leadership of the United Nations Food and Agricultural Organization for fishery development in the Indian Ocean.

In 1965–66, ships of the United States, the Soviet Union, France, Brazil, and other nations took part in "Equilant" cruises, even more tightly organized surveys of the tropical Atlantic. They obtained si-

The largest oceanographic vessel in the United States fleet, *Oceanographer*, arrives in Sydney, Australia, on a global expedition in 1968. (Robert S. Dietz)

multaneous data on temperature, salinity, and currents in an area little known but potentially extremely productive, off the west coast of Africa.

Eight countries have taken part in a cooperative study of the Kuroshio area, including the South China Sea; twelve countries are involved in cruises between Gibraltar and Cape Verde in Africa; in the Mediterranean and Caribbean, several countries plan cooperative research programs. It is certain that this trend will continue.

Scientific knowledge has increased more in the last few years than in all of the previous history of man, and it is estimated that ninety percent of all the scientists who have ever lived are active today. This remarkable situation is even more startlingly evident in oceanography than in nearly any other branch of science, so that it is impossible to summarize the recent history of oceanography with anything like the completeness or fairness that can be applied to its earlier development. Information about ocean currents, submarine geology, deep-sea creatures, the abundance of exploitable fish populations, and all the other specialized branches of oceanography is increasing at a dizzying rate, so that no man can keep up on one small area completely, let alone maintain a balanced view of the whole of ocean science.

The penetration of Africa in the last century and the conquest of Antarctica in the present one have led much of mankind to the mistaken view that human exploration and pioneering on the earth are over, and that such adventure can be experienced only in space. In this estimate man is once again allowing his prejudice as a land animal to influence him unduly—he has forgotten the seven-tenths of the world that lies under water. In the United States, a spirited and sometimes acrimonious debate is taking place over whether so much of the nation's wealth should be invested in space exploration. Thoughtful scholars, such as historian Arnold Toynbee, label space research a "dead end" and plead that this expensive effort be diverted to ocean science and the development of sea resources, which promise so much more for mankind.

Space research is enormously exciting and challenging; it satisfies man's intellectual hunger and adventurous spirit as no other activity in history has done. Whether oceanography is equally challenging and satisfying is futile to argue, but the material gains to be expected from ocean exploration are far greater than those from space. Space research promises little more in this regard than a better understanding of the earth and how to use *its* resources. A chunk of minerals from the moon will cost millions of dollars a pound—and will add nothing to the satisfaction of our material needs. By contrast, vast quantities of scores of minerals await harvest from the ocean. We are already getting close to a fifth of our oil and gas supplies from the sea—well over a billion dollars' worth a year—and the proportion is increasing steadily. It has now been firmly established that the moon is not made of green cheese, nor will any other kind of food come from space. But the sea promises the possibility of overcoming much of the world's protein shortage, even if it will not solve the whole hunger problem.

Thus the promises of oceanography include not only the satisfaction of intellectual curiosity and a better understanding of natural processes and the principles controlling the operation of the universe, but to an important degree the fulfillment of man's requirements for material things like water, minerals, and food. This guarantees that research in the sea will accelerate at an even faster rate in the future than it has in the recent past. Oceanographic historians a few decades hence will have even more exciting tales to tell than those who have written this book.

2 THE UNDERWATER

by Robert S. Dietz

Almost nothing was known about the distance to the bottom of the oceans until long after the vastly greater distances to the moon, sun, and Mars were precisely determined. Until just a few years ago, regions as large as Mexico existed for which there was not a single depth sounding. The blank areas on the charts are gradually being filled in, but the oceans are so vast that it will be another generation before man has even a reconnaissance knowledge of the deep-ocean floor. The deeper regions alone—those covered by 1,000 fathoms or more of water—total 120 million square miles, an immense area eight times larger than the surface of the moon, near and far sides combined.

A classic attempt to fathom the ocean is ascribed to Ferdinand Magellan in 1521. In the central Pacific Ocean, between St. Paul's Island and the Los Tiburones in the Tuamotu Archipelago, he spliced together six pieces of rope and then lowered this improvised sounding line to 2,500 feet without reaching bottom. He then proclaimed, so the story goes, that the Pacific Ocean was immeasurably deep. With grand disregard for logic, he wrote in his log that he was over its deepest depression. Modern charts tell us that 15,000 feet of water lay beneath his keel— a typical depth for the open ocean. Magellan had reached but one-sixth of the way to the bottom.

Not until the mid-eighteenth century did natural philosophers seriously speculate about oceanic depths. One astronomer calculated, on the basis of the speed of the tide wave, that the oceans were twenty-three miles deep! The eminent French astronomer Pierre de Laplace more conservatively computed the depth as twelve miles. Other savants were content with the view that the ocean was simply bottomless. Another prevalent view, based merely on the "fitness of things," was that the seas were as deep as the mountains are high. This last conjecture, although pure speculation at the time, has proved to

be a fair guess, although something of an underestimate. Entire mountain ranges as lofty as any on land lie completely submerged. Only exceptionally high peaks pierce the sea's vast blue expanse.

Sounding the Seven Seas

In his *Depths of the Sea*, published in 1873, Sir Wyville Thomson tells of the first sample ever brought up from 1,000 fathoms.

In the year 1818, Sir John Ross, in command of H.M.S. *Isabella*, on a voyage of discovery for the purpose of exploring Baffin's Bay and inquiring into the possibility of a Northwest Passage, invented a machine for taking up soundings from the bottom of any fathomable depth, which he called a deep-sea clamm. A large pair of forceps were kept asunder by a bolt, and the instrument was so contrived that on the bolt striking the ground a heavy weight slipped down a spindle and closed the forceps, which retained within them a considerable quantity of the bottom, whether sand, mud, or small stones. On September 1, 1818, Sir John Ross sounded in 1,000 fathoms, lat. 73°–37′ N., long. 75°–25′ W. The sample consisted of soft mud, in which there were worms, and entangled on the sounding-line at a depth of 800 fathoms was found a beautiful *Caput Medusae* [a basket star, a relative of the common starfish, whose arms are subdivided so many times that the writhing mass resembles the head of the mythical Medusa]. The clamms were used with strong whale line, made of the best hemp, 2½ inches in circumference.

One precise measurement, such as this, is more eloquent than endless philosophizing. Twenty-two years later, in 1840, Sir John's nephew, Captain James Ross, paused in his voyage of antarctic exploration aboard the H.M.S. *Erebus* to sound the depth of the abyssal ocean for the first time. Ross, from whose ship Mt. Erebus, Antarctica's only active volcano, took its name, made this sounding in the South Atlantic at 17° 30′ W. and 27° 24′ S., approximately

LANDSCAPE

halfway between the lower halves of Africa and South America. A line of yarn was made up and rolled onto an enormous spool. A 72-pound lead weight was attached, heaved into the sea, and allowed to sink freely to the bottom. It struck at 2,425 fathoms. A few days later, a second sounding was taken with similar dispatch, this time using a 540-pound weight. After fifty minutes of running free, the plummet struck bottom at 2,677 fathoms.

Ross's initial sounding is still found on navigational charts; rather remarkably, however, it appears as 2,426 fathoms, rather than the original reading of 2,425 fathoms, on modern American and British charts. This must be ascribed to creeping errors in transcription by chart compilers, who, like the medieval copyists of manuscripts, did not go back to the original source. In 1968, at this writer's suggestion, the U.S. Geodetic Survey's *Discoverer* briefly resurveyed the original site with modern position control and precise echo sounders. The Ross sounding was found to be located in an area of rough topography on the west flank of the Mid-Atlantic Ridge, but nowhere within five miles of the site was any sounding found as deep as Ross's. At the precise location, a sounding of 2,100 fathoms—325 fathoms less than recorded by Ross—was obtained. It would appear that Ross had not immediately recognized when his plummet struck bottom, and so he had overestimated the depth.

Just five years after Ross made his historic measurements, Lieutenant Matthew Fontaine Maury, founder of the agency that became the U.S. Navy Hydrographic Office, initiated a systematic program of sounding the oceans. In 1845 several Navy ships began to secure soundings in the North Atlantic by lowering a cannonball at the end of a free-running spool of twine. It was erroneously supposed at first that when the cannonball touched bottom, a shock would be felt and the line would go slack. Actually a sounding line continues to run out under its own weight, making early sounding reports more confounding than useful. Soundings of 34,000, 39,000, 46,000, and even 50,000 feet, which did not reach bottom, were obtained and duly transmitted to the Hydrographic Office, where they were entered on the charts. The ocean seemed truly bottomless. However, it would appear that the question actually being investigated was, "How long is a piece of string?"

Maury, aware of some plaguing Neptunian prank, conjectured that the dilemma lay in a strong deep-sea crosscurrent that acted "with a swigging force on the bight." A solution was soon found to his so-called "Law of the Plummet's Descent," about which he promptly instructed all Navy ships. He prescribed twine of 60-pound strength and a standard 32- or 68-pound cannonball for the plummet. The twine was marked with bunting every 100 fathoms, and the rate at which it reeled out was carefully recorded. When this sinking rate *abruptly* decreased—even though the line continued to pay out—it was assumed that the seabed had been struck. This technique gave reasonable results. By 1852 Maury was able to publish a depth chart, based on a few sparse soundings, along the proposed telegraph cable route between North America and Great Britain. By 1854 he was able to draw a generalized bathymetric chart of the entire North Atlantic, using 150 soundings deeper than 1,000 fathoms.

As with many pioneers who attempt to establish new programs and precedents, Maury's career in the Navy, although brilliant in retrospect, was not entirely happy. Early in his career, Maury was nearly cashiered out of the service, ostensibly because he was unfit for sea duty due to a slight disablement, the result of a fall from a horse. Later, after the serious stagecoach accident that did force him to take a shore post, he never rose above the rank of lieutenant, despite his great accomplishments in collecting,

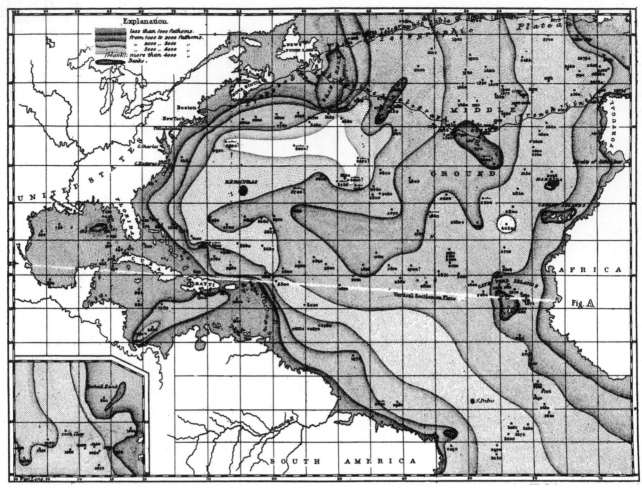

Maury's map of the basin of the North Atlantic, from his *The Physical Geography of the Sea*, exhibits the status of scientists' knowledge of the ocean floor as of 1855.

interpreting, and making generally available information about the ocean. Unfortunately for the progress of oceanography, Maury's career as the Navy's hydrographer was cut abruptly short by the outbreak of the Civil War, since his allegiance was to the South.

The question of the average depth of the Pacific Ocean was ingeniously solved in 1856 by A. D. Bache of the U.S. Coast and Geodetic Survey. A tsunami, or "tidal wave," set up by the great Japanese earthquake of 1854 gave the answer. A Russian officer's account of the destruction of his ship provided the basic data:

On the 23rd of December 1854, at 9:45 A.M., the first shock waves of an earthquake were felt on board the Russian frigate *Diana* as she lay at anchor in the harbor of Simoda, not far from Jeddo [now known as Tokyo], Japan. Fifteen minutes afterwards a large wave was observed rolling into the harbor and the water on the beach to be rapidly rising. The town as seen from the frigate

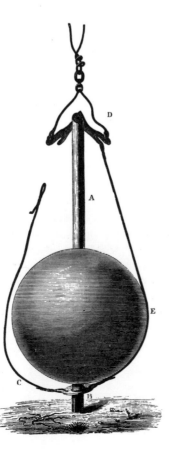

Brooke's deep-sea sounding apparatus played an important role in supplying the data for Maury's bathymetric chart. This illustration is from *The Depths of the Sea*, by Wyville Thomson.

24

appeared to be sinking. This wave was followed by another and when the two receded which was at 10:15 A.M. there was not a house, save an unfinished temple, left standing in the village. These waves continued to come and go until 2:30 P.M., during which time the frigate was thrown on her beam end five times. A piece of keel 81 feet long was torn off, holes were knocked in her by striking the bottom and she was reduced to a wreck. In the course of five minutes the water in the harbor fell, it is said from 23 to 3 feet and the anchors of the ship were laid bare. There was a great loss of life; many houses washed into the sea, and many junks carried up—one, two miles inland—and dashed to pieces on the shore. The day was beautifully fine and no warning was given of the approaching convulsion. It was calm in the morning and the wind continued light all day. ["Notes of a Russian Officer," *Nautical Magazine* (London, 1856), v. 25, no. 2, p. 97.]

Some twelve hours after the earthquake, a series of seven waves washed into San Francisco and San Diego—the first tsunamis ever to be recorded on tide gauges along the California coast. Bache used a newly discovered physical law that related the speed of very long waves such as tsunamis (but not ordinary swell or wind waves) to the depth of the ocean. He computed the average depth along the wave path to be from 14,000 to 18,000 feet—in full agreement with the depth of roughly 16,000 feet shown on modern charts.

Early soundings were laborious and took a whole day. H.M.S. *Challenger* (1872–76) used a stout hemp rope flaked out on deck and then allowed to run free overside. It was hauled in by hand over a steam capstan. With the advent of telegraphy in the mid-nineteenth century, plumbing oceanic depths for cable routes became of prime importance, and improvements in the art of sounding were made. In 1885 the U.S.S. *Tuscarora* revolutionized sounding techniques by using a single strand of piano wire, permitted several soundings to be taken each day along proposed alternate cable routes between America and Japan. By 1911, 5,969 oceanic soundings deeper than 1,000 fathoms had been laid down, with the deepest being 5,269 fathoms (31,614 feet) in the Nero Deep off Guam, where the bathyscaphe *Trieste* would one day commence her series of ultradeep dives to the bottom of the world.

Curiously, the term *sounding* in the depth-finding sense is derived from the verb *to sound,* meaning *to probe,* as to sound a gasoline tank with a metal rod or stick. It is fortuitous that the use of acoustic sound later became the standard method for taking soundings. Using acoustic sound, it is possible to obtain a depth reading within a matter of a few seconds by

The men of the *Discoverer* attempt a sounding in 1968 by the classical method of lowering a cannonball, at the site of Sir James Ross's original deep-sea sounding in the central South Atlantic in 1840. A sonic sounding survey of the area revealed that Ross's measurement of 2,425 fathoms was 325 fathoms too deep. (Harley Knebel)

remote sensing of the bottom with sound waves, which are transmitted in water even better than in air. This procedure is far simpler than lowering any sort of line two or three miles down to the bottom. Early hydrographers were aware of the advantages offered by timing echoes bounced off the bottom, and they tried beating the ship's hull with great hammers. Maury wrote of "exploding petards and ringing bells to attempt to hear an echo when the wind was hushed and all was still, but no answer was heard."

Echo sounders are a relatively modern development. They are a product of the invention of electronic devices for precise measurement of elapsed time, the magnetostrictive property of nickel, and the piezoelectric property of certain crystals. Magnetostrictive substances contract or expand under the influence of a magnetic field and so can be made to oscillate underwater and in turn generate a high-frequency sound impulse. Similarly, certain crystals, such as quartz, as well as some ceramic materials, such as barium titanate, also can be made to oscillate and send out an ultrasonic sound impulse, or "ping," when they are excited by an electric impulse. Racing through the water, the ping bounces off the bottom and returns as an echo to the source crystal, which reconverts the sound pulse into an electric signal for graphic registration. Since sound travels through

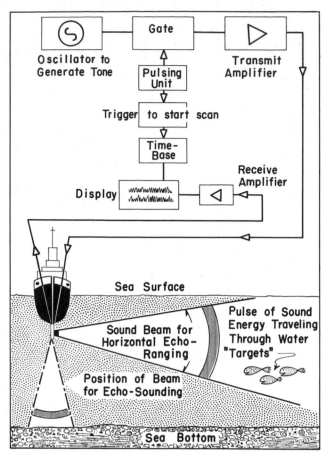

Modern soundings are obtained by bouncing echo beams off the ocean floor. The echo of an acoustic impulse, which travels five times as fast through water as through air, can be accurately timed to determine distance.

water about five times as fast as through air, the round-trip time, from vessel to deep-sea floor and back, usually is about five seconds.

The first really successful sonic soundings were made in 1920 by the German scientist Alexander Behm in the North Sea. Prior to that date efforts to electrically record echoes bounced off the bottom had been confounded by the high speed of underwater sound, which caused a merging of the outgoing pulse with the returning echo. Since timing was not accurate enough, or the ping short enough, to permit such discrimination, most people doubted the presence of any echo at all. Then, in the United States, Reginald Fessenden, former chemist-assistant to Thomas Alva Edison, was stimulated by the tragic collision of the *Titanic* with a fog-enshrouded iceberg to develop an echo-sounding device for navigational use. He obtained an echo return by bouncing sonic waves off an iceberg 2.5 miles away. This was the first *sonar* (an acronym from Sound Navigation and Ranging), which, roughly speaking, is simply an echo sounder tilted horizontally. More significant than showing how to avoid icebergs, Fessenden had demonstrated a technique with great potential for hunting sub-

Magnetostrictive substances, which contract or expand under the influence of a magnetic field, can be made to generate high-frequency sound impulses. Scripps's deep-towed instrument, "The Fish," operates with a sonar beacon network, controlled from a research vessel laboratory. (Scripps Institution of Oceanography, San Diego)

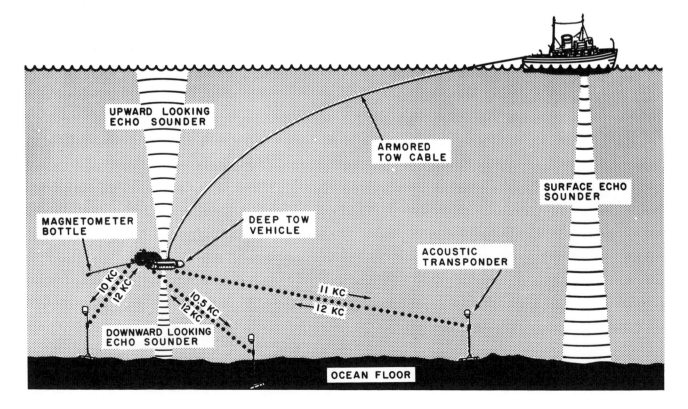

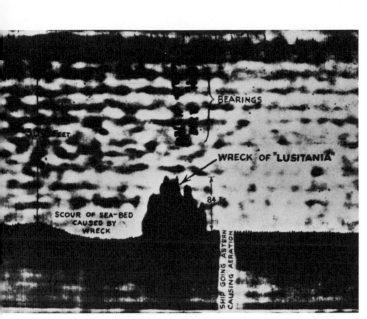

The sunken *Lusitania* was the first important wreck recorded graphically by echo sounder. (S. Smith & Sons, Kelvin Hughes Division)

marines. His simple experiment triggered the explosive development of a new field of science—underwater sound.

Even a cursory knowledge of the bathymetry of the deep-ocean basin is quite new. For the North Pacific this was achieved during the fifteen years before World War II, largely by the efforts of a single U.S. Navy supply ship, the U.S.S. *Ramapo*. Operating between the United States and the Philippine Islands, *Ramapo* was directed to follow a slightly different route on each great-circle crossing, acquiring more than sixty sounding tracks in all. Sonic soundings were made each ten miles, but with a stopwatch for timing, so that depths were determined to the nearest eighty fathoms only. During this time another extensive and precise survey of the Gulf of Alaska was made by ships of the U.S. Coast and Geodetic Survey en route to their summer station in Alaskan waters. By wisely routing the ships over well-spaced parallel tracks, the Survey was able to develop the bathymetry of the entire Gulf of Alaska; its investigations revealed a score of giant flat-topped seamounts rising from a seaward-sloping abyssal plain. This was the first oceanic region to become well known.

Prior to World War II, plotting the world's new deep-sea soundings was largely the job of one man who sat in a musty office at the International Hydrographic Bureau in Monaco compiling the *Carte générale bathymétrique du monde*. A nearly hopeless task from the start, it became exceedingly so with the advent of echo sounding. Now torrents of deep-

Ceramic spheres that send sound signals underwater are lowered into the Pacific as part of an antisubmarine-warfare research project. (Lockheed Aircraft Corporation)

oceanic soundings are being collected, but surveying the deep-ocean floor remains an enormous task, which will not be completed even in this century.

Precise to the fathom, modern echo sounders now are revolutionizing our knowledge of the ocean floor by taking soundings every second and graphically displaying a continuous profile of the bottom relief along the ship's track. Remarkable though it is, this technique still leaves much to be desired, for sound is "blurry-eyed," reflecting only the gross topographic highlights in a shadowy way. No information about fine structures is revealed, as these are too small to be resolved by sound waves. Whether or not seamounts are scarred by gullies for example, will remain unknown until man himself can descend to see with his own eyes.

Another valuable depth-measuring device, the narrow-beam echo sounder, is now replacing the old broadbeam sounder that recorded echoes off the nearest bottom—which was not necessarily directly beneath the ship. Still more vital are advances made in the technique of precise positioning, as accurate soundings are of little importance if the ship does not know where it is. A solution to this plaguing problem recently was found with the development of satellite navigation for the U.S. Navy by the Johns Hopkins Applied Physics Laboratory. Now four satellites that will respond to electronic interrogation are injected into polar orbit ninety degrees of longitude apart. An automatic computerized system aboard a

Bottom-scanning sonar equipment is recovered from the water. (Westinghouse Electric Corporation)

survey ship can now pick up these satellites every few hours and compute its own position within about two-tenths of a nautical mile. Since this is a type of radar system, positions can be fixed in a full overcast, for radar is independent of whether or not the sun shines or the stars are out. This is the most striking innovation in navigation since the compass for giving orientation, the sextant for determining latitude, and John Harrison's chronometer for determining longitude. It makes deep-sea surveying a feasible enterprise for the first time in history.

The hydrographic offices of the world all prepare sounding sheets on which are compiled information about the depths of their territorial waters. In addition, the major nations have broken away from the constraint of making nearshore continental-shelf surveys only, and they now maintain oceanic sounding sheets for the entire world. In the United States, the U.S. Naval Oceanographic Office and the National Ocean Survey are the repositories for this information.

A fine example of international cooperation is the GEBCO (General Bathymetric Chart of the Oceans) series of oceanic sounding sheets. In this scheme the world ocean has been divided into regions of chief responsibility. Each of the sixteen cooperating nations adds new depth data and corrects the old data within its assigned region. Up-to-date sounding sheets are then archived in a central repository at the International Hydrographic Bureau at Monaco.

In recent years some new maps of the ocean floor, based on new data, have appeared. Notable examples are the physiographic diagrams of various oceans of the world by the National Geographic Society, based on studies by Bruce Heezen and Marie Tharp of the Lamont Geological Observatory, which reveal the

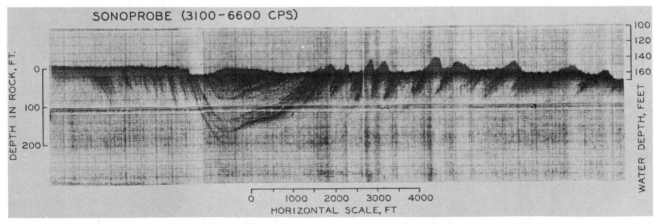

While most sounders simply bounce echoes off the bottom, more specialized devices have been developed to penetrate deep into the substrata. A high-powered sonoprobe that profiled the bedrock structure on the Anacapa Island shelf off southern California gave this "X ray" view of the ocean bottom. (D. G. Moore)

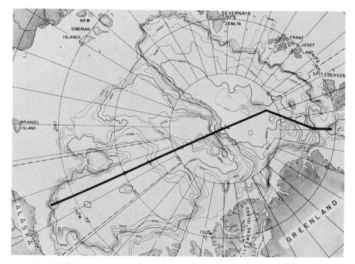

In 1958 the *Nautilus* crossed under the permanent ice pack of the North Pole, submerging north of Alaska and surfacing 1,200 miles later in the vicinity of Spitzbergen. Along the way, *Nautilus* obtained the first continuous bathymetric profile of the Arctic Ocean.

vast sweep of the mid-ocean ridges and the grandeur of lofty seamounts. Another series of bathymetric charts (1:12,000,000) and 1:25,000,000), which has been assembled in the U.S.S.R., utilizes much new data collected from the research vessels *Vitiaz* and *Lomonosov*. Still another set, that of the U.S. Naval Oceanographic Office, depicts the entire ocean floor in a series of eight charts. Undoubtedly many more such maps will be printed in the next few years as the morphology of the ocean floor becomes further illuminated by fathometers with their searchlight beams of sound.

Morphology of the Ocean Floor

Planet Earth is the "water planet," the only member of the solar system having an ocean. It is not so well appreciated that the earth is also the only planet with ocean basins. Even without the water, the ocean basins would be a remarkable part of our realm, for the two-level aspect of the earth's hypsographic curve (the continental level and the oceanic level) is fundamental to this planet. Photos taken on the recent missions to Mars by Mariner spacecraft reveal no oceanic depression on that red planet. The moon, which is closer to us, has "maria"—the dark depressions in the lunar rind—but despite their name ("mare" from the Latin for "sea"), these are not ocean basins. Galileo, looking through the first telescope in 1608, thought they were, and even assumed that they contained water. Of course they are dry, and it

seems quite certain that they are explosion basins caused an aeon (a billion years) or so ago by the impact of minor planets, or asteroids.

Such a genesis by cosmic cannonballs could hardly account for our ocean basins. Geologists have convincing evidence that they were evolved slowly by the earth's deep internal forces. If the moon were covered with a sea, the water would only roughly fit the maria, since they do not have a common depth, like the earth's oceans, nor are all the depressed areas on the moon located within the maria. And the maria are isolated basins in the overall lunar framework of "uplands," or *terra*, while the earth has an interconnected and all-encircling world ocean from which the continents rise like great islands. These are but two of the significant differences between the terrestrial ocean basins and the lunar "seas."

The scientists of the *Challenger* expedition expected to find the ocean floor a haven for ancient

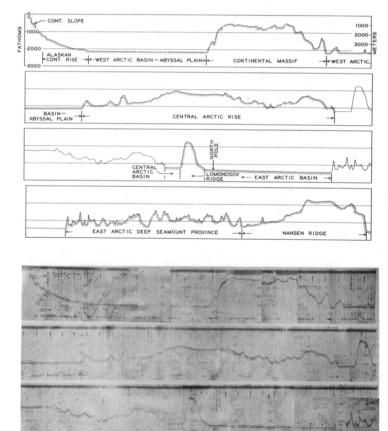

The bathymetric profile of the Arctic Ocean (top) was drawn from the echogram recorded by the *Nautilus* (bottom).

archaic life, but virtually no "living fossils" were found; instead, the abyssal life was modern and closely related to extant animals inhabiting continental-shelf waters. Similarly, geologists until very recently expected the ocean floor to contain Precambrian seamounts more than 600 million years old. It was argued that a seamount once formed would remain forever in the still abyss where no wind blows and no stream flows. But now the ocean floor does not seem to contain physiographic "museum pieces," preserved like the features of the moon over aeons without change. Although undersea erosion is sluggish, internal tectonic forces and volcanic effusions have vigorously shaped and reshaped the ocean floor. In some manner not fully understood, the ocean floor has undergone geologic renewal that has destroyed, or altered beyond recognition, the old physiography. The oldest seamount chain known in the Pacific Ocean is the deeply submerged Mid-Pacific Mountains, which pierce the surface only at Wake Island. But this mountain range is a mere 100 million years old, its birth being synchronous with the decline of the dinosaurs on land. These mountains are much younger than, for example, the 250 million-year-old Appalachian Mountains along the eastern margin of the United States.

Perhaps the most significant discovery made about the ocean basins after World War II is that they are crustless. Geologists still speak of an oceanic "crust," but the layers of rock that make up the ocean basins are quite unlike those of the continents. The continents are composed of granitic rock called *sial*, while the ocean basins are underlaid by dark, heavy rock known as *sima*. The expression *crust* really should be reserved for the continents, since the sial of the continents alone comprises the outer shell, or "epidermis," of the earth. The sea floor is underlaid by what can more correctly be termed a *rind*, a modified or chemically altered surface of the earth's mantle—like the rind on a head of cheese. To a reasonable approximation, the ocean basins are virtually exposures of the earth's interior mantle. And the mantle is where the geologic action is: The magma of volcanoes is born there, as well as the mountain-making forces that wrinkle and mold the earth's outer skin. One can appreciate the importance of the mantle by realizing that it comprises 80 percent of the earth, with the core making up 19 percent and the sialic continental crust a mere one percent. These ratios are similar to the distribution of the main gaseous components of the atmosphere—nitrogen, oxygen, and argon. The continental sial, then, is as unrepresentative of the earth as argon is of the atmosphere. On

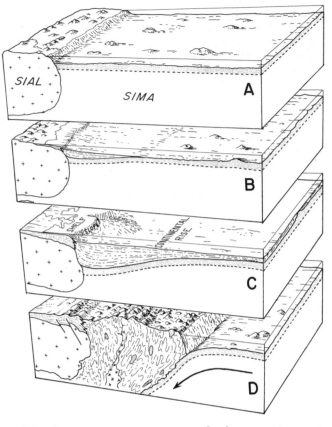

While the continents are composed of a granitic crust called *sial*, the oceans are underlain by *sima*, a chemically altered surface of the earth's mantle. The earth's surface is constantly reshaped by the geologic forces that are generated within the earth's mantle.

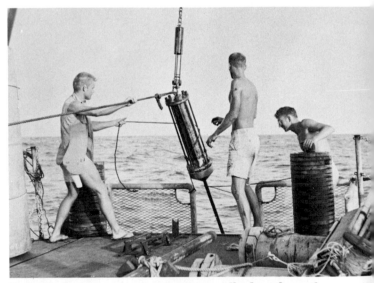

Probably because of vast convection cells deep beneath the ocean floor, the amount of heat reaching the earth's surface varies considerably. Scientists lower a temperature probe into the Pacific sea floor during the MidPac Expedition of 1950. (Scripps Institution of Oceanography, San Diego)

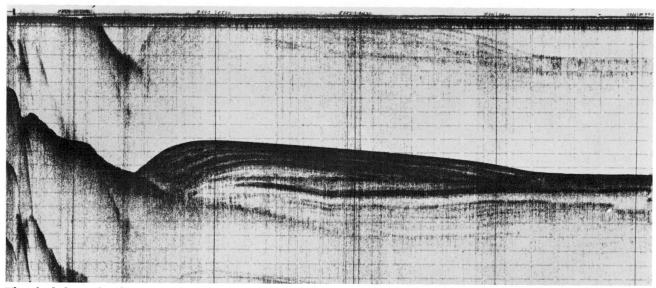

The thick lens of sediment in the Florida Strait east of Miami is revealed by seismic reflection profiling. The area shown is 10 miles across and 1,300 feet thick. (Richard J. Malloy)

the other hand, we can appreciate the importance of the mantle and, in turn, of marine geology, for in this pursuit we can learn about the earth's inner workings.

The land surface and the ocean floor are realms of contrast as unlike each other as each in turn is unlike the surface of the moon. The land has experienced strong erosive leveling, so that mountains have been reduced to nubbins. The land forms we usually see, whether they be folded mountains or volcanic blisters, have been largely etched out by erosion, and therefore they reflect the varying hardness of rock layers rather than the original features. Shielded from erosion, the ocean floor volcanoes, block-fault mountains, and rifts retain an almost pristine crispness compared to their land counterparts. Other major domains on the ocean floor—the trenches, the mid-ocean rises, and the abyssal plains—have no equivalents on the land. Even the volcanic seamounts attain a scale of grandeur that makes them quite unlike volcanoes we know on the continents. The land's folded mountains, like the Appalachians, appear to have no counterpart under the sea. Mountains there are, but of quite a different type, and not caused by accordion-like pleating of the crust. The ancient stable areas like the Canadian Shield also are unknown beneath the sea. So truly the abyss is a world apart.

The Continental Margins

The ocean floor is divided into two great domains, the marginal shallow seas and the deep-ocean floor. The continents seem to lie low in the water, partially submerged like so many hippopotamuses; their wet shoulders are termed the *continental shelves*. These fringe all the continents; nowhere do the margins plunge directly into the abyss. Typically the shelves are about fifty miles wide, but they may vary from as little as ten miles to as much as 800 miles—an extreme width found in the Arctic Ocean off Siberia. Geologically the shelves belong to the land—being simply extensions of the continental domain—and not to the oceanic realm. Their total area is enormous, comprising about eight percent of the world, or a region as large as the continent of Africa. When the United States recently proclaimed territorial rights to its adjacent continental shelves, it acquired a realm as large as the Louisiana Purchase.

Invaded by the shallow shelf seas, the continents are now twenty percent submerged, but this is drier than usual for geologic time generally. During the recent interglacial epochs of the Ice Age, when continental ice was largely melted, the sea rose high enough to cover another ten percent of the land surface. At other times in the far past, when the invertebrates were first evolving, North America was as much as sixty percent submerged. But during the peak of the recent glacial epoch, when ice covered most of Canada, the continental shelves were largely dry, and wooly mammoths wandered over these flat expanses. Archaeologists believe that many of the remnants of early man are now covered by the transgressing sea, so they must search the shelves to complete the record of man's past. The waxing and waning of the shallow seas has been one of the slow rhythms of the earth's history. And it has been in these shallow seas that the marine sediments that mantle our continents (limestones, sandstones, and shales) were laid down.

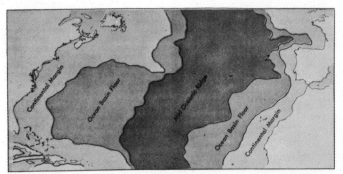

The deep areas of the North Atlantic lie between the shallow continental margins and the Mid-Oceanic Ridge, which forms a central area. (*McGraw-Hill Encyclopedia of Science and Technology*)

Well-defined terraces at Middleton Island, Gulf of Alaska, were cut by the surf in the recent geologic past. The island subsequently was uplifted in a series of steplike pulses, probably during great earthquakes. (S.R. Capps, U.S. Geological Survey)

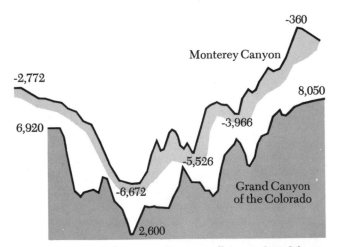

The Monterey Submarine Canyon off central California cuts an enormous chasm into the continental shelf. Its granite walls tower more than 6,000 feet above its winding axis, exceeding the depth of the Colorado River's Grand Canyon.

The outer edge of the continental shelf is marked by an abrupt brink, or shelf break, where the sea floor plunges in an unbroken fall of two or three miles to join the abyssal floor. The *continental slopes* are the longest, highest, and straightest boundary walls in the world. The lofty Himalayan rampart facing India is the only mountain front on the earth that attains the scale of a continental slope. If the waters were removed, the continents would appear much like huge mesas towering above the ocean floor.

The basic reason for the ocean slopes is evident. The continents are made up of light rock, which is not spread uniformly over the globe. The continental blocks are buoyantly supported by the earth's mantle so that they actually float. The continental slopes are the margins of iceberg-like continents.

Canyons Beneath the Sea

The continental slopes are the most imposing features of the earth, but their grandeur is not confined to their steepness and height, for they almost everywhere are dissected by vast clefts called submarine canyons. How and when they were carved no one can say for sure. If they were not sealed from view, these canyons would constitute the world's most spectacular scenery. Like those in mountains on land, the undersea canyons are deep, winding, V-shaped valleys cut in hard rock. The Monterey Submarine Canyon off central California is one such enormous chasm. Beginning but a stone's throw from shore at Moss Landing, it incises the shelf and plunges down the slope for 100 miles. Its granite walls tower 6,000 feet above its winding axis, exceeding the Grand Canyon in grandeur. Rich fishing grounds all, the deep submarine clefts have attracted the attention of fishermen, receiving picturesque names—"Swatch of No Ground" off the Indus River, or "Trou sans Fond" (Bottomless Hole) off Abidjan in West Africa.

These canyons, cutting the continental slopes, have bitten deeply into the minds of geologists, too, who have hotly contested their origin for decades. It was at first natural to explain these canyons as simply drowned river valleys, mainly because they often are found off the mouths of large rivers like the Columbia, Hudson, Congo, and Indus. It was argued that such rivers could easily have cut them when the oceans were drawn down during the Ice Age—a time when much water was locked up as continental ice caps. This seemed quite reasonable until surveys revealed that these canyons went much deeper than 400 feet below present-day sea level—generally considered the depth to which waters dropped during the Ice Age. Glaciologists were asked to revise upward their estimate of the thickness and extent of the continental ice caps. But this whole scheme collapsed when further surveys showed the canyons to continue down to the very base of the continental slopes—12,000 feet or more. Clearly the entire ocean could not have congealed on top of the continents. Submarine canyons with deep mouths are also found in the Mediterranean Sea and the Sea of Japan, which are isolated from the world ocean by shallow sills. If a great Ice Age withdrawal of sea level had occurred, these bodies of water would have remained isolated and perched above the world ocean, so the Ice Age scheme cannot account for these canyons.

One must look to the sea to explain things of the sea. The emerging view is that the canyons were almost wholly formed in Neptune's realm, with river cutting relegated to the minor role of modifying some canyon heads to a depth of 400 feet. Turbidity currents—swift torrents of mud—appear especially important. An earthquake, or some other stirring of the bottom, triggers a mud slide in a canyon head. The slide converts into a dense mixture of mud and water, and finally into a raging turbidity current, pouring down the canyon axis like a stream of mercury.

Unknown on land, and by their nature unseeable beneath the sea, the very existence of turbidity currents has had to be inferred by their effects. Through sad experience cable companies have learned the folly of laying their cables across submarine canyons. Every several years turbulent streams of mud carry away their cables precisely where they cross canyon axes and bury the snarled shreds far out at sea.

One of the most famous of mud slides occurred on November 18, 1929, when a great earthquake shook the continental slope off the Grand Banks south of Newfoundland, an area which has a remarkable concentration of transatlantic submarine telegraph cables. Like undersea avalanches, turbidity currents poured down the slope, snapping cables everywhere. At the time, the broken cables were thought to be casualties of the quake itself, but later studies by oceanographers of the Lamont Geological Observatory revealed a remarkable fact. From the evidence of the interrupted telegraphic services in London and New York, it was apparent that the cables near the initial shock point, or epicenter, broke immediately, while those farther downslope broke in a delayed sequence. For thirteen hours after the shock, cables farther and farther downslope con-

The rugged terrain in the submarine Oceanographer Canyon is viewed from one of *Alvin's* external 35-mm. cameras at 4,923 feet below the surface. The tools in the foreground are used for taking samples.

tinued to break one by one in a regular sequence until the last break took place 300 miles from the initial shock point. It was quite clear that this series of breaks was caused by a gigantic submarine mud flow as it cascaded downslope and onto the abyssal plain. Calculations as to the speed of the flow revealed that it attained speeds of up to fifty miles per hour on the steep portions of the slope and slowed down to fifteen miles per hour as it died out on the deep plain. Subsequent coring of the bottom has revealed that a layer of mud about three feet thick was laid down as the aftermath of this event.

In many parts of the world, piers have been built out to canyon heads to take advantage of the deep water that permits ships to approach close to shore. But the canyon heads are notoriously susceptible to landsliding. Piers have suddenly disappeared into the heads of the Abidjan Canyon off the Ivory Coast and the Hueneme and Monterey Canyons off California.

Since the advent of scuba diving, oceanographers have been able to descend and personally investigate the nearshore heads of submarine canyons like the La Jolla Canyon lying just off that resort city. I have enjoyed this experience frequently. Swimming over the sandy bottom at twenty fathoms, one suddenly encounters the brink of this sheer rocky chasm. Diving deeper, one sees vertical and even overhanging rock walls covered with heavy encrustations of life. Lobsters and crabs inhabit crevices, and myriads of fish dart about. An endless stream of bubbles rises out of the amphitheater-shaped head—seepage caused by erosion, which has cut deep enough to truncate strata rich in natural gas. The thought that invariably crossed my mind was: "This is no drowned and dead land canyon. Here is a domain of active erosion, even more so than in the land canyons in the cliffs behind the beach." Here and there are large blocks of sandstone marking recent rockfalls. Sand pouring into the canyon has cut smoothly scoured channels. On one dive I came upon a reflecting pool trapped in a scour-basin. My face reflected clearly in this strange quicksilver-like mirror, but when I stirred it, it proved to be a pocket of turbid water—the remnant of a turbidity current that had recently swept by. Such torrents of mud must frequently cascade down the axis of the canyon—cutting and eroding, and carrying sand and silt into the 4,000-foot-deep San Diego Trough, twenty miles out to sea. A great cone of sand and seaweed constantly pours into the canyon head, yet it never becomes filled. This mass of debris apparently slides down the canyon axis about once a year, making room for the head to trap more sand as it moves in from farther up the coast.

The presence of submarine canyons off river mouths is easily explained by the turbidity-current concept, since the river can be the source of the mud needed for propelling the current. This concept also better explains numerous canyons that have no relation to rivers. The Barrow Canyon is one of these; it lies immediately beneath the place where the North Alaska Coastal Current leaves the coast and strikes out to sea. Many other canyons, like the La Jolla Canyon, seem related to headlands. Littoral currents move sediments southward along the California shore until Point La Jolla forces the current, with its load of sediment, to turn out to sea. Each year about 100,000 cubic yards of sediment pour into the La Jolla Canyon; essentially all the new sand drifting into the area is siphoned down the canyon. It may strain credulity to believe that turbidity currents have cut such enormous gashes, but it is no more unbelievable than the cutting of the Grand Canyon by the running water of the Colorado River. Given time—and there are millions of years to spend—the tooth of running water is sharp.

The Mid-Oceanic Ridge

The continents and the ocean basins are the earth's first-order features. Second only to these fundamental earth levels is the Mid-Oceanic Ridge, the greatest mountain range on earth. It is a continuous feature that extends through the Arctic, Atlantic, Antarctic, Indian, and Pacific Oceans for a total distance of nearly 40,000 miles—or nearly 20,000 leagues under the sea. The ridge is a broad, low swell in the ocean floor, with an elevation above its surroundings of one or one and a half miles and with a width in most places of about 1,000 miles. It is a curious paradox that the continents are highest near their margins while the ocean basins are highest in their centers.

The Mid-Atlantic Ridge has been sketchily known for a century. Only recently have oceanographers realized that it is but one section of a continuous range of mountains extending around the entire world. Only in Iceland does the arched back of the Mid-Oceanic Ridge breach the surface. Elsewhere its crest is fully submerged, although occasional volcanoes lying along the ridge do pierce the surface to form such islands as the Azores, Ascension, Tristan da Cunha, Kerguelen, and Easter Island. This mountain range is unlike anything on land; it does not consist of folded strata but is rather a broad bulge in the earth's mantle.

Mountain ranges on land are created by compression and crustal shortening, but just the opposite seems true for the Mid-Oceanic Ridge. The crest of the ridge shows evidence of crustal stretching, consisting of a central rift valley, rift mountains, and a high fractured plateau. The median rift valley is the ridge's most striking feature, appearing in profile like a deep notch or cleft 1,000 fathoms deep. This central fossa is remarkably narrow, ranging between fifteen and thirty miles. The ridge descends in a series of steps, and farther out on the flanks one finds

A drawing from the *Challenger Reports* shows the historic brig moored at St. Paul's Rocks, one of the few places where the Mid-Atlantic Ridge pokes above the surface.

a chaotic topography of abyssal hills. There is little sedimentary cover mantling the ridge, indicating that it cannot be very old.

The central rift valley is an earthquake belt, and the effusion of much lava occurs there. This is seen clearly in Iceland, which exudes more lava than any other place on land—with the possible exception of the Hawaiian Islands. In 1964 and 1965 vast eruptions of hot molten rocks occurred there, forming the new island of Surtsey. Basalt (hardened lava) is the most prevalent type of rock dredged from the ridge, and underwater photos commonly reveal freshly frozen flows. It is clear that volcanic activity plays an important role in the construction of the ridge.

The term "mid-ocean" is something of a misnomer for the Pacific branch of the ridge, for it lies along the eastern margin of that basin, striking northward near the western margin of South America. It then enters the Gulf of California and seems to underrun western North America, apparently accounting for the overall uplift of the Rocky Mountains. There, it is not even an oceanic ridge at all. It would seem that the entire world ridge system is somehow tied up with the thermal regime of the earth's mantle and that it marks a hot zone where thermal expansion has taken place. Quite certainly the Atlantic Ocean is a rift ocean, which was split asunder 180 million years ago by the formation of this ridge. This would account for its mid-ocean position as the cicatrix, or leaf scar, along which the Atlantic continents parted company.

The Oceanic Trenches

The greatest depths on the face of the globe hold a special fascination. Early geographers supposed that these would be found in the middle of the ocean basin. Curiously, this quite logical presumption has turned out to be wrong. Virtually all the greatest depths lie along the margins of the Pacific, and not as ovoid depressions but as long, linear gashes termed trenches. The ocean basins seem to lie at depths ranging from 15,000 to 18,000 feet. Less than one percent of the ocean floor is deeper than 20,000 feet and virtually all of these "hadal" (that is, ultradeep) depths are confined to trenches.

Until the last decade the discovery of great deeps seemed to be an incidental byproduct of war. During World War I the German raider *Emdem* located a hole with a depth of about 34,000 feet off Mindanao in the Philippine Trench. This depth exceeded the Nero Deep off Guam by half a mile. Further oceanographic exploits by *Emdem* were cur-

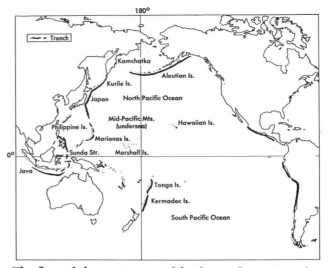

The floor of the sea is gouged by linear depressions that plunge almost seven miles down. Surprisingly, most of them are close to land, as shown by this map of the great trenches of the Pacific. (Edwin L. Hamilton, U.S. Navy Electronics Laboratory, San Diego)

tailed by her capture a short time later by the Allies at Clipperton Island, and the Emdem Deep remained as the deepest known spot in the world until World War II, when the U.S.S. *Cape Johnson* discovered an even greater depth, 34,430 feet, in the same trench. The discovery of the Cape Johnson Deep was not purely accidental, since the navigation officer aboard was the Princeton oceanographer Harry Hess, who was vitally interested in this quest. Hess commenced a detailed survey of the area, but this was quickly broken off by the sighting of Japanese torpedo bombers.

While the ashes of war were still smoldering, Britain, with her traditional zeal for scientific endeavor, in 1949 dispatched an oceanographic research ship on a round-the-world cruise that lasted three years. The ship was christened *Challenger II* after her famous predecessor. Her principal mission was to learn about the crustal layers beneath the sea by recording sub-bottom echoes from the firing of war-surplus antisubmarine depth charges, which provided a strong explosive sound source. An unexpected dividend was the discovery of the Challenger Deep —the nadir of the earth in the Mariana Trench not far from Guam. As British geophysicist Thomas Gaskell tells it:

As we approached the southern end of the Mariana Trench our Hughes echo sounder went off soundings— that is, the water was too deep to return an echo. Undismayed we hove to and tossed a 1¼ pound charge of TNT off the fantail. Using a hydrophone for the echo detector, we recorded the echo from the jarring explosion with our seismic apparatus. To our amazement the travel time for the echo was over 14¼ seconds and corresponded to a depth of 5,900 fathoms, nearly 1,000 feet deeper than the previously known greatest depth in the Philippine Trench. We spent the rest of the day carrying out one of our standard seismic experiments but that night returned to the deepest part of the trench to check our sounding with a sounding line of thin steel piano wire. At 1710 we let the reel run free, dragged down by a 140-pound iron weight. As though a marlin had struck, the reel spun crazily as the sinker dropped ever deeper— 1 mile down, 2, 3, 4, 5, 6 and finally 7 miles down! At 1840, or 1½ hours later, our weight had finally struck bottom. The depth reading was 5,944 fathoms. After making a 45 fathom correction for probable deviation of the wire from the vertical, we signaled the Admiralty that a depth of 5,899 fathoms had been attained only 50 miles east of where the old Challenger had found her deepest hole.

A decade later I was privileged to be over this same Challenger Deep aboard a heaving and wallowing destroyer, the U.S.S. *Lewis*. Our task was to send two men down to personally "sound" this deepest of all depths. Standing off and tethered to a fleet tug, the U.S.S. *Wandank*, was the bathyscaphe *Trieste*, ready to challenge the Challenger Deep. Like the *Challenger II*, we too were "off soundings." Even with the most careful tuning, our echo sounder was quite unable to bounce the slightest whisper of an echo from the distant bottom. As a last resort we undertook to create our own "ping." We made up homemade depth charges by taping together a fuse, a detonator, and TNT blocks, and we heaved them overside. Every few minutes a jarring explosion sent a plume of water into the sky—and a strong sound pulse racing toward the bottom. Faint echoes now came back, so for a day and a night a crude survey was made of the Challenger Deep until we delineated an area from which, by stopwatch measure-

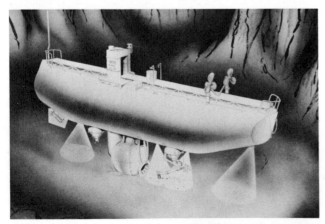

Built to withstand underwater pressures of eight tons per square inch, the bathyscaphe *Trieste* descended approximately 35,600 feet into the Challenger Deep in 1960. (U.S. Navy)

ment, it took more than fourteen seconds for the echo to return. This had to be the deepest hole in the deep. We could be satisfied with nothing less. We wanted the *Trieste* to hang up a record that would last for all time, but we knew also that it would be highly dangerous if the thin-skinned bathyscaphe struck the jagged side walls of the deep. Delaying the dive in these rough seas at this remote site entailed much risk, but the survey had to be carried to completion. Finally, at first light on the morning of January 21, 1960, flares and dye markers were spread on the precise dive spot. The *Trieste* was then towed in. Jacques Piccard and Lieutenant Donald Walsh put out in a small rubber dinghy, climbed aboard the wallowing *Trieste*, and at 8:30 A.M. began their five-hour descent to the nadir of the earth.

As of today, the Challenger Deep remains as the deepest hole in the deepest ocean, and a spot much deeper is not likely ever to be found. Robert Fisher of the Scripps Institution of Oceanography, who has made an especially careful study of the greatest depths, considers that, when all corrections have been made, the true depth of the Challenger Deep will be found to be approximately 5,933 fathoms, or 35,600 feet. Several different maximum soundings have resulted from subsequent probes of this area, but since the depression is flat-bottomed, Fisher's corrected depth is likely to stand. According to him the second deepest known spot is the Horizon Deep, discovered by Scripps' R.V. *Horizon* in the Tonga Trench, with a depth of 5,096 fathoms, or 35,440 feet. The third deepest is that discovered by the Soviet Union's *Vitiaz* in the Kurile Trench, with a depth of 34,584 feet. The Cape Johnson Deep in the Mindanao Trench now is only the fourth deepest spot, with its depth of 34,430 feet. All of these, and the axes of several other trenches as well, far exceed Mount Everest's 29,028-foot height.

Abyssal Plains

Until recently the ocean floor was commonly believed to be everywhere monotonously flat like a lake bottom floored with mud. This has been proved to be wrong; the ocean floor is generally more rugged than the land. But there are extensive linear or ovoid depressions covering about ten percent of the ocean floor that are exceedingly flat. These are the abyssal plains, which are completely devoid of irregularities other than an occasional seahill poking through the smooth bottom. Abyssal plains slope no more than one part in 1,000, or five feet per mile; generally they are even flatter, and so it is quite beyond the ability of the human eye to recognize them as being other than perfectly flat. Sediment cores and sub-bottom seismic reflection profiles (which give an "X-ray" vision of the layers beneath the bottom) reveal thick, flat-lying fills of sands and silts, which drown any older topography.

These sediments were apparently laid down by turbidity currents, as abyssal plains are always found in topographic localities favorable for receiving the deep-sea deposits of such currents, which cascade down submarine canyons cut into the continental slope. The settling of pelagic sediments—the dead husks of planktonic life that rain down from the surface water and were poetically described by Rachel Carson as an "endless snowfall"—apparently plays no more than a minor role. The Pacific sea floor lacks large, open abyssal plains, except in the Gulf of Alaska and in the peripheral seas, for this hemispheric basin is rimmed by a nearly continuous line of deep-sea trenches that act as traps for any detritus shed from the land. The DRV's (Deep Sea Research Vehicles) of the future will find smooth, if not monotonous, sailing across these vast flats in the deeper parts of the oceanic depressions.

Mountains Beneath the Sea

Undersea mountains, or *seamounts,* are mostly isolated features, and although distributed through all of the oceans, are most characteristic of the Pacific Ocean. There echo soundings have already revealed about a thousand seamounts higher than 3,000 feet (isolated rises of less relief than 3,000 feet are called "sea-knolls"). By the time this ocean is completely surveyed, ten thousand such seamounts probably will have been discovered.

The few Pacific seamounts whose summits form islands are mostly coral atolls. These are strung in festoons from northwest to southeast, like strings of pearls across the tropical Pacific. As hypothesized by Charles Darwin, and recently demonstrated by drilling and seismic studies, these atolls are great "coral" (actually mostly algal) caps, several thousands of feet thick, resting on extinct volcanoes as pedestals. Reef-forming corals, depending as they do upon the presence of algae for their very life—since the algae play a vital part in their metabolism and respiratory process—can only grow to depths that are illuminated by strong light (about 180 feet). Hence, coral reefs living on the top of several thousand feet of dead coral have obviously grown up on a platform that has subsided. Atolls are monuments to departed volcanic islands.

Not all Pacific islands are atolls. A few of them, like the Hawaiian and Society Islands, are the tops of great basaltic volcanoes built up from the seabed. Even nonscientist Herman Melville recognized this at Nuka Hiva in the Tuamotu Archipelago, or Low Islands, for in 1842 he wrote, "The origin of the Island of Nuka Hiva cannot be imputed to the coral insect secretions for indefatigable as that wonderful creature is, it would hardly be muscular enough to pile rocks one upon the other more than 3,000 feet above the level of the sea. That the land may have been thrown up by a submarine volcano is as possible as anything else." An impressive aspect of these sea-floor volcanoes is their enormous size as compared to those on land, suggesting a type of volcanism quite different from that with which we are commonly acquainted. Hawaii, for instance, has a relief of 32,024 feet above its base, making it by far the greatest single mountain on the earth's crust.

Nearly all seamounts are isolated peaks, oval in plan, having slopes of ten to twenty-two degrees—all strongly suggesting that they are volcanoes. To date, about fifty seamounts have actually been sampled by the scientist's rock dredge. Without exception they yield basic volcanic rocks, although those in the Gulf of Alaska also yield a few "continental" rocks apparently rafted out by icebergs of the past. The Pacific Basin appears, therefore, to be a vast volcanic province, confirming one of the broad generalizations made by the *Challenger* expedition scientists nearly one hundred years ago.

During World War II the surprising discovery was made that many seamounts, instead of rising to an abrupt peak, were truncated cones with extensive flat summits 300 to 1,000 fathoms beneath the sea. The original, roughly conic volcanic form had obviously been modified by erosional decapitation. But this cutting could not have happened at the present great depths of these summits, for although currents persist to all depths, they are not of a nature that could cause abrupt cutting of some particular level. Actually this truncation could only have occurred at the surface, where breakers and surf provide the sea's only "knife edge." These *guyots,* or *tablemounts* as they are sometimes called, must be deep-drowned ancient islands that for some reason have grown no coral caps to mark their graves.

This supposition was verified several years ago with some dredgings from the tops of five one-mile-deep tablemounts of the Mid-Pacific Mountains—a newly discovered, completely submerged range, stretching for 1,000 miles between Hawaii and Wake Island. Large hauls of olivine basalt were recovered, showing the general volcanic nature of these table-mounts. Some rounded cobbles were dredged up as well. The latter could only have been so smoothly fashioned by being rolled about in the surf at sea level. But the most significant result came from two of the tablemounts—the recovery of a good collection of fossil reef-forming corals, rudistids, and related atoll fauna. The recovery of the fossils, which could only have lived originally in water at less than 180 feet, was positive proof of great drowning. The fossils also enabled geologists to tell when the seamount was an island, because they included extinct middle-Upper Cretaceous forms, which lived about 100 million years ago.

Incidentally, these fossils are about the oldest ever recovered from the Pacific Ocean, or from any ocean for that matter. And, although now deeply drowned, the seamounts from which they were dredged nevertheless have a relief of nearly two miles, so that the Pacific must have been a deep ocean in the Cretaceous period. When they were islands, the Mid-Pacific seamounts stretching for a thousand miles across the central Pacific must have provided convenient "stepping-stones" for the migration of plants and animals over the ocean.

Although the explanation of how the great drowning could have taken place is unknown, there are many feasible conjectures. New, or juvenile, water is constantly being added to the oceans from the earth's interior, but at a very slow rate. Reasonable calculations suggest that only a few tens of feet could have been added to the sea's level since Cretaceous time by this mechanism. The shedding of sediment from the continents and its deposition in the ocean basins also causes a relative sea-level rise with respect to seamounts, but this too must be extremely small.

The belief that the guyots were decapitated by great oscillations of sea level, owing to water becoming locked up in icecaps or to some other cause, seems to be definitely ruled out, since the tops are not concordant. In other words, the seamounts are not all decapitated at some particular depth, nor does there seem to be any preferred depth. This lack of concordance argues strongly against the oscillation theory, for it would have been necessary for a great lowering of sea level to take place in order to decapitate the deepest seamounts, followed by a rise of sea level and the birth of new seamounts that were subsequently cut by another lowering not quite so great as the previous one, and so on. This is a most unlikely sequence of natural events.

Many bands of seamounts are situated on great linear rises of the sea floor—and there are cogent reasons to believe that these are comparatively transitory features of the crust rather than permanent deformations. It appears likely that such rises form above ascending thermal convection cells in the earth's mantle. These cells apparently are created in the earth's interior by heat released by the decay of radioactive substances. As sections of rock are heated, they expand and become relatively lighter than surrounding rock masses. As a result, the heated rock tends to rise, setting up an overturning motion like that in a slowly boiling pudding. Molten rock, or magma, ascends and breaks through the crust to create volcanoes, and because of the expansion, the crust is arched up as well. If the heat source is eventually withdrawn, the arch, of course, would subside once again to its original level. Thus, a volcanic seamount developing on one of these swells could have been given a temporary boost to the surface of the sea, only to be later drowned once again.

The principal reason for the subsidence is probably that the earth's crust is just not sufficiently strong to withstand the enormous load of a superposed volcano of the dimensions attained by many seamounts—far larger than land volcanoes. Mountain ranges "float" in the earth's mantle, and their high relief is due to the light weight of the rock of which they are composed; thus they are not real loads on the earth's crust. Volcanoes, on the other hand, are *real* loads on the crust imposed by the piling up of lava. The crust must either support them or fail. Volcanoes are as precariously situated as a small boy standing on thin ice, and they are in constant danger of foundering. Seismic studies show that the earth's crust is in fact depressed beneath some of the great seamounts. Also, some seamounts are surrounded by moatlike depressions—good evidence that our so-called terra firma actually has failed.

Another mystery concerns the comparatively youthful age of the seamounts. There are good reasons for believing that the Mid-Pacific Mountains are one of the oldest chains in the Pacific, but geologically speaking, their Cretaceous age makes them fairly youthful. The Cretaceous was the last period of the Mesozoic era, and this era spanned from 70 million to 225 million years ago. The question arises: What has become of the chains of volcanic seamounts that must have been formed in the early Mesozoic, the Paleozoic (between 225 million and 600 million years ago), or before that? Unlike mountains on land, seamounts are subjected only to extremely slow erosion, so that once formed they should persist through an enormous span of geologic time. When studying the Pacific, one finds that reasonable geological history can be pieced together subsequent to the beginning of the Cretaceous period, but that prior to that time the record is completely erased— a blank page. Possibly, the record has been wiped clean by drifting of continents through the sea floor, or by enormous outpourings of lava combined with crustal buckling. More likely the answer lies in *sea-floor spreading*, whereby the ocean bed acts like a conveyor belt, carrying seamounts along from the mid-ocean rises and down into the trenches, where they are destroyed as the sea floor descends to become a part of the mantle.

Two seamounts in the Gulf of Alaska provide some evidence of having been shifted toward the Aleutian Trench as though on a conveyor belt. One of these is a guyot now lying on the flank of the Aleutian Trench. Its flat top, quite remarkably, is tilted eight degrees toward the trench. It would appear that this guyot owes this tilting to its being carried toward the trench. Another guyot nearby is situated in the axis of the trench and appears about to be destroyed. This one now has the unusual summit depth of 1,400 fathoms, far deeper than is normal for guyots. Here, then, may be evidence for a slowly shifting sea floor.

Drifting Continents

In the past few years theories about the origin of ocean basins have undergone drastic changes. A new theoretical model has evolved to replace the "doctrine of permanency," which had prevailed since being laid down by a Yale professor, James Dwight Dana, in 1864. Briefly, that doctrine stated that "the continents and the ocean basins had their general outline or form defined in earlier geologic time." Sir John Murray, geologist of the *Challenger* expedition, wrote in 1891: "With some doubtful exceptions it has been impossible to recognize, in the rock of the continents, formations identical with those of pelagic [deep sea] deposits. . . . It seems doubtful if the deposits of the abyssal areas have in the past taken any part in the formation of the existing continents."

Alfred Wegener challenged this viewpoint in 1912 with his theory of continental drift, which hypothesized that the continents were once all together in a universal land mass he termed Pangaea. But Wegener was regarded by most scientists as a zealot, whose ideas were bizarre and unacceptable. One Chicago University professor regarded drift with such scorn that it is said he became incoherent for

several days if the concept was even mentioned in class. A prominent Stanford professor published a paper entitled "Continental Drift: A Fairy Tale"— hardly a suitable title for a scientific dissertation.

According to Wegener's theory, the Atlantic and Indian Oceans are new basins that did not exist prior to 150 million years ago (during the age of the dinosaurs), while the Pacific Ocean is the old ocean basin upon which the drifting continents have encroached. This does not mean that the Pacific sea floor is primitive, for it, too, has been renewed, even though the basin itself has not been destroyed in the process. Thus Wegener's theory is in accord with the later observation that no very old sediments are to be found anywhere in the deep ocean basins.

In fact, it now appears that Wegener was basically right. The continents, indeed, seem to have drifted to their present position as the result of an episode unique in earth history that commenced 150 million years ago. Earth scientists are now rapidly turning to what is termed the "new global tectonics," which integrates a group of concepts including sea-floor spreading, transform faulting, and the rigid-plate rotation of segments of the earth's 60-mile-thick outer shell. Continental drift, in turn, becomes a consequence of this new model of global evolution. According to this concept, the earth is divided into about eight major, rigid sectors, which are bounded by the mid-ocean rift, the oceanic trenches, major mountain chains, and great shear zones. These plates are separating along the mid-ocean rift zone, and new mantle-rock is rising into the gap to create new crust. But this does not cause the earth to grow larger, for an equal amount of crust is destroyed by descending into the trenches and returning toward the earth's deep interior. The earth's crust seems to be broken up into a group of conveyor belts, which are ultimately mobilized by the thermal overturn or convection within the earth's mantle. The tectonic model is an improvement upon Wegener's theory, since it does not require that the continents plow their way through the earth's mantle—a movement which, as Wegener's critics were quick to suggest, is actually impossible mechanically. Instead, the new theory holds that the continents are carried along passively like so many icebergs, being moved by the currents beneath them.

The study of rock magnetism has been critical in the proof of continental drift. When volcanic rocks solidify, they take on the earth's ambient magnetic field, so that they in effect become fossil compasses. By studying such rocks scientists have been able to determine that continents are not in the same

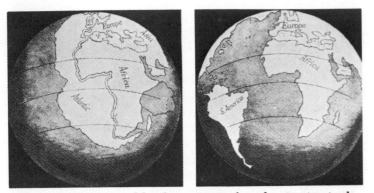

Although only accepted by the majority of earth scientists in the past few years, the concept of continental drift is not new. Reproduced here is the first depiction of the drift opening of the Atlantic Ocean by A. Snider in 1859 in *The Creation and Its Mysteries Explained*. His bizarre conclusion was that the New World was really Atlantis, which had been split from the Old World by the biblical flood. (Thomas Nelson and Sons, Ltd., London; the Ronald Press, Co., N.Y.)

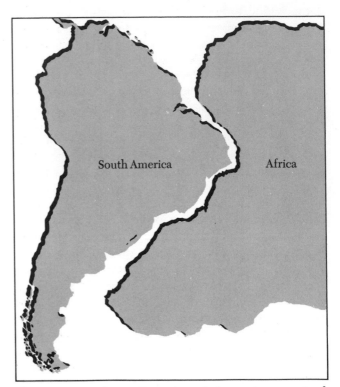

positions they were formerly. It is not possible to tell precisely the original positions of the continents, since the magnetic imprints reveal only the latitude and orientation of the rocks at the time they solidified. As with any compass, longitude remains indeterminate. The problem is the same as attempting to discover a ship's position prior to the invention of the chronometer—and it has no solution. The magnetic

The jigsaw-puzzle configuration of the eastern coast of South America with the western coast of Africa lends credence to the theory that the land masses were once joined.

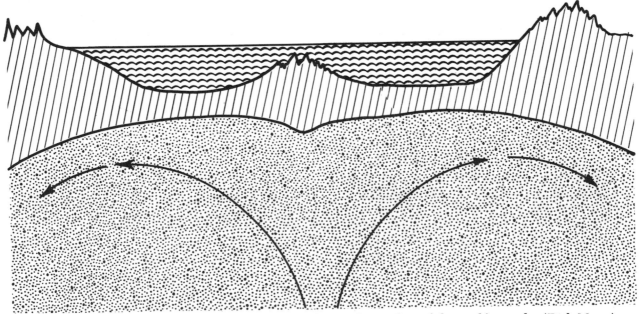

Continents on the move are caused by the upwelling and spreading of the earth's mantle. (Dick Marra)

studies, however, show that continents certainly have shifted; and the reconstruction of the land masses into the single universal continent of Pangaea, as suggested by Wegener, or into two supercontinents of Laurasia (a Northern Hemisphere group) and Gondwana (a Southern Hemisphere group) is now virtually certain.

The studies of rock magnetism have also proved that new oceanic crust is being generated constantly along the mid-ocean ridges, further supporting the tectonic model. Because the earth's poles reverse every half million years or so, with south becoming north and vice versa, the polarity of the rock on the ocean floor varies according to when the lava solidified. The effect is to create a series of readily detectable magnetic anomalies so that a magnetic chart of the ocean floor takes on a zebra-like pattern. With this information, it is possible to show that new zones of crust a thousand miles or more wide have been added to the ocean basins in the past 100 million years. It is easier, of course, to show that something new has appeared on earth than that something has disappeared. However, evidence is growing which suggests that an equal amount of old crust is disappearing into the trench zones of the world, which are located mainly around the Pacific Ocean.

With the establishment of the theory of continental drift, the doctrine of permanency has been violated in many respects but not all. For example, while the continents have changed in outline and position, their area has been roughly preserved over the aeons. Also, despite drift, the continents have maintained *vertical permanency*. Continents have remained as high blocks, and the ocean basins have remained as low depositional basins throughout earth's history. The continents have been repeatedly transgressed by shallow-shelf seas a few tens or hundreds of feet deep, but never covered or drowned by deep oceanic waters.

The Last Frontier

Although the oceanic depths are awesome to mere fathom-high man, on the geographer's globe they are no thicker than the layer of blue printer's ink that marks their presence. In these days of advanced technology it seems anachronous for the deep-sea floor to remain an unknown and shadowy frontier. The world today is experiencing an information explosion that applies to oceanographic data as much as anything else. More oceanographers and more oceanographic vessels ply the seas than ever before. Even so, the task of surveying the oceans is enormous. The extent of their area exceeds their depth a thousandfold, and it will take several decades to develop anything but crude maps of their depths. Today, when all mountains have been climbed and even the icy wastes of Antarctica have yielded their secrets, it is refreshing to know that the ocean floor remains as an endless frontier for human challenge and exploration in the heroic tradition.

3 BIOLOGY OF THE

by Charles E. Lane

The sea is the only environment on earth that provides all the chemical ingredients of the most primitive living organisms. Of course nothing definite can be known about the events that culminated in the first assemblage of the different components of living matter, but recent research offers some insight into possible mechanisms. In 1953 biochemist Stanley Miller, and, somewhat later, Sidney Fox, reproduced in the laboratory most of the chemical conditions that must have been present in ancient seas and then flashed experimental bolts of lightning through their flasks. After a time the treated solution contained new, complex chemical substances such as amino acids, the essential building blocks out of which proteins are constructed, and nucleic acids, the chemical substances of heredity that insure that modern animals and plants resemble their parents. These experiments strongly suggest that life must have originated in the shallows of warm primitive seas, perhaps aided by bolts of lightning from the turbulent atmosphere.

The seas offer several advantages for animal life that are not shared by other life zones. Among these are uniform concentrations of major chemical elements, relatively stable temperature, protection from excessive ultraviolet radiation, and uniform reduction of light intensity with depth. Large-scale current systems mix the mass of water, making all the oceans of the world similar in composition. Seawater contains a wealth of dissolved substances that provide marine animals with the ingredients for shells, skeletons, protective coatings, and general life processes. The slight alkalinity of seawater, combined with its other attributes, creates an optimal environment for living matter.

All major groups in the animal kingdom are represented in the seas. Certain of these, such as the starfishes and their kin, contain only marine representatives. These animals have never developed the capacity to cope with an environment as variable as fresh water, or even the brackish water in estuaries. Other kinds of animals, such as sponges and jellyfishes, contain mostly marine forms, with occasional representatives inhabiting fresh or brackish water. Still other major groups, such as that including the crabs, crayfishes, and shrimps, and that embracing the relatives of the oysters and snails, include animals with a considerable capacity to adjust to different environments. These are represented not only in the sea, but in fresh and brackish water and even on land.

The First Fishermen

The earliest human remains found in Africa suggest that the human species probably originated far from the sea. Man became a social being with sizable permanent settlements in the Middle East. During these earliest stages of cultural evolution (about 650,000 to about 8000 B.C.) all the basic tools to be used in the future were developed. Man had partly solved the problem of climate with clothing and fire. He perfected the barbed spear point and the bow and arrow, and learned to catch fish with the harpoon and the gorge. The gorge was a stick pointed on both ends, with a line attached about its middle. It was implanted lengthwise in bait. When it was swallowed by a fish, a pull on the line turned the gorge, perforating the gut and allowing the easy capture of the victim. Just before the dawn of the Neolithic, or New Stone, Age man developed the recurved, single-pointed fishhook made of bone. By 5000 B.C. early man was using copper pins for fastening his clothing and copper hooks for catching fish. These findings suggest that he was familiar with the use of fish for food. His familiarity with other edible aquatic animals is shown by the heaps of freshwater clam shells found among the kitchen middens that mark the location of temporary settlements of ancient

SEA

peoples along the shores of inland lakes and streams. Thus, when the hunting travels of early man brought him to the seashore he was already prepared to exploit some of its resources.

Man's first experience with marine organisms was probably limited to edible shallow-water species that could be easily collected. Thus, sessile or slow-moving invertebrates such as mussels, clams, oysters, crabs, lobsters, and their relatives were the first sea animals to be exploited by man. In coastal regions simple but effective fishhooks were fashioned from bone or shell. Somewhat later, as this primitive technology developed, the abundant shore fisheries became accessible. Contemporaneously with his early tentative efforts at fishing for sustenance, man must have acquired, through painful experience, firsthand information about certain kinds of animals that were either poisonous, venomous, or otherwise unsuitable as food. On the tomb of the Egyptian Pharaoh Ti, Fifth Dynasty, who lived nearly 5,000 years ago, there is an inscription that shows the poisonous puffer-fish, *Tetraodon stellatus*, together with a hieroglyphic description warning of its toxicity. This fish, and some near relatives, occur in the eastern Mediterranean and in the Red Sea. Somewhat later—in the second century B.C.—similar information was published in the first Chinese pharmacopoeia, *The Book of Herbs*. Only in 1968—thousands of years later—have biochemists and marine biologists finally determined the nature of the puffer's extremely virulent poison (tetrodotoxin) and described its mode of action.

Commercial exploitation of the sea for trade and conquest was first highly developed in the Mediterranean. The small size of boats, and lack of navigational equipment that could provide guidance at night, or out of sight of land, limited early sailors and merchantmen to coastal and inshore waters, so there was very little opportunity to investigate the animal life of the sea.

The Origins of Marine Biology

Development of stable societies in the eastern Mediterranean provided man both leisure and opportunity to examine the nature of his environment. Xenophanes, about 500 years before Christ, noticed fossilized marine animals high up in the mountains of southern Italy and concluded that the mountains must have been underwater at one time. By the time of Aristotle (384–322 B.C.) there was sufficient familiarity with littoral marine animals to permit him to attempt a classification. His writings include descriptions of about 520 species—over a third of them marine forms—recognized by present-day zoology. These forms all occur in Greece and its seas. Marine fauna interested him almost as much as land animals; fishes, bivalves, and crabs are better represented in his works than are the common land animals. Aristotle is chiefly remembered for his philosophical writings, but his boundless curiosity and great energy led him to study the gills of fishes, which he recognized as their breathing apparatus. His detailed description of the anatomy of the cuttlefish suggests that he dissected one or more specimens. His curiosity also impelled him to make detailed observations on the eggs of animals and their development. For his studies of marine animals, Aristotle might well qualify as the first marine biologist.

Gaius Plinius Secundus (Pliny the Elder), who is remembered as one of the foremost natural philosophers of ancient Rome, was born in A.D. 23. Principally from the works of others he assembled lists of animals, together with comments on their distribution and use to man. These are found in his only treatise that has survived, the 37-volume *Natural History*. Understandably, his lists are heavily weighted in favor of the large exotic land animals, such as the elephant, the aurochs, and the lion. He did, however, include common cold-blooded land animals, as well

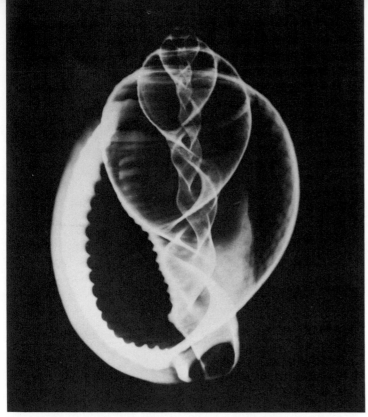

The dissolved substances in seawater provide marine animals with the ingredients for shells, skeletons, and protective coatings. An X ray of the marine snail *Phallium granulatum* reveals the considerable internal complexity of the shell. (Charles E. Lane)

All major groups in the animal kingdom are represented in the sea. Certain of these, such as the brittle starfishes, are found only in salt water. (Charles E. Lane)

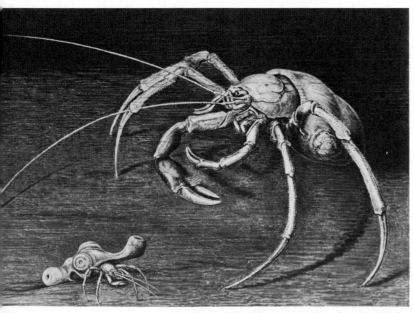

Some major groups, including the exotic deep-sea hermit crabs shown here, have the capacity to adapt to different environments. They are found in all types of water and even on land. (Paul Winther, *Galathea* Expedition)

as fishes and bivalve mollusks among marine forms.

During the Middle Ages in Europe the spirit of inquiry was oddly suppressed. Scholarship was the domain of the Church. Curiosity about biological matters was largely satisfied by reference to the written works of the ancients—Hippocrates, Plato, Pliny, and Galen. These authors were principally concerned with medical matters, giving only incidental comments on marine organisms. Arabian philosophers, during this same period, devoted most of their writings to interpretation and explanation of the works of the ancients. No systematic study of marine organisms was attempted before the early years of the nineteenth century.

During the late eighteenth and early nineteenth century many scientific disciplines burgeoned. Geographic exploration extended over the entire globe after the pioneering voyages of Captain James Cook and Alexander von Humboldt. Biology shared in this splendid flowering through the work of such men as Jean Lamarck and Georges Cuvier, the celebrated French comparative anatomists; the Austrian monk Gregor Mendel, whose pioneering studies initiated the science of genetics; Charles Darwin; Alfred Wallace, an English biologist whose astute observations supported Darwinian theories of evolution; and Thomas Henry Huxley, an early defender of Darwinian theories.

The appearance of Darwin's *Origin of Species* in 1859 generated a philosophical revolution that pervaded most scientific disciplines, affected social theory, and generated discord within the Church of England. There have been repercussions even within twentieth-century religious organizations in the United States. The publication of this work stimulated scientists in many countries to renewed efforts to understand the nature of living things, both in the sea and in the laboratory. Before the manifold adaptations of animals and plants normally living in the oceans could be properly appreciated, however, it was essential to understand the physical and chemical characteristics of the seas.

In the early years of the nineteenth century, most scientists agreed with Edward Forbes' theory that plants and animals could not survive in the numbing cold and stygian darkness of the ocean depths. Evidence to the contrary, however, gradually accumulated. In 1858 a transatlantic telegraph cable linking

---

Although usually found in the sea, sponges and jellyfishes will occasionally inhabit fresh or brackish water. This Portuguese man-of-war dangles its long, stinging tentacles beneath a brightly colored float. The tentacles sometimes extend to 100 feet. (Charles E. Lane)

Charles Darwin's five-year voyage on the H.M.S. *Beagle* provided many of the raw materials for his *Origin of Species*. This monumental work, published in 1859, stimulated scientists to renewed efforts in understanding the nature of living things.

Professor Thomas Henry Huxley, who defended and popularized Darwin's theories, was pictured in the March 1881 issue of *Punch* riding the back of a fish.

the United States and England was completed. The cable route traversed regions of the Atlantic more than a mile deep. After a brief service life, the cable failed and had to be retrieved for repair. When it was brought to the surface, it was found to be covered with creatures that no one had ever seen before. Clearly these animals had flourished at depths far greater than Forbes had thought possible. This disquieting event, together with other deep-sea finds by English and Norwegian scientists, stimulated investigators in several countries to undertake a series of dredging efforts to determine whether other regions of the deep-ocean floor also supported animal life.

In 1868, the *Lightning* dredged between Scotland and the Faroe Islands at depths of 3,000 feet and found not only primitive, simply organized animals but more highly developed creatures, including even some bony fishes. The success of this cruise convinced the British Admiralty of the desirability of continuing these explorations with the *Porcupine,* a ship that was better equipped for deep-sea study than was the *Lightning.* On cruises during the summers of 1869 and 1870, the *Porcupine* dredged many animals from depths ranging to 14,610 feet, or more than two and a half miles, including representatives of species that were unknown up to that time.

Wyville Thomson, co-leader with Dr. W. B. Carpenter of the *Lightning* and *Porcupine* expeditions, also collected samples of the myriad small floating organisms, many invisible to the naked eye, that were later to be known as "plankton." This term was first used by Victor Hensen in 1887 to include all those wanderers of the oceans—animals and plants—that float or drift passively in the sea. Prior to this time, biologists had devoted very little attention to these organisms, which are the initial links in the complex food chain of the oceans. During the past century the nutritional interrelationships of marine organisms have been clarified by the studies of marine biologists. Tiny plants, often of only microscopic dimensions, trap some of the abundant energy of sunlight when they make carbohydrates in the process called photosynthesis. Starches and sugars resulting from this process are then available to animals. Marine plants, like their terrestrial relatives, are the ultimate synthesizers of amino acids, the building blocks out of which proteins are constructed. In general these cannot be made by animals. Certain fatty acids that occur abundantly in the fats of bony fishes are also manufactured by planktonic plants. A continuous cyclic transfer of matter and energy from microscopic plants through smaller to larger animals characterizes the complicated food web of the oceans. These exchanges are effected as grazing animals feed on the plants and are, in their turn, preyed upon by larger animals. On the death of the ultimate predators the energy-rich materials of their bodies are returned to the microorganisms of the seas to begin a new rotation of the food cycle.

Largely based upon the results of his expeditions on the *Lightning* and the *Porcupine,* Wyville Thom-

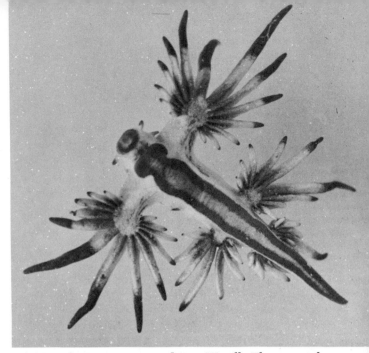

While on the *Porcupine* expedition, Wyville Thomson collected samples of the myriad small, floating organisms that were later known as "plankton." These tiny organisms, many invisible to the naked eye, were discovered to be the early initial links in the complex food chain of the oceans. (Charles E. Lane)

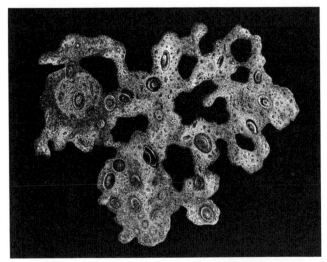

Wyville Thomson described what he thought was a primitive living creature, "Bathybius," as the "mother of protoplasm." This supposed deep-sea creature, illustrated in *The Depths of the Sea,* had no organs but seemed to react to stimuli and be capable of assimilating food.

son wrote *The Depths of the Sea,* which proved to be immediately popular. This book, still widely read, may be regarded as the first textbook of oceanography in English.

As a consequence of Thomson's results and the wide interest generated by his book, the British Admiralty organized the great *Challenger* expedition

In a three-and-a-half-year voyage, H.M.S. *Challenger* brought back more information about the seas than had been amassed in the previous eighteen hundred years. This drawing of the contents of the dredge being emptied appeared in the fifty-volume *Challenger Reports*.

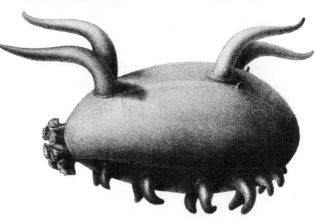

The abyssal sea cucumber *Scotoplanes globosa*, discovered by the *Challenger* expedition, probably spends most of its time buried in the mud. It has been caught in the Philippines Trench at depths of 32,800 feet.

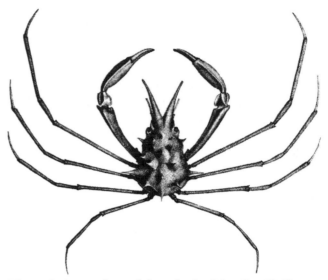

The crab *Anamathia pulchra*, dredged by the *Challenger* from 2,250 feet of water off the Philippine Islands, has long legs that help it stride over the soft sediments of the deep sea.

to study "every phase of the sea." The itinerary of the *Challenger's* three-and-one-half-year voyage included Portugal, Spain, the Azores, the Cape Verde Islands, Madeira, the West Indies, Bermuda, South America, the Straits of Magellan, Cape Horn, Australia, New Zealand, the Indian Ocean, New Britain, Japan, the Aleutians, Vancouver Island, and California. Before she returned to England in May 1876, the *Challenger* was to go back and forth across the Atlantic, Pacific, Indian, and Antarctic Oceans. Her homecoming was properly triumphant, for she brought back more information about the seas than had been learned in the previous eighteen hundred years.

Later expeditions, with more modern equipment and better ships, could only add to and amplify the nascent discipline of the biology of the sea that the *Challenger* expedition had launched. Other workers were to discover many new and exciting animals; indeed an entirely new major group of marine animals, the Pogonophora, has been described during the past decade. Knowledge of the physics and chemistry of the sea would be continuously expanded by later workers. But the small band of men on the *Challenger* established the broad foundation of our knowledge of the oceans and the creatures that live in them; they founded a new scientific discipline. Among their pioneering discoveries may be mentioned the unexpectedly uniform chemical composition of the seawater around the globe, the absence of regions in the sea devoid of life, the presence of some form of animal life at all depths, and an astounding array of adaptations that permit marine

animals to colonize even the most unpromising regions of the sea.

The Voyages of Agassiz

In 1871, while Wyville Thomson was assembling the results of deep-dredging operations of the *Porcupine* and *Lightning,* he was visited in Edinburgh by Alexander Agassiz, a young staff member of the Museum of Comparative Zoology at Harvard. The son of Louis Agassiz, Alexander was a brilliant young man who had been admitted to Harvard at the age of fifteen. After graduating in 1855, he took a degree in engineering, and he later served at various times as president of a coal-mining concern in Pennsylvania and of the Calumet and Hecla mining operation in northern Michigan—all the while holding a part-time position at the Museum of Comparative Zoology, which had been founded by his father. Alexander Agassiz had maintained a lively interest in biology from his earliest youth and had missed no opportunity to collect and study marine animals accessible to him. His association with Wyville Thomson was to be a means of sustaining his preoccupation with marine animals during his long and productive life. On his return to the United States he sought every opportunity to participate in biological oceanography. He was shortly afterward invited by the director of the United States Coast and Geodetic Survey to supervise the dredging operations conducted by the survey steamer *Blake*. His first cruise, in 1877, was to the waters surrounding Cuba, the Yucatán Banks, Key West, and the Dry Tortugas. Agassiz's training in mining engineering stood him in good stead when the dredging equipment of the *Blake* required modification. He also redesigned some of the trawls and substituted wire cable for the hemp rope that had been used for dredging and deep sounding on the *Challenger*. During two subsequent summers, Agassiz directed the scientific operations of the *Blake,* returning to his post at Harvard between cruises.

It was during the second cruise of the *Blake* under his direction that Agassiz was dramatically reminded of the extreme pressure and low temperature prevailing at great depths in the ocean. On a particularly bright, hot day in the tropics, after working on deck to examine the contents of a deep dredge, he was struck by the refreshing chill of the animals in the dredge. To derive a practical benefit from this observation, he had a bottle of champagne lowered to 14,400 feet for one hour. When it was re-

trieved, the bottle and its contents were suitably chilled, but the cork had been driven directly into the bottle and the champagne was now mostly cold seawater. At this depth the weight of the water column exerts a pressure of more than three tons on each square inch of surface.

During his second season on the *Blake* Agassiz collected animals from two hundred separate localities and dredged animals from all depths between 600 and 14,000 feet. He found that his richest hauls, i.e., the greatest abundance of animal material, came from 1,200 to 12,000 feet, while the largest number of different species was found between 1,800 and 6,000 feet. His collections were so extensive, when added to those of the previous year, that for many animal groups his material was almost as rich as that from the *Challenger*. Indeed, Agassiz wrote to Wyville Thomson to ask whether Thomson would like some of his material to supplement what had already been collected by the *Challenger*.

In the summer of 1880 Agassiz was engaged in still a third cruise aboard the *Blake*. This time he probed the deeper waters of the Gulf Stream, a portion of the Atlantic Ocean that had not been investigated by the *Challenger*. Agassiz was impressed by the similarity between the fauna of the West Indies and that of the Pacific coast of Central and South America. He concluded from this that at one time the Caribbean Sea must have been an arm of the Pacific Ocean, or at least closely connected with it. He noted the scanty animal life in the area of the Gulf Stream and attributed this in part to the swift current of the stream that carried everything along its course off the bottom. Today we know that the Gulf Stream is a relatively shallow system, occupying only the upper layers of the water column.

Agassiz called attention to the color of animals from different depths in the sea, with brilliant colors predominating in the surface layers—the blues, greens, and yellows being replaced by blacks and deep-red pigmentation in deeper waters. He related this to the absorption of different wavelengths of light at different depths in the water column. He said:

In contrast to the brilliantly colored animals of the shallow water the deep sea floor is a monotony only relieved by dead carcasses of animals which find their way from the surface to the bottom, and which supply the principal food for the scanty fauna living there. . . . It requires but little imagination to notice the contrasts as we pass from shallow water regions of the sea, full of sunlight and movement and teeming with animal and plant life, to the dimly lighted but richly populated Continental zone, and further to imagine the gradual decrease of this

While conducting voyages aboard the steamer *Blake* (above), Alexander Agassiz invented an improved dredge that would work equally well regardless of how it fell on the bottom. He was the first to use steel cables for dredging. (U.S. Coast and Geodetic Survey)

Continental zone as it fades into the calm, cold, dark, and nearly deserted abyssal regions of the oceanic floor at distances from the continents.

In 1887 Agassiz was placed in charge of a deep-sea expedition, sponsored by the United States Commission of Fish and Fisheries, to the Pacific side of the Isthmus of Panama on the new survey ship *Albatross*. On this and subsequent cruises of the same ship, he made extensive collections of Pacific fauna from regions that had not been previously studied. His Pacific studies renewed his interest in the problems of the distribution of reef-building corals and in the growth of coral reefs—an interest that was to occupy much of his attention for the rest of his life.

Coral reefs occur in all the oceans of the world where the water temperature remains above 70° F. the year round. These massive limestone structures are constructed by tiny individual animals that resemble, and indeed are related to, sea anemones. During the voyage of the *Beagle*, Darwin studied these structures in the Pacific and observed that coral animals grew on the sloping shores of oceanic islands down to depths of about 180 feet. He was trained as a geologist and was familiar with the fact that the surface of the earth has suffered many changes—now rising in mountain-building episodes, at other times subsiding. He theorized that the rate of subsidence of islands might be matched by the upward growth of coral reefs, a process that would ultimately produce the familiar Pacific atoll. Agassiz disagreed with Darwin's subsidence theory of coral-reef formation, contending that local elevation of the sea floor, caused perhaps by volcanic eruptions or movements of the earth's crust, created the platforms on which reef-building corals could grow. While Agassiz was right to the extent that the atolls had formed on the tops of extinct volcanoes, Darwin's theory that the reefs grew upward as the islands subsided appears today to be the more nearly correct explanation.

Exploring Arctic Waters

The *Challenger* expedition, the voyages of the *Blake* and *Albatross*, the German *Valdivia* expedition, and the series of collecting voyages made by the prince of Monaco on his succession of oceangoing yachts provided basic information on the biology of the middle latitudes of the world's oceans. The higher arctic and antarctic latitudes were studied by two remarkable scientists, Fridtjof Nansen and Alister Hardy.

Nansen, a Norwegian sportsman and zoologist, was inured to cold and hardship by his early training

Alexander Agassiz watches the crew pull in the deep-sea trawl on the *Albatross*. From this survey ship Agassiz made extensive collections of Pacific fauna. (Sir William Herdman, *Founders of Oceanography*. London: Edward Arnold & Co.)

Coral reefs, constructed of limestone by tiny individual animals, occur in all oceans of the world where the year-round temperature remains above 70° F. A coral skeleton is shown here. (Charles E. Lane)

and field studies in his native northland. Soon after winning his doctorate, he was able to implement an ambition that originated early in his career—to study the current systems of the polar seas. The North Magnetic Pole, which had never been reached by man, appeared to Nansen to be attainable if he could design a vessel that would survive the enormous stresses she would suffer while being frozen in the polar ice.

49

He proposed to enter the ice where the winds and currents would drift the ship across the top of the world from east to west. He expected to pass over the North Pole, or very close to it. For this purpose he designed the *Fram* (meaning "to push forward"), a smoothly rounded hull 128 feet long, whose stout framing timbers and resilient planking together were twenty-four to twenty-eight inches thick. The crew consisted of thirteen carefully selected, healthy, compatible men, experienced in the Arctic.

On June 23, 1893, the *Fram* expedition departed from Christiania (now Oslo). The ship was frozen in the ice for thirty-five months, during which slow and erratic polar currents moved the ice in an average drift of less than a mile per day. Often the drift was in the wrong direction, i.e., away from rather than toward the magnetic pole. During their long imprisonment in the ice, the crew conducted regular oceanographic and meteorologic observations, recording air and water temperatures, determining the salinity of the water at various depths, determining the depth by deep sounding through holes in the ice, and making regular collections of plankton and other creatures. These observations showed that the North Polar Sea, contrary to beliefs then current, was a deep body of water. Some soundings showed depths of 12,000 feet, or more than two miles. The expedition also documented for the first time the extraordinary plankton bloom of early summer in the Arctic, which feeds one of the world's greatest fisheries. Nansen correctly attributed this explosion of growth to the presence of abundant nutrients.

Perhaps the most impressive aspect of the *Fram* saga was the result of Nansen's impatience with the dreary pace of the drifting ship. Together with a single companion, Hjalmar Johansen, Nansen left the *Fram*, still locked in the ice nearly five hundred miles from the Pole, and set off by dog sled across the ice, carrying food and supplies for a journey he estimated would take fifty days.

During the early days of their travels the average daily temperature was −45° F. Although their heavy clothing protected them from the cold, it trapped sweat resulting from their exertions. This quickly froze, imposing an additional burden. Sleeping bags froze during the day, only thawing briefly at night when the weary explorers sought shelter in them. These unexpected problems reduced the daily distance traveled from the planned twenty miles to two or three. After four and a half weeks, Nansen's position-fix showed that he was still more than two hundred miles from the North Magnetic Pole. The men were exhausted; some of their dogs had died. They

Norwegian explorer Fridtjof Nansen studied the movement of polar ice aboard the 128-foot *Fram*. While the vessel was locked in the ice for thirty-five months, the crew documented for the first time the early summer plankton bloom. (Norwegian Embassy Information Service)

had seen no bear, walrus, or any other edible game animal, and their food supply was running low. So Nansen decided to turn back. Four months later, in early August 1895, they stumbled ashore on Franz Josef Land. Winter snows and darkness were already upon them, so Nansen and his companion built a rude hut on the shore in which they spent the long winter of 1895–96, subsisting on seals and walrus. In June 1896 they were discovered by members of the British Jackson-Harmsworth polar expedition and returned to Norway. Meanwhile the crew on the *Fram* had managed to free the ship from the ice and return her to Norway, thus ending what must be one of the monumental oceanographic expeditions of all time.

Nansen continued to be active in oceanography until World War I, after which he turned to politics and problems of refugee resettlement. For these efforts he was awarded the Nobel Peace Prize in 1923. He died in 1930 at the age of sixty-nine.

Commander Robert E. Peary of the U.S. Navy finally reached the North Magnetic Pole on April 6, 1909, leaving the South Magnetic Pole the last of the earth's unexplored major geographic features. Nansen's ship, the *Fram*, carried the Norwegian explorer Roald Amundsen to the antarctic continent on his successful south polar expedition—surely a testament to the soundness of her design and the skill of her builders. Amundsen reached the South Pole on December 14, 1911.

The Biology of the Whale

Nansen's achievements in clarifying the biological oceanography of the north polar seas were matched by those of Alister Hardy for high southern latitudes. Hardy's contributions arose from his studies of the biology of the whale and of its principal food organisms.

Whales, the largest animals in the world, have challenged man's skill and ingenuity, and aroused his cupidity, since earliest times. They were originally sought primarily for their oil, which was used for heat and light, as an essential ingredient in paints, and in soap. During World War I there was an enormous demand for whale oil as a source of the glycerol used in the manufacture of explosives. Readily accessible stocks of whales were rapidly decimated, and whale hunters were forced to journey farther and farther for their prey, finally beginning to exploit even the great antarctic stocks.

The evolution of factory ships on which all the processing steps could be conducted, from retrieval of the whales killed by the catch boats to packaging of the final products of commerce—whale oil, baleen, pet food, and pharmaceuticals—greatly accelerated the decline in worldwide whale stocks. The British government, in an effort to place the harvest of whales on a sound biological basis, organized a systematic study of the antarctic whaling industry and placed the young zoologist Alister Hardy in charge of its biological aspects. His assignment included study of the reproduction, growth, and nutrition of the whales, and the optimal oceanographic conditions for their survival. A five-year study was planned, and the three-masted wooden steam bark *Discovery* was bought for this purpose. The ship was 198 feet in length and specially reinforced for service in the antarctic ice. It had been built shortly after the turn of the century and had been used as a base by the unsuccessful Scott-Shackleton South Pole expedition in 1901–1903.

For its service in the South Atlantic and the Antarctic, the *Discovery* was extensively rebuilt in 1923 and equipped with the most refined oceanographic and biological collecting gear available. It was early recognized that the study should be conducted along two general lines. First, the biology of the whales themselves should be studied at an established whaling station on shore, where numbers of animals would be available. Here their anatomy could be clarified and detailed records of breeding season, length of gestation, numbers of young, and rate of prenatal growth could be assembled. Meanwhile, another equally important phase of the study would concern itself with water masses, salinity, temperature, food, and the general ecology of the whales. The latter portion of the study was to be conducted on the *Discovery*.

Five species of commercial whales are found in some abundance in the South Atlantic and the Antarctic. The sperm whale is the only important rep-

The sperm whale is part of the toothed-whale group that includes porpoises and killer whales. Its oil, formerly a source of heat and light, is today used for lubricating and for making soap, glycerine, and oleomargarine.

resentative of the toothed whales (Odontoceti), a group that includes the porpoises and the killer whales. The fin, blue, humpback, and sei whales constitute the whalebone whales (Mysticeti), so called because of the development of a complex straining apparatus from the roof of the mouth. This consists of plates of fringed, horny whalebone known as baleen. The overlapping plates constitute a filter to retain the euphausid shrimps that are the food of these creatures. During the era when fashion dictated tiny waists for ladies, corsets were reinforced with strands of whalebone, creating an important market for this product.

Some measure of the food requirements of these creatures—the largest animals that have ever evolved on earth—is provided by their rate of growth. The blue whale grows at a rate of 200 pounds per day. The whale is a mammal and like all other mammals it nurses its young. The newborn whale is about twenty-five feet long and weighs about two tons. It suckles for about seven months, during which time it doubles its length and may add more than twenty tons to its mass. The nursing mother produces over a ton of milk per day—milk that may contain ten times more butterfat than cow's milk. The young whale is adapted to swallow and breathe simultaneously. Basically this is the same structural refinement that permits horses to drink without drowning—a separation between respiratory and digestive portions of the pharynx that converts this common chamber into two functional parts.

In addition to fundamental studies of whale biology, Hardy unraveled the life history of the krill *Euphausia superba*. This shrimplike creature, scarcely more than an inch in length, is the principal, if not the sole, food of the whalebone whales in the Antarctic. Clearly, any decrease in the abundance of these animals would seriously limit the population of whales that any given area of ocean could support. As Nansen had earlier reported for the Arctic, in high southern latitudes Hardy found that an abundance of nutrients supported enormous populations of diatoms and dinoflagellates—the basic elements of the food chain. These were eaten by small crustacea, which in their turn nourished the important euphausids.

Hardy's early studies of the plankton were conducted with traditional gear. Nets of various sizes, with about two hundred meshes per inch, were attached to a wire cable at intervals. These were lowered quickly to a predetermined depth, towed for a standard time, and were then closed when weighted messengers attached to the cable struck a spring-loaded release mechanism. This system provided a series of samples of the planktonic animals and plants that were unable to avoid the net. At best, the method gave a discontinuous impression of the distribution of plankton in the water column and allowed only a limited sampling in a horizontal direction, because the speed of the ship had to be reduced during the towing period to avoid damage to the nets. Hardy was prompted by these disadvantages to design a continuous plankton sampler—a device that permits the vessel to operate at normal cruising speed and yields a continuous sample of the plankton in the water through which it is towed. Basically the continuous plankton sampler is a hollow bronze housing, tapered and faired to a streamlined shape. The inlet opening of the rectangular tunnel is smaller than the outlet. A long strip of fine-meshed silk gauze is arranged to wind slowly across the water in the tunnel from bottom to top. The gauze sieves out the plankton, which remains pressed against it by the flow of water. As it emerges from the tunnel the plankton-loaded strip is joined by a second similar strip winding off a spool placed above the tunnel. The two strips sandwich the plankton between them and pass together to a storage spool. All the moving parts of the machine are driven by gear trains activated by an external propeller. Thus the rate of movement of the gauze strip is related to the speed at which the sampler is towed. The collecting strip is ruled transversely at intervals equal to the distance from the bottom to the top of the water tunnel. By selecting the appropriate gear train, each numbered rectangle of collecting gauze can be made to represent any desired horizontal distance, for example one nautical mile. It will then contain a sample of all planktonic

creatures living at its depth in that mile of ocean. Early models of the recorder could accommodate strips sampling continuously for two hundred miles. This device greatly facilitated the plankton survey of the antarctic whaling ground. Various later modifications of this principle have been employed in studies of the fisheries of the North Atlantic and the North Sea.

After two years of labor in the South Atlantic the *Discovery* expedition returned to England. All hands recognized that the many achievements of the expedition merely emphasized the vast amount of work that still needed to be done in the area before the complex interrelationships between the ocean water, the plankton, and the whales could be satisfactorily described. The results of the first *Discovery* expedition required more than five years to organize and publish. A newer, bigger *Discovery II* was planned and built to continue the investigation of the biology of the whale and to extend the plankton studies initiated by Hardy on Scott's old *Discovery*.

Alister Hardy accepted a university appointment and has devoted himself to the education of generations of biological oceanographers and to continuing his early studies of the interrelationships between plankton and larger marine animals. He has written numerous scientific papers and several books describing the relations between marine organisms and their environment. He was knighted for his contributions to biology and has been a professor of zoology at Oxford for the past twenty years.

The Quest for Causes

Biological oceanography, like other sciences, has matured through a succession of stages. The earliest of these sought to systematize the bewildering array of events and beings in the ocean. This quest for order in general biology is expressed in the descriptive, or systematic, era that extended from medieval times through the end of the nineteenth century and established the foundations for the modern classification of animals and plants. In biological oceanography the latter years of the nineteenth century were especially fruitful. This was the period of the great expeditions, when a few scientists endured the rigors of the sea together with the frustrations and inadequacies of small ships, in order to collect representative specimens of marine organisms from all oceans of the earth. When these were returned to the expedition base at the conclusion of the voyage, the real work began. The preserved specimens were sorted and dispatched to experts for classification and description. The oceanographic data describing the environment from which a particular specimen was taken gave some insight into the adaptations demanded of the specimen in nature. Such conclusions were largely speculative, and most emphasis was placed on descriptive aspects of biology.

As the nineteenth century drew to a close some rumblings of dissatisfaction with purely descriptive marine biology began to be heard. Scientists were no longer content merely to know "what" occurred in the sea—they were curious to determine "why" and "how" marine organisms adjusted themselves to the special demands made by their environment. The cramped, wet, and dark "laboratory spaces" that were available on most oceanographic ships of the time afforded neither the physical room nor the facilities needed for physiological studies. For these, minimal requirements are tanks where recently captured marine animals can be maintained for extended periods. Some means of renewing the water in the tanks must be provided. Temperature, as well as light, must be controlled. Having provided for the healthful survival of his experimental material, the physiologist then can concern himself with housing the recording, stimulating, sensing, and testing equipment he needs. Much of this equipment employs electric or electronic components whose efficiency is not enhanced in a wet, salty environment.

By the late nineteenth century better facilities were needed not only for physiologists but for scientists representing other aspects of functional biology, such as biochemistry and the nascent discipline of biophysics. Furthermore, although oceanographic ships had steadily increased in size and scientist capacity since the days of the *Challenger*, they were still limited in the number of scientists they could sustain for long periods at sea. The establishment of marine laboratories on land satisfied many of these deficiencies.

One of the earliest and most successful of these institutions was the Stazione Zoologica established on the Bay of Naples in 1872 by the German zoologist Anton Dohrn. From the Italian government he obtained a grant of land in a park on the city side of the bay; from German sources and from scientific societies throughout the world, and with supplements from his own fortune, Dr. Dohrn constructed the first laboratories of the station. The station is presently directed by Dr. Peter Dohrn, a grandson of the founder. During nearly a century of service to marine biology, the Stazione Zoologica has been host to most of the great zoologists of Europe and of the United States. Many fundamental studies completed there have altered the course of the development of biology. Among these may be mentioned studies by Alexander Kowalewski, Ernst Haeckel, Karl Ernst

The first permanent headquarters of the Bureau of Commercial Fisheries (above) was adjacent to the Marine Biological Laboratory at Woods Hole, Massachusetts. (Paul S. Galtsoff)

von Baer, and others that revealed the complex changes experienced during development of certain invertebrate and primitive vertebrate animals native to the Bay of Naples. These studies not only clarified developmental processes but they also shed some light on the broader problems of the interrelationships and evolution of major groups of animals. These studies alone would have insured the historic importance of this station. Modern investigations at the Stazione Zoologica reflect the orientation of contemporary biology toward molecular biochemistry, while continuing to stress the relationship between the total organism and its environment. Thus, J. Z. Young and his students have studied the brain and central nervous system of the octopus, not only as a relatively simple model for the organization of more complex types, but also as the central coordinating agency for the complex behavior of this animal. Very few major developments in modern biology have occurred without significant contributions from this international research center on the Bay of Naples.

The Marine Biological Laboratory at Woods Hole, Massachusetts, the oldest marine laboratory in the United States, was established first in 1873 by Louis Agassiz on Penikese Island off the Massachusetts coast. Here he was host to some sixty scientists during the summer. After Agassiz's death, later in that year, the Penikese project was transferred to the neighboring mainland at Woods Hole, at that time a small fishing village at the base of Cape Cod. The establishment of the Marine Biological Laboratory (MBL) and of the facilities that it came to possess was influential in the early location of the Bureau of Commercial Fisheries Laboratory and Aquarium on adjacent property. This physical arrangement has permitted considerable exchange of information between the staffs of the two institutions. The Woods Hole Oceanographic Institution (WHOI), with the specialized mission to describe the physical, chemical, and geological characteristics of the oceans, was constructed adjacent to the biological laboratory. The MBL operates intensive summer courses for graduate students from inland universities. For many this program provides a first opportunity to see living marine creatures in their natural environment. The MBL has afforded many biologists the opportunity to continue their research programs while enjoying a summer holiday in a pleasant seaside community. The MBL library houses one of the most complete collections of ocean-oriented books and periodicals in the Western Hemisphere.

During the latter part of the nineteenth and early twentieth centuries an outstandingly productive gen-

Aerial view of the Woods Hole Oceanographic Institution, as it appears today. (Woods Hole Oceanographic Institution)

eration of biologists, including such names as E. G. Conklin, E. B. Wilson, Jacques Loeb, C. M. Child, and Robert S. Lilly, among others, firmly established the preeminence of the Marine Biological Laboratory in American biology. Early work at the MBL generally did not seek to clarify the interrelationships between marine organisms and their environment so much as to employ readily accessible marine materials to investigate problems fundamental to general biology, such as the role of the cell nucleus in the life of the cell, and the development of diverse organs of animals by successive divisions of an apparently homogeneous fertilized egg cell. To a considerable degree this philosophy has persisted to the present day.

During the first half of the present century, several American universities established marine stations, usually as facilities of the department of biology. These were generally small, and served more as collecting stations than as coordinated research centers. Since World War II, and particularly during the last decade, many of these have grown markedly and now possess broad capabilities in both research and teaching. Among these are the Scripps Institution of Oceanography of the University of California, the Institute of Marine Sciences of the University of Miami, the University of Washington's Marine Laboratory, and the Duke University Marine Laboratory.

In 1879 the Marine Biological Association of the United Kingdom was founded at Plymouth. With additions and extensions built since that time, the Plymouth Laboratory now provides facilities for the study of most aspects of ocean science.

In addition to the Stazione Zoologica, other European marine laboratories include the imposing museum, aquarium, and laboratory constructed in Monaco by the distinguished oceanographer Prince Albert I. This structure houses an outstanding collection of marine organisms from all the oceans of the world, many collected by the prince in his succession of yachts: the *Hirondelle I* and *II*, and the *Princesse Alice I* and *II*. The present scientific director of this institution is Captain Jacques-Yves Cousteau, whose development of scuba gear has made the shallow waters of the world available to thousands of avocational divers. Other noteworthy marine laboratories are those at Concarneau on the Atlantic coast of France and at Arcachon on the Mediterranean. These, together with the host of other laboratories found in most countries of Europe, are chiefly devoted to marine biology in the classical sense.

High-seas oceanography in the latter half of the twentieth century requires large ships and expensive, complicated equipment, supported by extensive shoreside facilities that generally include electronic computers and other sophisticated data-processing aids. The costs of maintaining programs of this magnitude require massive governmental subsidies and clearly exclude all but the largest marine institutions.

How Animals Survive in the Sea

Pioneering studies in marine biology supplemented by twentieth-century investigations have considerably improved our understanding of the general biology of creatures that live in the sea. All recent studies show that the biology of marine organisms does not differ radically from that of creatures occupying other environments. All animals must solve the fundamental problems of living and reproducing their kind. Basic solutions for these common problems are shared by the simplest one-celled animal and the most complex beings. From this point of view it is difficult to justify "marine biology" as a separate specialty. All animals and plants that live in the sea, however, have had to adjust to its demanding nature; unraveling the precise mechanisms of these adjustments continues to challenge the ingenuity and dexterity of biologists throughout the world.

Animals can survive in the sea by adopting one of two strategies: they either tolerate free exchange between the medium in which they live and their blood and body fluids, or they evolve mechanisms such as impervious shells or cuticles to insulate their special internal chemical composition from their environment. Most primitive animal groups adopt the former and easier solution to this problem; thus the body fluids of single-celled animals, sponges, jellyfishes, and starfishes and their relatives are nearly identical in composition with seawater, with certain quantitative differences. Other marine animals—the more active predatory types such as lobsters, worms, and fishes—evolved mechanisms for preserving a

The steamer *Hirondelle II* was second in a succession of oceanographic yachts outfitted by Prince Albert of Monaco. The marine organisms he collected became the basis for Monaco's laboratory and museum. (Musée Océanographique, Monaco)

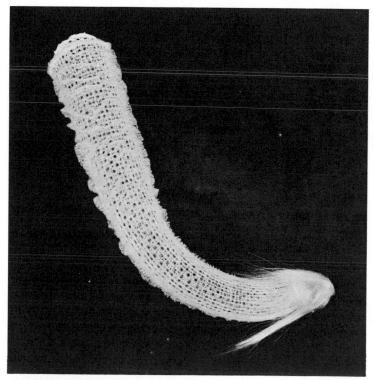

The glass sponge and other primitive animals survive in the sea by freely exchanging their body fluids with the medium in which they live. (Charles E. Lane)

More complex organisms, such as lobsters, worms, and fishes, have evolved mechanisms for preserving a stable internal composition different from that of seawater. Since their body fluids contain less salt than the surrounding water, they must compensate for continual loss of water by osmosis. (Walter Starck)

stable internal chemical composition different from seawater. Thus the total concentration of various mineral salts in their bodies may be only a half to a third that of the medium in which they live, and concentrations of major chemical elements may be quite different from those found in seawater. These concentration differences impose a severe stress upon animals in the groups affected by them. Unlikely as it may seem, these marine animals share with inhabitants of arid deserts the hazard of death by desiccation. Since their body fluids are less saline than their medium, they tend constantly to lose water by osmosis through all permeable body surfaces, and to survive they must conserve water in the midst of plenty.

Ingenious water-conservation mechanisms have evolved in response to this requirement. Among these may be mentioned the progressive reduction in size and number of the glomeruli, the kidney structures most important in excreting water. This reduction leads to the disappearance of the glomeruli from the kidneys of certain groups of marine bony fishes, such as the anglerfish, the sea horse, the pipefish, and their relatives. Many fish imbibe seawater and then excrete the salts, retaining the water. Since the excretory capacity of the kidney is reduced to conserve water, it is advantageous for these animals to exploit some other salt-excreting mechanism. Many fishes have specialized cells in the gill membrane that apparently extract sodium chloride from blood and

transfer it into the surrounding water. This process of transport against a concentration gradient is only poorly understood at the present time. Clearly, when the phenomenon is explained, however, it will have vast industrial potential, for it might be used to desalt seawater or to convert the bitter alkaline water found in many desert areas of the world to potable quality.

Marine mammals—the whales, seals, walruses, and their kin—are able to drink seawater without hazard because their kidneys produce urine that is more concentrated than seawater; that is to say, they can excrete all the salt they have swallowed in a smaller volume of water. The human kidney lacks this ability; when man drinks seawater he actually must excrete more total volume than he has imbibed, so he suffers a net water loss. Certain desert mammals—the little kangaroo rat, for example—have so developed the concentrating ability of the kidney that they can live indefinitely on a diet of dried cereal grains without ever drinking water. Their small water requirements are satisfied by the water that is liberated in metabolism of their foodstuffs.

Marine birds and reptiles eliminate the salt they swallow when they drink or feed, through a special set of glands located between the eyes. These glands drain into the nasal cavity of the birds, or into the conjunctival sac of most marine reptiles. Secretions of these glands contain a concentrated salt solution that has only small amounts of other dissolved materials.

These secretions explain the tendency of the female marine turtle to shed copious "tears" while laying eggs. The turtle is not suffering in the process, nor is she saddened by the almost certain fate of most of her progeny—she is simply disposing of the salt that she swallowed with her last drink or with her latest meal. The same phenomenon accounts for the clear fluid often seen dripping from the beak of pelicans, gulls, and other marine birds.

Another adaptation by marine organisms occurs in the visual system. Since light penetrates only the uppermost levels of the total water column in the sea, those animals that inhabit the deeper levels show several interesting modifications of their equipment for seeing. Some animals have greatly enlarged eyes, the better to respond to what little light exists at their levels. Others have abandoned sight as a means of orientation in favor of other senses, such as those of taste, touch, or ultrasonic vibration. Most animals that live in the deep sea develop prominent organs for light production, presumably employing the same biochemical processes used by the firefly and some single-celled marine organisms. These photophores may be of several different colors and kinds. Sometimes they occur as rows of lights arranged symmet-

In 1901 Prince Albert of Monaco caught this abyssal *Grimaldichthys profundissimus* in 19,800 feet of water off the Cape Verde Islands. For half a century it held the record as the deepest-dwelling fish ever brought to the surface. (*Monaco Reports*)

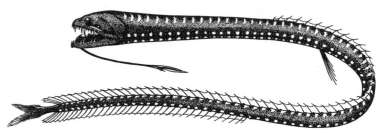

Since light penetrates only the surface regions of the sea, animals that inhabit the deeper levels must modify their equipment for seeing and being seen. The *Idiacanthus panamensis*, found by the *Dana*, has a luminous organ at the end of its chin barbel. Its body is studded with light-producing photophores.

The bottle-nosed dolphin, like other marine mammals, is able to drink seawater because its kidneys produce urine that is more saline than the ocean. (J. W. LaTourrette, Wometco Miami Seaquarium)

rically along the body; in other cases, as in some relatives of the anglerfish, the luminous organ is located at the tip of a stalk that overhangs the mouth. Smaller fishes that may be attracted to this structure are trapped and swallowed when the capacious mouth is abruptly opened. Some invertebrates, such as deep-sea squids, are hosts to luminous microorganisms that are cultured in the photophores. Various lenses and filter mechanisms have evolved to intensify and to regulate the color of the light that is emitted.

Floating and Diving

Since marine animals occupy a medium that has a tremendous vertical dimension in relation to their size, they experience hazards that terrestrial animals rarely know. Most pelagic marine animals are more dense than the medium in which they live; hence they tend constantly to sink. Various flotation devices have evolved that minimize this hazard. Most fishes have gas-filled swim bladders that can be inflated by active secretion of a gas gland, or deflated by reabsorption or release of the contained gas to achieve neutral buoyancy. There is still considerable disagreement among physiologists about the mechanism by which this gas secretion is achieved. At 3,000 feet in the ocean the oxygen in the swim bladder of many deep-sea fish exerts a pressure of more than 1,500 pounds per square inch. (The familiar oxygen cylinders of industry or medicine are generally filled to produce a pressure of about 2,000 pounds per square inch.) A gas-filled bladder provides a sensitive pressure gauge that may prevent the fish from either ascending or descending too rapidly. In the former situation the expanding gas would painfully distend the sensitive swim bladder, and this distension, like any other painful sensation, would cause the animal to alter its behavior. During rapid descent the increase in density, caused by compression of the gas in the swim bladder, would require the fish to swim more vigorously to keep from sinking.

Various animals, such as some of the deep-sea squids (Cranchiidae), secrete large volumes of fluid into the body cavity containing salts that are lighter than those of seawater. The luminous protozoan *Noctiluca* employs a similar mechanism, replacing considerable quantities of sodium in its intracellular fluids with the lighter ammonium ion. (These animals originated the flotation mechanism that was exploited in principle by Auguste Piccard in the deep-diving vehicle *Trieste*. In this vessel the passenger compartment is suspended beneath a flexible bag containing gasoline. Being less dense than seawater it tends to make the vehicle float in the water column. This tendency is opposed by iron shot held by electromagnets. To descend, gasoline is jettisoned; to return to the surface, some of the electromagnets are turned off and shot ballast is released.) Other cephalopods achieve neutral buoyancy by secreting gas into the interstices of their internal skeleton—the cuttlebone. Some pelagic fishes, such as mackerel, avoid sinking in the water column as a result of an inherent inadequacy of their respiratory system that requires them constantly to swim to insure sufficient ventilation of their gills. Other animals, such as the nudibranch mollusk *Glaucus marinus,* reduce their tendency to sink into the depths by extravagant expansion of surfaces that may at the same time serve as gills. A similar expansion of appendages occurs in the phyllosoma stage in the development of many marine crustacea. It seems likely that the ribbonlike leptocephalus stage in the development of the eel and certain other marine fishes may represent a similar evolutionary effort.

Hydrostatic pressure is another characteristic of the environment to which marine animals must adapt. Pressure in the sea increases approximately one atmosphere (fifteen pounds per square inch) for each thirty-three feet of vertical descent. Since most invertebrates and many vertebrates in the seas lack significant gas-filled spaces, the pressure that increases with depth is transmitted equally throughout their bodies and no parts are subjected to any unusual stress. Bony fishes with swim bladders, and marine mammals with air-filled lungs, however, suffer the consequences of compression of trapped gases as they sink in the water column. Moreover, gases tend to dissolve in water in proportion to the pressure of the gas. These two phenomena account for some of the well-publicized hazards that limit human divers to the upper levels of the water column and emphasize the essentially terrestrial nature of man.

Self-contained underwater breathing apparatus (scuba), designed during and since World War II, has freed the human diver of cumbersome "hard-hat" diving gear and has made him independent of a surface source of air. This equipment provides the diver with air from a high-pressure cylinder worn on his back, supplied through a regulator that adjusts delivery pressure to the surrounding water pressure. Using standard safety precautions, divers experience little difficulty if they limit their dives to approximately thirty feet. Deeper than this, increased pressure forces larger amounts of atmospheric gases into solution in the blood plasma. The most abundant atmospheric gas is nitrogen, constituting approximately four-fifths of the atmosphere. At depths greater than a hundred feet, enough nitrogen may be dissolved in the blood to interfere with the normal functioning of the brain and central nervous system; this produces the condition known as nitrogen narcosis or "rapture of the deep," which is characterized by poor coordination, confusion, errors of judgment, and even unconsciousness. Enhanced amounts of dissolved nitrogen constitute a serious hazard when the diver returns to the surface from a deep dive. With abrupt reduction of pressure, the blood and body fluids may effervesce like a bottle of soda when the cap is removed. Gas bubbles so liberated may lodge in joints, causing painful "bends," or they may block the smaller blood vessels in the heart, lungs, or brain, with much more serious consequences.

Marine mammals are notoriously successful divers. Whales are known to descend more than a thousand feet and remain submerged for more than an hour. Observers in the deep submersible *Aluminaut* have reported seeing porpoises at depths greater than six hundred feet. Moreover, these animals regularly prolong their dives far beyond man's capability. Recent physiological studies have clarified some of the adaptations contributing to the success of these interesting animals.

All diving animals, whether they be birds, rep-

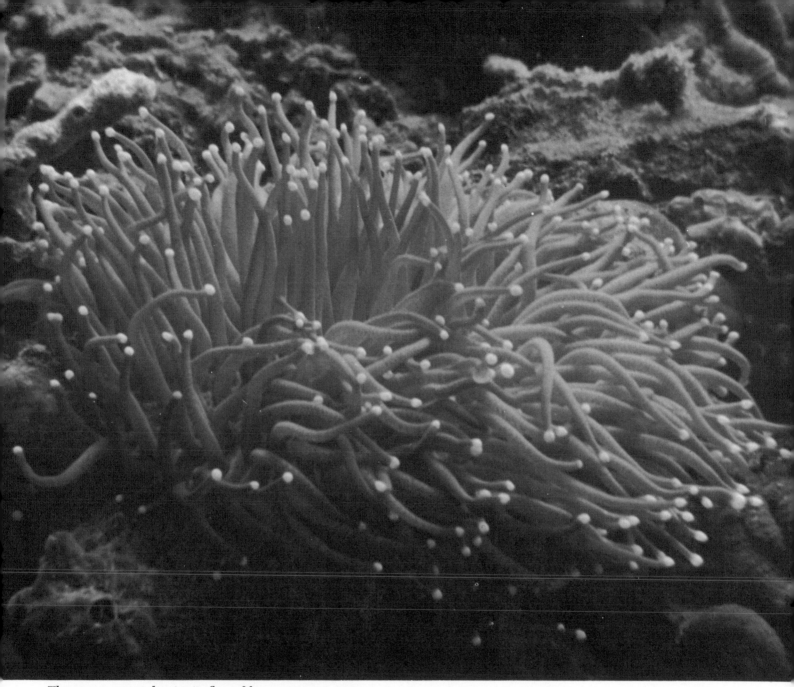

The sea anemone, despite its flowerlike appearance, is an animal, related to the corals and the jellyfishes. (F. M. Bayer)

tiles, or mammals, depress the heart rate dramatically upon beginning a dive. In the seal the rate may drop from a normal seventy-five beats per minute to only five on submergence. Various shunting pathways in the circulatory systems of seals and whales are activated early in the dive to deliver blood preferentially to the brain, sense organs, and diving muscles, at the expense of the abdominal viscera. These animals have more red blood cells than terrestrial animals and thus have a higher oxygen-carrying capacity. Of equal significance is the insensitivity of their brain

centers to the accumulation of carbon dioxide in the circulating blood. Thus the protracted dives of the whales and seals are explained by greater oxygen capacity, more economical circulation, and reduced sensitivity to accumulated carbon dioxide.

The ability of the great whales to dive very deep without incurring nitrogen narcosis on descent or the bends on ascent is probably due to the construction of their respiratory system. Their diaphragm is oblique instead of transverse as in most mammals, and their rib cage is relatively flexible. As a great whale

Modern marine biologists at work: Dr. Robert H. Williams (right) and Dr. J. F. Walker examine the catch of a thirty-minute plankton tow aboard R. V. *Gerda*, research ship of the Institute of Marine Sciences, University of Miami. (Charles E. Lane)

dives, its rib cage squeezes the contained respiratory system, collapsing the lungs and driving the air up into capacious head sinuses. Since gas is only absorbed from the terminal portions of the tubular respiratory system and not from the trachea and bronchi, this mechanism effectively turns off gas absorption and spares the animal the consequences of enhanced absorption of nitrogen.

Flora of the Sea

In sharp contrast to the wide variety of animal life in the sea, the marine flora is quite limited. Among plant phyla, only the thallophytes and spermatophytes (flowering plants) occur in the sea; the other two major divisions of the plant kingdom, the bryophytes (the mosses and liverworts) and the pteridophytes (the ferns and their allies) are not represented there. Nearly all marine plants belong to the phylum Thallophyta. These are primitive plants with no differentiated true root, stem, or leaf. This group includes marine algae, marine fungi, and marine bacteria. The common plants of the seashore, the sea lettuce *Ulva,* and the brown rockweed *Fucus,* together with the large kelps *Nereocystis* and *Macrocystis,* comprise only a tiny portion of the plant life of the ocean. Far more important, both numerically and in organic productivity, are the diatoms and dinoflagellates. These microscopic plants are called the primary producers of the sea because they account for most of the carbon fixed by photosynthesis

in the pastures of the ocean. They form the bottom level of the food web that supports all the organisms in the sea.

The only marine flowering plants are a few species that have secondarily invaded the seas from adjacent fresh water. The eel grass *Zostera marina* is an outstanding example of this group. This plant of worldwide distribution flowers underwater, and its threadlike pollen is distributed by water currents.

The marine seaweeds first seen by a visitor to the seashore are classified as either brown algae or red algae. On the shores of North America and Europe the most conspicuous seaweeds are brown algae. The giant kelp, *Macrocystis pyrifera,* is an impressive brown alga that occurs in massive beds off the coast of California and Baja California. Brown algae were formerly widely used in the production of iodine, iodides, and potash. These substances are no longer commercially extracted from seaweeds, but the brown algae have assumed new industrial importance as sources of alginic acid and its derivatives. This organic acid may constitute up to 40 percent of the total dry matter in many seaweeds. Sodium alginate—the sodium salt of alginic acid—is readily soluble in water and is used to size and stiffen fabrics. It is sometimes used in place of starch, since it fills the cloth with a tougher and more elastic film than starch.

Alginates are widely used in pharmaceutical preparations such as emulsions, tablets, ointments, and dental-impression materials. Since alginates are nontoxic and may be partially digested, they are used in food industries as stabilizers for ice cream, chocolate milk, sherbets, and the like. Alginates are often employed in many bakery products such as icings, fillings, and glazes, and in packaged whipped-cream substitutes.

Red algae, the second major kind of seaweed, occurs in great variety in tropical waters but is also abundant in boreal regions. Cells of the red algae are enclosed by thick walls containing complex carbohydrates that may be extracted and that readily form colloidal solutions when heated; on cooling, a firm gel develops. Common examples are agar and carrageenin.

Seaweed gels are widely accepted, particularly in warm climates, because they set readily at elevated temperatures and do not melt during warm weather, as do gels based on gelatin. On the northeastern coast of North America several million pounds of Irish moss (*Chondrus crispus*) are harvested annually from York, Maine, to the maritime provinces of Canada. Carrageenin, readily extracted from Irish moss, is used in prepared foods, in cosmetics, and in pharmaceutical preparations. Its value in these products is based on its remarkable thickening, suspending, filling, and stabilizing powers. Perhaps its most important use in the United States is as a stabilizing agent for chocolate milk.

Agar is another common commercial product derived from red algae. Before World War II most of the world's supplies of agar came from Japan, where annual production was approximately 2,000 tons. Beginning in 1941, when Japanese supplies were abruptly reduced, agar production was begun in California, and in North Carolina and Florida on the eastern seaboard.

Agar is indigestible but nontoxic, and is widely used by the food industry. Bakeries use agar to reduce the tendency of icings to become sticky in summer and to adhere to wrappings. It is used as a stabilizer in pie fillings and meringues. Agar also is used in place of gelatin in many jellied candies and marshmallows, and as fillers in candy bars. Jellied desserts, aspic, salads, and puddings generally owe their gel strength to agar, and mayonnaise and salad dressings are usually stabilized either by agar or carrageenin. Total food uses in the United States amount to several hundred thousand pounds per year. Agar is the chief gel in bacteriologic culture media and is important in the manufacture of dental-impression compounds.

Few seaweeds are eaten as such in the United States, but they have long formed important constituents of the diets of many oriental countries. Although many of the more complex components of seaweeds are indigestible and pass unchanged through the human digestive system, various inorganic salts, trace minerals, and vitamins that have been concentrated by the life processes of the plant are readily absorbed. Since marine algae are good to excellent sources of vitamin C and riboflavin, as well as trace minerals, they are often used as components of animal and poultry feeds.

New Directions in Research

The variegated fabric of man's present knowledge of the sea and of its inhabitants has been woven by the patient efforts of many generations of investigators. The number of living scientists occupied in these endeavors probably equals, and may even exceed, the total number who have ever labored at this loom. It would be impossible to mention all those now contributing to the growth of biological oceanography.

Much contemporary research in marine biology seeks to clarify the fundamental problem of how a marine organism conserves an internal chemical composition different from that of the seawater bathing it. The laboratories of W. T. W. Potts and J. D. Robertson in Great Britain, and those of A. W. Martin, P. F. Scholander, and K. Schmidt-Nielsen in the United States are particularly active in these studies. In addition to their intrinsic theoretical interest, these investigations may suggest practical and more efficient methods for desalting seawater to satisfy the growing water requirements of the world's population.

Fundamental studies of gas-exchange mechanisms by Hermann Rahn, P. F. Scholander, R. J. Harrison, N. B. Marshall, and numerous others have clarified many of the relationships between diving animals and their environment and have suggested how man can extend his limited diving capability. Progress in this direction encourages those who propose the establishment of semipermanent work stations at depths of 600 feet or more on the world's continental shelves. These stations would be occupied by divers who would spend weeks or months at a time on the bottom, engaged in mining, mariculture, or scientific studies, surfacing only for reassignment.

Greater understanding of the dynamics of populations of marine organisms and more general use of the storage and retrieval capability of computers will allow more intelligent exploitation of marine resources for food and fiber, and will insure that existing stocks will not be dangerously depleted.

4 PHYSICS OF THE

by James B. Rucker

The world's oceans contain some 330 million cubic miles of water distributed over the earth's surface, with an average depth of approximately two miles. In spite of these colossal figures, the mean ocean depth actually represents only one seventeen-thousandth of the earth's radius. This ocean depth across the face of the earth may be considered analogous to the thickness of a layer of heavy dew on an average American automobile.

In spite of the thinness of this layer of "dew," its presence on our planet controls the destiny of mankind. The thin veneer of seawater that is distributed over 70 percent of the earth's surface has provided maritime nations with food and wealth, with avenues of commerce, communication, and cultural exchange, and with natural defense. It is not surprising that many of the most powerful and prosperous nations have been maritime civilizations. Neither is it surprising that for thousands of years man has been interested in understanding the sea—an interest that has within the last century evolved into the science of oceanography.

Ancient Charting and Navigation

The forebears of the modern oceanographers were mariners: men who earned a living from the sea, and who gradually learned to understand her ways. Traveling under sail, they were keenly aware of the effect of changes in weather on the behavior of the sea. They also observed many of the ocean's unique characteristics: its temperature, waves, currents, tides, and its saltiness. Their main concern, however, was to know their exact geographic location, and the depth of water beneath their keel. The birth of the science of oceanography had to wait its turn. Mariners first required methods of navigation, charts, and soundings. The development of navigation and charting, in turn, was based on scientific and technical developments in astronomy and mathe-matics, on the work of instrument makers and cartographers, and on the catalytic effect of commercial needs and national defense requirements.

The development of navigation preceded that of charting. Navigation has gradually evolved from an art into a science over the past six to eight thousand years. Although many epic voyages were made in the period prior to 500 B.C., little data exists to substantiate the methods of navigation that were used. Noah on his ark supposedly "navigated" by releasing a dove to find a landfall. Early Norsemen, on their voyages from Spitsbergen, used ravens and auks in a similar way. The Norsemen and Polynesians left little in the way of recorded information about their navigational methods.

Until seamen began to voyage out of sight of land they were, in effect, navigating by piloting; that is, they found their position merely by identifying surroundings, or by recognizing landmarks from the oral or written descriptions of other mariners. As they began to venture away from land, however, other signposts were utilized. In addition to birds, the sailors were guided by wave and cloud formations and by scents. Land has distinctive smells that breezes can carry as far as thirty or fifty miles out to sea. The scent of orange blossoms or of burning wood, carried by the winds, is easily recognizable as potential landfall. Micronesian and Polynesian navigators could read eddy and wave patterns formed by islands many hundreds of miles distant. In addition, and most important, were the conclusions drawn by early navigators from observations of the astronomical bodies during their travels.

As civilization progressed, it became evident—particularly with the advent of increased exploration by man—that navigation would not suffice as an art; an organized and reliable science of navigation was needed. The mariner required a greater understanding of the physical character and geometry of his world, as well as some understanding of its place in

SEA

the system of the universe. During the four centuries before Christ and in the early Christian era, certain notable scientific achievements had significant effects on the development of navigation, and consequently on charting. Pythagoras, in the sixth century B.C., gave shape to the earth when he proclaimed it to be spherical. Two centuries later Aristotle and Plato reasserted his belief. In the third century B.C., Eratosthenes, the brilliant scholar of Alexandria, Egypt, gave the spherical earth its dimensions, calculating the circumference to be 25,000 miles. (Today, twenty-three centuries later, our measurement is 24,902 miles.) Between 300 and 350 B.C. Pytheas of Massilia, a Greek astronomer, made one of the best-documented voyages up to that time. He sailed from the Mediterranean to England, following an established trade route. From there he ventured north to Scotland, Norway, and the rivers of Germany. He did this without the aid of mechanical devices, using only the sun, stars, wind, and possibly some form of sailing directions. About 150 B.C. the branch of mathematics known as trigonometry was devised by Hipparchus, laying a foundation for the development of many types of map projections during the ensuing centuries. This period was indeed one of awakening, during which the basis for scientific navigation and charting was firmly established. Along with these more scholarly achievements of the pre-Christian era, other developments of more immediate practical use were introduced: the establishment of lighthouses in 600 B.C. by the Cushites and Libyans along the coast of Egypt; the use of the lead line (perhaps the world's oldest navigational instrument) for depth soundings; and the accumulation of geographical knowledge in the major sailing centers of the world, principally around Alexandria.

As the science of navigation developed, mariners gradually extended their commercial and military expeditions beyond the Mediterranean, where tidal ranges are slight, into regions where the rise and fall of the tide, and tidal ebb and flow in harbors were more noticeable. During the four centuries before Christ, several notable scholars—Herodotus, Aristotle, Selencus of Babylon, Pliny the Elder—observed and recorded the tidal phenomenon outside of the Mediterranean, and discovered some of the effects of moon declination (angle between the position of the moon and the plane of the earth's equator) on tidal variations. It must be realized, however, that during this period there was little real understanding of the tidal phenomenon, and virtually no systematic observations were made.

The First Charts and Tide Tables

From the first to the thirteenth century A.D. the developments in charting exceeded those in navigation. The outstanding contribution during this period was made by Claudius Ptolemy, an Egyptian, who is best known for his geographic compilations, atlases, and map projections, and for a description of the motion of the heavenly bodies that is known as the Ptolemaic system. His map of the world was a considerable achievement for the time; it was based on conic projection, listed more than eight thousand places by latitude and longitude, and served as a general basis for further cartographic work. It had however, one major flaw: Instead of using Eratosthenes' much more accurate figure, Ptolemy accepted the value of only 18,000 miles for the earth's circumference that had been calculated in about 150 B.C. by Poseidonius. This gave erroneously short distances, which affected navigation up to the time of Columbus' eventful voyage to "India." Because of the general use of Poseidonius' figures, Columbus died thinking he had reached the eastern shores of the Asian continent.

During this span of centuries notable contributions to charting were also made by Arabian cartographers. They developed many maps of the known

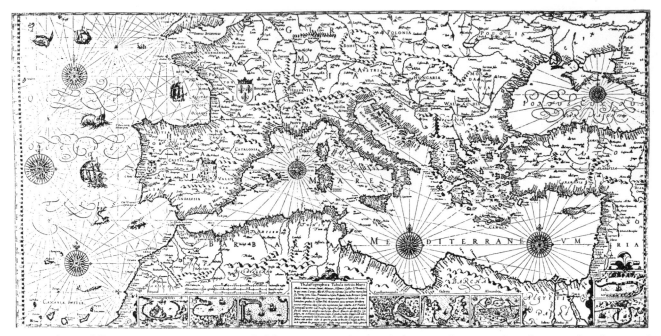

During the Middle Ages navigators were guided by portolanos, or harbor-finding charts. This copper engraving from 1595 is an almost unaltered copy of a fourteenth-century portolano. (Map Division, Library of Congress)

world and its nations; perhaps most importantly, they introduced a celestial globe.

The earliest known nautical charts date from this period—the portolanos, or harbor-finding charts. These portolanos were characterized by a group of rhumb lines intersecting at a common point, each rhumb line representing a constant and true direction between points along the line. Although, over a large area, the rhumb lines will "curve" with the curvature of the earth's surface, within a small area (as on a harbor chart), the rhumb lines appear to be straight. The portolanos carried a mileage scale and indicated known hazards to navigation. They were not marked with parallels of latitude or meridians of longitude, but later portolan charts had a *rose dei venti* ("rose of wind," i.e., compass directions), the forerunner of the modern compass rose. The appearance of the rose of direction on nautical charts indicates the introduction of the use of the magnetic compass by mariners during the Middle Ages.

Throughout the early Middle Ages there was little improvement in man's understanding of the tides; although some of the portolan charts warned of ports having high tidal ranges, no tidal predictions were available. Even the Anglo-Saxons, who were closely associated with the sea, and whose harbors experienced a great tidal range, seemed to have no understanding of the tidal phenomenon. They knew that the tide came earlier to the northern shore of England's east coast than to its southern shore, but they could neither predict its height nor explain its cause.

The first information regarding tide prediction is found in *Codes Cohonianus, Julius DVII*. These tidal observations were collected by the British Abbot Wallingford of Saint Alban's Monastery (today a suburb of London) in about A.D. 1200. In a table titled "Flod at London Brigge," Wallingford documented the times of high water and of the moon's zenith. A tidal lag of forty-eight minutes following the moon's zenith passage was determined; this is very close to today's value. No written evidence of other tide tables from this period has been found. Tide tables, like early charts, were kept as carefully guarded secrets in the steadily growing and competitive merchant-marine trade.

The Age of Exploration and Charting

Emerging from what has been considered the Middle Ages into the fifteenth and sixteenth centuries of exploration and navigation, charting began to develop at a rapid pace due to the necessity of finding safer and more accurate ways of crossing the oceans. During this period several major events took place: printing by movable type was invented, and Ptolemy's writings were multiplied by the printing process; the compass was introduced into general use as a navigational aid.

In addition, the voyages of many great explorers contributed enormously to the science of navigation and charting. Bartholomeu Diaz rounded the Cape of Good Hope in 1488, and Christopher Columbus made the first of his several voyages to the Americas in 1492, erroneously thinking he had made landfall along the easternmost shores of Asia. In 1497 Vasco da Gama circumnavigated Africa—west to east—en route to India. Amerigo Vespucci made several voyages to the New World (1499–1502) for Portugal, exploring some six thousand miles of South American coastline. He also developed an original method of determining longitude by comparing the known hour of the moon's conjunction with a planet in Spain with the hour this conjunction was observed along the coast of the New World. The passage between Antarctica and Tierra del Fuego, the southern tip of South America, was discovered in 1582 by Sir Francis Drake.

Of all early voyages of the period, the epic three-year journey (1519–21) of the Portuguese explorer Ferdinand Magellan probably was the most outstanding single contribution to the charting and navigation of the oceans. His instruments exemplified the

tools of fifteenth- and sixteenth-century navigation, including sea charts, a terrestrial globe, theodolite and quadrants, magnetic compass, hour glasses, and a log towed astern to determine speed. It seems that the only thing lacking was an instrument or method for accurately determining longitude. In addition to successfully circumnavigating the globe and discovering new lands, Magellan gathered data that established the length of a degree of latitude, and the circumference of the earth.

Many of the voyages conducted during this period, however, owe at least part of their success to Henry the Navigator (Prince Henry of Portugal), who in 1416 established a school where mariners and craftsmen could develop and improve instruments, charts, and ship designs. Most famous mariners of the period were students at this school or benefited indirectly from the knowledge it offered. Oddly enough, Henry himself never went to sea.

Chart development during this period progressed rapidly, due to the advent of printing. The portolan, developed much earlier, was still in use, primarily in the Mediterranean. Cartographers of Germany and the Netherlands drafted most of the small-scale

Chart development progressed rapidly with the advent of printing. The world map published by Abraham Ortell in Antwerp in 1570 shows Australia covering a considerable

portion of the Southern Hemisphere. (Map Collection, New York Public Library)

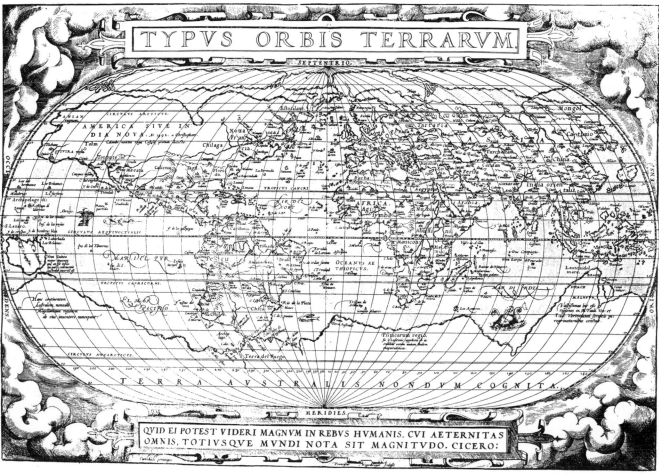

charts of the world. Most notable of these was the world map constructed by Abraham Ortell, which appeared in his atlas *Theatrum Orbis Terrarum,* published in Antwerp in 1570. In 1569 Gerhard Kremer (or Mercator), the Flemish geographer, published a world map constructed on the projection that now bears his name. Mercator projection makes possible the plotting of rhumb lines as straight lines even on small-scale charts covering large areas—a tool indispensable to the modern navigator. Mercator's efforts were correct in theory but somewhat inaccurate in computation. Some thirty years later Edward Wright published a study, including tables, revising the Mercator projection, thus enabling chart makers throughout Europe to utilize it.

Of Time and Tides

Supplementing the developments in charting was the extensive publication of sailing directions. The most notable were published by the Dutch pilot Lucas Waghenaer in 1584; they served as a standard for the next two hundred years.

The seventeenth and eighteenth centuries were marked by developments in the field of instrumentation, by widespread activities in cartography, by the first understandings of tide-generating forces, and by tide prediction.

The pendulum clock, Sir Isaac Newton's mathematical development of the spheroid, the work on longitudes done by the French Academy of Science —all contributed to advancing navigation. It was a series of ship disasters around 1690, however, that prompted governments to intervene in nautical science.

England, realizing (as did many countries) that the problem of longitude had to be solved, finally established the Board of Longitude. To know his position at sea, a mariner must be able to determine both latitude and longitude. Latitude, or angular position north or south of the equator, can be determined with relative ease by measuring the angle between the horizon and the known position of a celestial body. The determination of longitude, or angular position east or west of some north-south reference meridian, is not as simple because of the rotation of the earth. The earth rotates through 15 degrees of longitude each hour; therefore, to refer one's angular position to a celestial body, the mariner must know the precise time at which his observation is made. Between 1735 and 1764 John Harrison, a Yorkshire carpenter, developed five portable clocks, or chronometers, for use in the accurate determination of longitude.

Sir Isaac Newton applied his universal law of gravitation to the behavior of tides in his *Principia Mathematica* in 1684. His Equilibrium Theory was the basis of all subsequent work on tides.

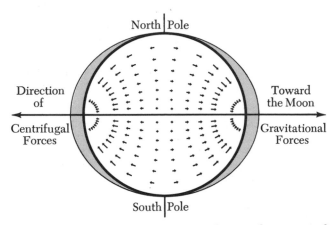

The gravitational force of the moon distorts the waters of the earth into an ellipsoid. The long axis remains directed toward the moon on the near side of the globe and away from the moon on the far side. The waters of the ocean do not move, but remain tidally deformed, with the earth rotating beneath them.

In the eighteenth century, at the time of Captain James Cook's explorations (between 1768 and 1779), the navigator had use of chronometers, charts of various projections, an accurate compass (and increased knowledge of its variations), improved versions of all known instruments, texts that simplified the mathematics of navigation, and official almanacs and sailing directions. Cook's voyages represented the beginnings of modern navigation.

The mariner's knowledge of tides up until the seventeenth century was essentially restricted to records of local tidal observations. In the early 1600's a few scientists, such as Johann Kepler and Galileo Galilei, recognized that the tidal phenomenon was

Swiss scientist Ferdinand Hassler was appointed by President Jefferson as the first superintendent of the Coast Survey. The agency later became the United States Coast and Geodetic Survey. (U.S. Coast and Geodetic Survey)

Louis M. Goldsborough was the first officer in charge of the U.S. Naval Depot of Charts and Instruments, established in 1830. The Depot, now the U.S. Naval Oceanographic Office, is one of the leading agencies in modern oceanographic studies. (U.S. Naval Oceanographic Office)

Lieutenant Charles Wilkes, Goldsborough's successor as head of the Depot of Charts and Instruments, is best known for his explorations in the Pacific and Antarctic Oceans between 1838 and 1842. (U.S. Naval Oceanographic Office)

causally related to the orbital and rotational motions of the earth, moon, and sun; however, no general concept of tide-generating forces was advanced until 1684, when Sir Isaac Newton's *Principia Mathematica* was published.

In this work Newton described his universal law of gravitation and applied it to the behavior of tides, and this gave precision to tidal theory. He assumed the earth to be a spherical body uniformly covered with water. Newton's "Equilibrium Theory," derived from this model, was the basis of all subsequent work on tides.

The most significant factor in Newton's theory of tides is that the change in level of the ocean surface varies directly with the mass of the disturbing body and inversely with the cube of its distance. The earth and moon revolve around a common center of gravity, which, because of the earth's far greater mass, lies within the earth's surface. The total gravitational attraction must be the same in magnitude and opposite in direction to the centrifugal effect of the revolving earth-moon system. The tidal bulges of the oceans are caused by gravitational force on the side of the globe facing the moon, and by centrifugal effect on the opposite side. By adding the gravitational force and the centrifugal effect at any point on the earth, we obtain a resultant, which is termed the differential force. This gives the magnitude and direction of forces acting on a "parcel" of water, moving it toward or away from the moon. This differential force is the tidal force.

The tidal force can be resolved into vertical and horizontal components. The vertical is of little significance, while the horizontal component causes the water masses to move toward the axis between the earth and moon. The tidal force tends to distort the earth and its waters into an ellipsoid; the long axis remains directed toward the moon and toward the opposite direction of centrifugal effect. In this condition of equilibrium, the waters of the ocean do not move, but remain tidally deformed, with the earth rotating beneath them.

In 1738 the French Academy of Science offered a prize for essays on the tides. The prize was awarded to Daniel Bernoulli, who refined the equilibrium theory so that tidal characteristics could be predicted for a particular area if observations of long-range tidal behavior were made there.

The next great advance in understanding tide-generating forces came in 1775, when the French mathematician Pierre Simon de Laplace approached the problem in another way. Thinking of tides as water in motion, he developed his "Dynamic Theory." Like Newton, Laplace based his theory on a model of the earth—a sphere entirely covered by a continuous ocean—and on the assumption that the vertical component of the tidal force is insignificant: With a water depth of 16,000 feet, the vertical component causes a sea-level rise of less than half an inch. According to the Laplace concept, the horizontal forces have a maximum effect at approximately 45 degrees from a line joining the center of the earth and the moon. Laplace's theory led the way to the harmonic analysis of tides, which assumes the tidal curve to be the sum of various tide-producing forces.

Oceanography Emerges as a Science

The nineteenth century was a period of extensive activity in the marine sciences. Not only were enormous strides made in the sciences of navigation and charting, and the theory of tide predictions firmly established, but worldwide information on winds, waves, ocean currents, and sea temperature and salinity was compiled. Oceanography finally began to emerge as a science distinct from the maritime and exploratory activities that were its forebears.

During the 1800's the United States Government, concerned with advancing the science of navigation and charting, founded the forerunners of our present Coast and Geodetic Survey and the Naval Oceanographic Office. These agencies were originally patterned after similar versions in foreign countries, such as France's Depot des Cartes, Plans, Journaux, et Mémoires Relatifs à la Navigation, which was established in 1720, and the Hydrographic Department of the British Admiralty, established in 1795.

The U.S. Coast and Geodetic Survey, originally known as the Coast Survey, was established in 1807 by directive of President Thomas Jefferson. Jefferson appointed Swiss scientist Ferdinand Hassler, who had founded the Geodetic Survey of his native country, as first superintendent of the new agency. Under the direction of Hassler, an expert in geodetic work, the Coast Survey engaged in hydrographic surveying, tide investigations, and chart compilation and production, as well as geodetic surveying.

The U.S. Naval Depot of Charts and Instruments was established in 1830 under the command of Lieutenant Louis M. Goldsborough. The Depot was to serve as a storehouse for charts, sailing directions, and navigational instruments to be issued to Navy ships when required. The Depot's first chart was produced from data obtained in a survey made by Lieutenant Charles Wilkes, who succeeded Goldsborough as head of the unit in 1833, and who later earned fame as the leader of a United States exploring expedition to Antarctica.

Supplementing developments in navigation and charting during the 1800's was an improvement in manuals for the navigator. Until the Renaissance any books or manuscripts that might be of help to the navigator were written by and for astronomers. The navigator was forced to make use of these, gleaning what little was directly applicable to his own profes-

68

In 1840 an expedition aboard the brig *Porpoise*, commanded by Lieutenant Wilkes, discovered that Antarctica was a continent. (U.S. Navy)

Nathaniel Bowditch's *The New American Practical Navigator* helped pave the way for Yankee supremacy on the seas during the Clipper Era. Since the original, which appeared in 1802, more than 70 editions have been printed. The manual is still published and sold by the U.S. Government. (U.S. Naval Oceanographic Office)

THE NEW AMERICAN

PRACTICAL NAVIGATOR;

BEING AN

EPITOME OF NAVIGATION;

CONTAINING ALL THE TABLES NECESSARY TO BE USED WITH THE

NAUTICAL ALMANAC,

IN DETERMINING THE

LATITUDE;

AND THE

LONGITUDE BY LUNAR OBSERVATIONS;

AND

KEEPING A COMPLETE RECKONING AT SEA:

ILLUSTRATED BY

PROPER RULES AND EXAMPLES:

THE WHOLE EXEMPLIFIED IN A

JOURNAL,

KEPT FROM

BOSTON TO MADEIRA,

IN WHICH ALL THE RULES OF NAVIGATION ARE INTRODUCED:

ALSO

The Demonstration of the most useful Rules of TRIGONOMETRY: With many useful Problems in MENSURATION, SURVEYING, and GAUGING: And a Dictionary of SEA-TERMS; with the Manner of performing the most common EVOLUTIONS at Sea.

TO WHICH ARE ADDED,

Some GENERAL INSTRUCTIONS and INFORMATION to MERCHANTS, MASTERS of Vessels, and others concerned in NAVIGATION, relative to MARITIME LAWS and MERCANTILE CUSTOMS.

FROM THE BEST AUTHORITIES.

ENRICHED WITH A NUMBER OF

NEW TABLES,

WITH ORIGINAL IMPROVEMENTS AND ADDITIONS, AND A LARGE VARIETY OF NEW AND IMPORTANT MATTER:

ALSO,

MANY THOUSAND ERRORS ARE CORRECTED,

WHICH HAVE APPEARED IN THE BEST SYSTEMS OF NAVIGATION YET PUBLISHED.

BY NATHANIEL BOWDITCH,

FELLOW OF THE AMERICAN ACADEMY OF ARTS AND SCIENCES.

ILLUSTRATED WITH COPPERPLATES.

First Edition.

PRINTED AT NEWBURYPORT, (MASS.) 1802,

BY

EDMUND M. BLUNT, (Proprietor)

FOR CUSHING & APPLETON, SALEM.

SOLD BY EVERY BOOKSELLER, SHIP-CHANDLER, AND MATHEMATICAL-INSTRUMENT-MAKER, IN THE UNITED STATES AND WEST-INDIES.

sion. After the printing press was introduced in the 1500's, the need for books on navigation resulted in the printing of a series of manuals that became increasingly valuable to the mariner throughout the next four hundred years. A landmark among these was Nathaniel Bowditch's *The New American Practical Navigator,* first published in 1802.

Bowditch vowed to put in his book nothing that could not be understood by the average seaman. His simplified methods were easily grasped by the seamen of the day, and his manual paved the way for "Yankee" supremacy of the seas during the Clipper Era. When Bowditch died the newly established U.S. Naval Depot of Charts and Instruments bought his copyright; a century and a half after its first appearance, the manual is still published and sold by the Government. More than seven hundred thousand copies have been printed in about seventy editions. Generations of navigators have considered Bowditch as their final authority.

Harmonic Analysis of Tides

In England during the early 1800's scientists were working on the problem of predicting tides. Among them was Sir John Lubbock, who analyzed a nineteen-year cycle of tide observations and identified most of the tide-producing component forces. In 1833 he published his first table for four English ports, based on analyses of the tidal history of each port.

In 1867 the British Association for the Advancement of Science appointed a committee to improve methods of harmonic analysis for tidal observations. A report on the subject prepared by William Thomson was published in 1868. Thomson, who later became Lord Kelvin, contributed many papers devoted to the theory of tides, but he is most prominently associated with the harmonic method of analysis and prediction. In 1872 he invented the tide-prediction machine, although the first such machine for practical use was not constructed until a few years later. He was supported in his work by Lord Rayleigh (John William Strutt) and young George Darwin, son of Charles Darwin, the famous biologist.

The harmonic method was greatly refined by Sir George Darwin in the late 1800's and early 1900's. Modern methods of analysis may appear to differ greatly from those he standardized, but they are essentially the logical development of his earlier work.

Harmonic analysis of tides is based on the assumption that the rise and fall of the tide in any locality can be expressed mathematically as the sum of a series of harmonic components related to astronomical conditions. For example, the relative positions of the earth, moon, and sun are continuously changing. This alters the direction and magnitude of the sum of tide-producing forces.

Consider some of the harmonic component forces produced by the earth, sun, and moon. When all three are nearly aligned—a condition known as syzygy—the independent tidal bulges produced by the sun and moon are superimposed on each other. The tidal range, or difference between maximum (high) water and minimum (low) water, is at its greatest; the tides produced at this time are called "spring tides," and they occur when the moon phase is new or full. During periods known as quadrature, axes between the earth and moon, and between the earth and sun, form nearly a right angle. Then the sun's tidal bulge is not superimposed on the moon's, but occurs where the moon's tidal forces produce low water. The resulting effect is reduction of the total tidal range, a condition termed "neap tide." During any lunar month (twenty-eight days), a tide station will experience two spring tides and two neap tides.

The orbit of the moon around the earth is not circular, but somewhat elliptical. The distance between the centers of the two bodies is therefore continually changing, and so are the tide-producing forces. When the moon is nearest the earth (perigee), the tidal range is increased and perigean tides occur. Smaller, apogean tidal range is experienced when the moon is farthest from the earth, in the apogee of its orbit.

There are also variations in the semidiurnal tides (two highs and two lows) generated by the continually changing declinations of the sun and moon. When the moon is at its maximum semimonthly declination (either north or south of the equator), tropic tides occur, and the inequality recorded in the tidal highs and lows is at a maximum. When the moon crosses the equator, the inequality in tides is at a minimum and equatorial tides occur.

These examples illustrate some of the variations in the separate tide-producing forces that produce the total tidal response. Harmonic constants are derived for each force at a given locality; these are used for predicting tides at that station by reuniting the constituents in accordance with known astronomical conditions prevailing at the time for which predictions are being made.

Finding the Gulf Stream

During the 1800's early investigations were undertaken to discover the major patterns of ocean

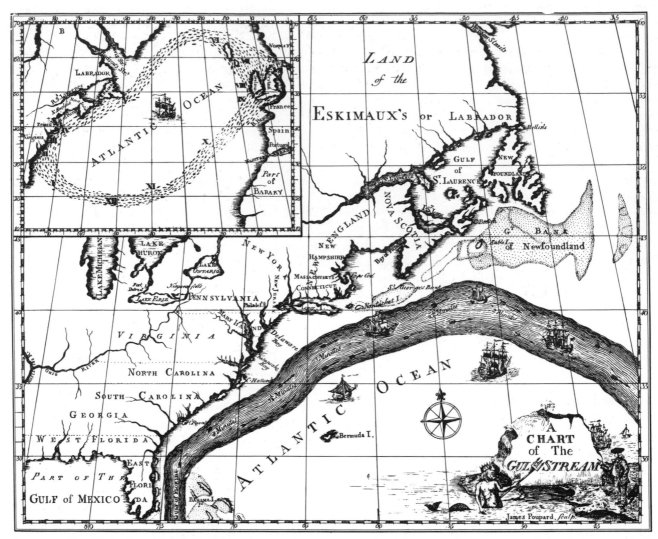

Benjamin Franklin's chart of the Gulf Stream records velocity in "minutes," which correspond roughly to knots, or nautical miles per hour. Franklin advised mariners to use thermometers to determine when they entered the current, thus taking advantage of it when sailing east and avoiding it on their return.

wind and current circulations, and the causal relationships between them.

Benjamin Franklin was one of the early pioneers in the field of observing currents. The chart that he compiled while serving as Postmaster General for the American colonies is believed to be the first to have shown the great Gulf Stream. It was issued to masters of vessels carrying mail packets between the Colonies and England. Franklin advised ships' officers to use thermometers to determine when they entered the warm Gulf Stream waters so as to take advantage of the favorable current when they were steering eastward and to avoid it when sailing west. Franklin attributed the Gulf Stream current to the action of the trade winds. This was the prevalent view among mariners for nearly a century until Lieutenant Matthew Fontaine Maury, in 1860, believing that the winds were insufficient to produce the current, advanced another hypothesis.

Maury was put in charge of the Navy's Depot of Charts and Instruments in 1841, relieving Charles Wilkes of command. During his nineteen-year tenure Maury's contributions to oceanography were truly remarkable. He systematically went through old ship logs and extracted such information as the daily position of the vessel by latitude and longitude, the force and direction of the currents, the temperature of the air and water, the prevailing force and direction of the wind, the height of the barometer, and the state of the weather. The persistent efforts of Maury in the collection of worldwide data pertaining to the oceans led to the compilation of his first wind and current charts of the North Atlantic in 1847.

Maury was one of the major promoters of the Brussels Maritime Conference in 1853, at which he issued a plea for international cooperation in data collection; he was answered by a deluge of ships' logs. From the logs, he correlated huge volumes of

70

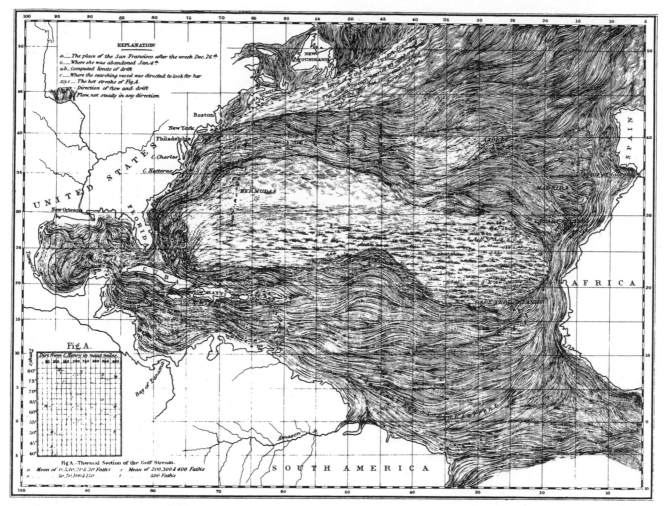

Matthew Fontaine Maury, who followed Wilkes as head of the Navy's Depot of Charts and Instruments, systematically studied wind systems and ocean drifts. This chart of the Gulf Stream and North Atlantic surface currents is from his *Physical Geography of the Sea,* published in 1855.

information, which he then reduced to further improve and refine his wind and current charts. Seasonal wind and current variations evident in the observations were included in his charts. Mariners saved many days of sailing time and much money—perhaps $40 million a year in the late 1800's—by following his sailing directions. In 1854, for example, the British estimated that Maury's sailing directions shortened the passage by thirty days from the British Isles to California, where gold had recently been discovered. In addition, on the basis of Maury's research and recommendations, the first telegraph cable was laid between England and the United States in 1853.

For all his observations, Maury appears to have had a much better understanding of the wind systems of the earth than of its ocean current systems. He wrote, "winds have little to do with the general system of aqueous circulation in the ocean." He felt, on the basis of temperature observations, that deep-ocean currents circulated in such a way as to replace the water moving in the surface currents. Maury theorized that the principal driving force of ocean currents was the difference in density between equatorial waters and the cooler waters of the polar region. Because of greater solar warming in the equatorial regions, their waters were expanded and stood "higher"; they would then move "downhill" from the equator toward the poles.

The Scottish geologist James Croll took issue with Maury's view and championed a rival hypothesis, which had been introduced nearly one hundred years earlier by Benjamin Franklin. Croll contended that the force of gravity on a slope of the sea's surface must surely be too slight to produce the existing circulation system of ocean currents. He said that frictional drag of the winds across the water surface caused surface ocean currents, as evidenced by the coincidence of the major wind and current systems at the ocean-air interface. The Franklin-Croll theory seems more nearly correct than Maury's, although neither hypothesis is adequate to fully explain the

71

Exploring the Oceans

Unlike any other part of our solar system, the earth is truly a water planet. Its five oceans, covering more than 70 percent of the globe's surface, average two miles in depth and contain some 330 million cubic miles of water. Aeons ago life on earth probably began in the warm shallows of primitive seas, and the oceans support life today. Without water the earth would be as desolate as the moon or Mars.

The oceans have always fascinated mankind, but they have terrified him, too. For most of recorded history, man has had to stay very close to shore. Neither his maps, nor his ships, nor his navigational instruments were good enough to enable him to voyage safely beyond the sight of land. Even the ancient Phoenicians, who traded throughout the Mediterranean and sailed as far away as Britain (one Phoenician captain may have circumnavigated Africa), usually kept to coastal waters.

The depths of the ocean have held an even greater mystery than its surface. In the fourth century B.C. Alexander the Great, conqueror of a large portion of what was then known of the land world, is reported to have had himself lowered into the Mediterranean in a crude diving bell. There he extended his sovereignty to the underwater world. Except for divers who relied on the strength of their lungs to search for sponges, pearls, or goods from sunken ships, Alexander's alleged venture was not to be duplicated for nearly two thousand years.

The ocean frontier was first explored systematically in the fifteenth century by Portuguese captains sailing under the direction of Prince Henry the Navigator. Under his direction, Vasco da Gama rounded the Cape of Good Hope and found a water route to India. Christopher Columbus, sailing for Spain but a product of Prince Henry's school, rediscovered the New World. In 1519–21, Ferdinand Magellan, a Portuguese naval officer who had transferred his allegiance to Spain, captained an expedition that circumnavigated the globe. Thus Prince Henry, in addition to extending man's knowledge of the world ocean, was largely responsible for the founding of the empires of Spain and Portugal.

The beginnings of modern navigation date from the voyages of Captain James Cook between 1768 and 1779. In three scientific expeditions to the Pacific, Cook made soundings to depths of 200 fathoms, and frequent and accurate observations of winds, currents, and water temperatures.

While Cook was exploring the South Seas, a Connecticut Yankee named David Bushnell was developing the first operational submarine—a true ancestor of the deep-diving vehicles that today enable man to plumb the oceans' greatest depths. With a one-man crew, Bushnell's *Turtle* attacked a 64-gun British ship-of-the-line in New York harbor in 1776. Although it failed to sink its target, the *Turtle* managed to evade counterattack and return safely to its base.

During the nineteenth century the British took the lead in the scientific exploration of the oceans. A series of great voyages, beginning with those of the *Lightning* and the *Porcupine* in the 1860's, culminated in the epic three-and-a-half year voyage of the *Challenger* in 1872–76. The *Challenger* scientists took hundreds of depth soundings, discovered and described many new species of marine life, and studied currents, temperatures, and salinities throughout the world. The expedition's reports, filling fifty bulky volumes, set the foundations for every major branch of modern oceanography.

The science of oceanography has matured rapidly in the 100 years since the *Challenger* expedition. While enlarging his ability to explore and probe the oceans from surface vessels, man has greatly expanded his ability to explore the depths. Scuba—self-contained underwater breathing apparatus—invented in 1943 by Jacques-Yves Cousteau and Emile Gagnan, now enables divers to literally immerse themselves in their work.

Small research submarines are conducting increasingly sophisticated investigations. Many are equipped with articulated arms that pick up objects on the ocean floor. Some carry elaborate sampling bottles and temperature-probing apparatus. Bathyscaphes, specially built to withstand high pressures, now descend to the oceans' deepest trenches.

Recent experiments are examining man's ability to live and work in the ocean environment. The U.S. Navy project Sealab, initiated in 1964, and similar studies being carried out in other countries, notably in France, are furthering man's knowledge and explorations of the seas.

The only planet with oceans in our solar system, the earth is a spectacular sight when viewed from space. This photograph of the Western Hemisphere was taken from a satellite at an altitude of 22,300 miles. (NASA)

According to legend, Alexander the Great descended into the Mediterranean in a diving bell lowered from a ship. This painting appeared in a Hindustan manuscript from the period of Akbar, 1556–1605. (Metropolitan Museum of Art, gift of S. Cochran)

Prince Henry the Navigator, who prepared the way for Portugal's golden age of exploration, never went to sea. But he encouraged exploration by establishing a school for mariners and map makers and sending his ships out into the Atlantic and down the west coast of Africa. (National Museum of Art, Lisbon)

On three voyages to a vast and little-known region, Captain James Cook carefully surveyed the Pacific Ocean from the Bering Strait to the Antarctic. On the third trip he discovered Hawaii, where he was killed in a fight with the islanders. (National Maritime Museum, London)

The first operational submarine, David Bushnell's *Turtle*,
attacked a British ship-of-the-line in New York harbor in
1776. Its one occupant cranked the propeller with one
hand while he steered with the other. (Matthew Kalmenoff)

The converted navy ship H.M.S. *Challenger* was assigned by the British Admiralty to investigate "the conditions of the Deep Sea throughout the Great Oceanic Basins." Out-fitted with the best equipment of the time, the *Challenger* logged 68,890 nautical miles over an area of 140 million square miles of sea surface in her epic voyage around the world. (Matthew Kalmenoff)

Submarines and diving equipment have enabled man to explore the sea from its depths as well as its surface. Here a scuba diver examines the anchor of the *Bounty,* the ship on which the famous mutiny took place. (© National Geographic Society)

The bathyscaphe *Trieste* explores the ocean floor. Built in France and purchased by the U.S. Navy, this remarkable vessel descended seven miles to the bottom of the Challenger Deep in 1960. (U.S. Navy Electronics Laboratory)

Sealab II plows through the water at La Jolla, California. Its Man-in-the-Sea experiment was conducted for 30 days in 205 feet of water off San Diego. Scripps Institution of Oceanography is in the background. (Ron Church)

The three-man crew occupies crowded quarters inside the small research submarine *Deepstar 4000*. Equipped with stereo cameras for topographic measuring, *Deepstar* is capable of eight-hour dives to 4,000 feet. (Ron Church)

circulation systems of the oceans. It is only within the last few decades that theories which quantitively describe ocean circulation have been advanced.

Although Maury's theories on currents may have been inadequate, his compilation of empirical data on these phenomena and on wind circulation was truly monumental. In 1855 he published *The Physical Geography of the Sea*, a comprehensive volume commonly considered the first text devoted to the science we now call oceanography.

Upon the outbreak of the Civil War, Maury resigned his post and returned to serve his native state of Virginia. Soon after the war, in 1866, Congress divided the Depot of Charts and Instruments into two activities, the Naval Observatory and the Navy Hydrographic Office. The functions of the Hydrographic Office were broadly increased along lines proposed by Maury. It was not to serve only as a "storehouse for charts and sailing directions" but was additionally authorized to carry out surveys, collect information, and print every kind of nautical chart and publication "for the benefit and use of navigators generally." The name of the Hydrographic Office remained unchanged for nearly a century. In 1962, as a result of a gradual shift of emphasis from charting to oceanography, it was redesignated the U.S. Naval Oceanographic Office.

Early Sea-State Observations

Ocean waves, the undulations of the surface water of the sea, are perhaps the most widely observed phenomenon of the oceans. Mariners under sail have long recognized that waves are caused principally by wind. While the generating wind blows, the resulting waves are called "sea"; waves that continue after the generating forces are reduced are termed "swell." "Sea state" is a term that relates to the appearance, or general degree of roughness, of the sea surface, usually including sea and swell, which commonly coexist. The mariner armed with a working knowledge of the interrelationship between winds and waves could use it to his advantage and operate with a minimum of discomfort and danger. Until the 1800's, however, systematic observation of the wind-wave relationship was not undertaken.

One of the first to investigate the relationship between winds and sea state was a British admiral, Sir Francis Beaufort. While a midshipman, Beaufort made short notes on wind and weather conditions every twenty-four hours; gradually his observations became more comprehensive, and records were made at two-hour intervals.

First mention of Beaufort's efforts is found in his log book of 1806. In it he ordinated the strength of winds on a scale from 0 to 12. Winds of zero force indicated a becalmed state, and wind-force 12 was designated a hurricane state.

Later, as an admiral, Sir Francis served as hydrographer of the Royal British Navy, from 1829 to 1855. The Beaufort Wind Scale was accepted for use on the high seas by the British Board of Admiralty in 1838. International Meterological Congresses through the nineteenth and twentieth centuries have further refined and improved Beaufort's original Wind Scale, modifications of which are still in use today.

The Beginnings of Modern Oceanography

The nineteenth century was an era marked by a succession of great scientific voyages—primarily under the direction of British naturalists—which established the framework of modern oceanography. Outstanding among these voyages was the expedition of H.M.S. *Beagle* (1831–36), renowned for the biological and geological data gathered by Charles Darwin. The scientific cruises of H.M.S. *Lightning* and H.M.S. *Porcupine* in the late 1860's sparked additional interest in marine studies and may be regarded as catalytic events leading to the great expedition of the *Challenger,* which began its three-and-a-half-year odyssey from Portsmouth, England, in December 1872 and returned in May 1876. The scientific and technical leader of this expedition was Wyville Thomson, who, upon his return, was knighted and appointed director of the *Challenger* Expedition Commission. He was charged with distributing to various specialists for analysis the many samples that had been collected and with overseeing publication of the results. Thomson did not live to see his monumental work completed; he died in 1882. Sir John Murray, his fellow scientist, shipmate, and friend, brought to a successful conclusion the analyses and the publication of results of that memorable voyage.

In addition to describing hundreds of newly discovered species of marine life, the *Challenger* scientists made hundreds of depth soundings and determined the exact positions of numerous islands and rocks. Ocean currents, both on the surface and at various depths, were studied. Oceanic temperatures were measured, and their independence of seasonal variation below the depth of 100 fathoms was demonstrated. Constant bottom temperatures over large areas of the oceans were observed. It was further noted that differences in bottom temperatures in

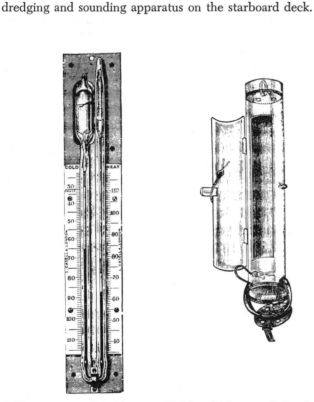

Current drags were used on the *Challenger* expedition to measure currents beneath the surface. The small surface float was the only portion of the apparatus that was influenced by surface winds and currents.

Challenger scientists obtained data that progressively denser and cooler water was found at increasing depths. This illustration from the *Challenger Reports* shows dredging and sounding apparatus on the starboard deck.

Self-registering thermometers (left), which recorded only the maximum and minimum temperatures to which they were exposed during any lowering, were used for most of the *Challenger*'s temperature observations. A series of such thermometers enclosed in cases (right) were spaced along a line and lowered to determine the temperature profile at a given ocean station.

On the *Challenger* expedition, the water piezometer (left) was attached to an oceanographic line to determine pressure of the seawater at a given depth. The hydrometer (right) determined the density or specific gravity of ocean-water samples at surface temperature and surface atmospheric pressure. A free-swinging suspended pan insulated the hydrometer apparatus from the rolling of the ship.

certain ocean areas were due to bottom waters being separated from each other by submarine ridges. For example, the temperature in the North Atlantic at depths exceeding 2,000 fathoms was found to be nearly constant at 36.5° F.; bottom temperature in the North Pacific was constant at 35° F.

It was also found that the positions of surface isotherms (lines connecting points of equal temperature at the surface) are governed primarily by intensity of solar radiation, which varies with latitude, but are modified by the mixing of surface current systems. Positions of isotherms are commonly subject to seasonal variations, especially in the temperate zones. Surface temperatures in the tropics may exceed 80° F., while in polar regions they may fall to 29° F. Data from the *Challenger* expedition showed that progressively denser and cooler water was generally found with increasing depths. Some variation from this rule was observed, however. For example, off the Norwegian coast, warmer water was found beneath denser and cooler water. It was recognized that this unstable arrangement of the water masses could generate a slow-moving current.

In addition to temperature and salinity (the total amount of dissolved salts), the density of seawater is dependent upon the pressure to which it is subjected. The pressure at a given depth was measured with a water piezometer, and seawater salinity was approximated by measuring the specific gravity of an ocean sample with a hydrometer. Later, in 1884, William Dittmar carefully analyzed seventy-seven samples of seawater collected during the *Challenger* expedition to determine the major dissolved constituents. These, together with temperature, govern seawater's specific gravity.

Temperature and salinity are characteristics of the seawater acquired through solar warming, evaporation, precipitation, and dilution by river discharge or the melting of polar ice masses. Both temperature and salinity are considered conservative properties of seawater; once acquired they change only gradually through mixing. Dense water masses with temperatures and salinities characteristic of surface antarctic waters are found along the bottom of the South Atlantic basin in the tropics, thousands of miles from their origin. It was known by scientists of the late 1800's that temperature and salinity, together with some pressure effects (since water is slightly compressible), govern the characteristic densities of seawater, and that an unstable arrangement of seawater densities can generate deep-ocean currents. It remained for the twentieth-century scientists— Vilhelm Bjerknes, Vign Wolfrid Ekman, and others— to develop the principles of ocean-current dynamics.

When the scientific methods of the pioneer investigators aboard the *Challenger* are compared to our modern techniques, we marvel at the scope and quality of their accomplishments. The *Challenger* reports laid the scientific foundation for most aspects of modern oceanography. It is doubtful whether ever again an oceanographic cruise will be made which can compare in scope with the *Challenger* expedition.

In the last quarter of the nineteenth century numerous less extensive, but nevertheless important, expeditions were undertaken by various nations and private individuals. Ships such as the *Tuscarora* (American), *Travailleur* and *Talisman* (French), *National* and *Valdivia* (German), *Vettor* and *Pisani* (Italian), *Ingolf* (Danish), and *Siboga* (Dutch) launched national explorations to investigate further the basic findings of the British *Challenger* expedition. Also, the private efforts of Alexander Agassiz of the United States and Prince Albert of Monaco made notable contributions to the fledgling science of oceanography.

Alexander Agassiz made numerous voyages on the Coast Survey's *Blake* and *Albatross* and chartered steamers for additional surveys. His studies in marine geology and biology and his charting cover more than 100,000 miles in the Caribbean, the Indian Ocean, and the tropical Pacific.

Prince Albert, with ample funds at his disposal, equipped and conducted numerous scientific expeditions from the late 1800's on. In 1885 he began to investigate the surface circulation systems of the North Atlantic. He dispensed some two thousand drift bottles in the Atlantic; sealed instructions in nine languages requested that the finder return the card and indicate location and date of recovery. From these drift-card data, he drafted a fairly accurate chart of surface-current circulation in Atlantic waters and became a leading authority of his time on surface currents. Although the drift-bottle method of studying such currents was old even in the nineteenth century, the technique is still used in some modern current studies.

Navigation in the Electronic Age

Toward the close of the nineteenth century, Heinrich Rudolph Hertz proved James Maxwell's theory of electromagnetic waves and their reflection; in 1895 Guglielmo Marconi transmitted the first wireless message. With these events navigation entered into the electronic age.

The applications of electronics to navigation have been many. The Radio Direction Finder (1907), a practical echo sounder (1922), practical radar (1937), and the development of distance and direction finders such as Hydrodist, are but a few of the electronic advances that aid the modern mariner.

Electronic navigation systems, such as Loran and Decca, have been perfected. Using these systems, the navigator may accurately fix his position by measuring the time interval, or phase difference, between signals received from synchronized transmitters at two different points on land. Such navigation networks as these cover extensive areas of the world's oceans. They consist of transmitters, usually located 200 to 400 miles apart, and charts with hyperbolas connecting all points of equal time (or phase) difference between any two transmitters. Twentieth-century navigation has progressed to a point where the basic problems involved in finding one's position at sea have been solved. The advent of electronics has made possible faster, more accurate solutions to most navigational problems. Much of the detailed work of navigation has been simplified; integrated with automated equipment and experimental satellite navigation systems, it will be simplified further still.

In addition, the navigator is no longer limited to a few charts—he has available a wide variety, in different scales and covering a major portion of known water masses, to back up most of his electronic aids. Photographic coverage of topographical areas adjacent to the coastline, when combined with precise marine survey data, gives a more accurate portrayal of coastal areas. Generally speaking, automation and computerization have made available to the modern navigator an enormous volume of information. High speed, multiple-color printing and the use of standard symbols have assured volume production of easily readable charts that portray information required by the navigator more accurately than ever before.

Modern Tidal Analysis

Although the fundamentals of tidal theory and of harmonic analysis were fully developed nearly a century ago, oceanographers are still unable to predict the tides at a particular location solely from positions of the earth, moon, and sun. Additional complex variables, such as irregular configuration of the shoreline, varying slope of the sea floor, and tidal refraction, have prohibited the construction of adequate mathematical models.

In order to predict accurately the tide at a given locality, tidal observations must be made and recorded for at least half a lunar month (fourteen days), preferably for an entire year. The record is then analyzed, and the various harmonic constants are determined for the principal tide-generating forces.

At the close of the nineteenth century the art and science of tide prediction was advanced by an American, William Ferrel, who was associated with the U.S. Coast and Geodetic Survey. Ferrel conducted researches on tidal theory and independently developed the harmonic method of analysis. In 1884 he devised a mechanical analog tide-prediction machine, which could handle up to nineteen harmonic constituents. Another American mathematician of the time, who was also with the Geodetic Survey, was Rollin A. Harris. Harris advanced the hypothesis that in each ocean there are regions capable of oscillations with a period close to that of a principal tide-producing force. He issued world charts showing co-tidal lines, i.e., lines joining points where high water occurs at the same hour. Together with E. G. Fischer he devised, for the Geodetic Survey, a mechanical analog tide-prediction machine that could handle up to thirty-seven tidal constituents; it replaced the old Ferrel machine in 1914. Since 1965 the work of these analog machines has been accomplished by computer.

Not only have tide analyses and prediction been computerized, but the world network of tide-observation stations has been greatly expanded, and methods of obtaining tidal information have changed rapidly during this century. The old mechanical gauge staff recorded water levels by drawing a stylus across a revolving drum that was driven by a clock mechanism. This is being replaced by more sophisticated electronic recorders, which encode tidal data for radio transmission and direct computer input.

Study and prediction of tides and tidal currents continues to expand, because of their many important applications. Tidal data is necessary for safe navigation, especially in harbor areas; dredging and maintenance costs may be reduced by constructing new harbor facilities to take advantage of the flushing action of tides. Detailed tide prediction is required for ocean engineering problems, such as construction of sea-level canals and protection of shore and harbor areas. From a military point of view, the need for tide predictions in amphibious operations is obvious.

Additionally, since tidal ranges are affected by the configuration of the shoreline and the slope of the bottom, tides funneled into some restricted estuaries may build up to fifty feet and tidal currents of as much as ten knots may occur. Some areas of great tidal ranges are the Bay of Fundy, Bristol Channel,

William Ferrel, of the U.S. Coast and Geodetic Survey, developed a mechanical analog tide-predicting mechanism, which could handle as many as nineteen tidal harmonic constituents. (U.S. Coast and Geodetic Survey)

The exact times of high and low water and the approximate values of tidal heights were read directly from the dial of the Ferrel Maxima and Minima Tide Predictor. The "Ferrel" was retired from service in 1914, and it is now displayed in the Smithsonian Institution. (Environmental Science Services Administration)

and the Sea of Okhotsk. The famous tidal phenomenon in the Bay of Fundy, between Nova Scotia and the Canadian mainland, moves more than 100 billion tons of water a day, and displays normal tidal ranges of fifty feet.

Surface Waves

Building upon the empirical observations of waves and wind initiated by Sir Francis Beaufort, observers in the early twentieth century continued to study the relationship between generating winds and the resulting waves. Scientists such as Thomas Stevenson and Vaughan Cornish developed empirical formulas relating wave height to wind velocity, duration, and fetch (the distance of open water over which a wind blows). Fetch, velocity, and duration are involved simultaneously, and diagrams were developed that showed this interrelationship. Sir George Gabriel Stokes developed a wave theory for two-dimensional oscillatory waves; he assumed a frictionless, homogeneous, incompressible fluid of uniform depth. In deep water the velocity of the wave's form is directly related to its length: The greater the length, the greater the velocity of the form.

Since a long wave with a large period (i.e., a relatively long time between the passage of that wave and its successor), has greater speed, swell generated in a storm area tends to become sorted, as waves with large periods outdistance the others. Swell dissipates very gradually, and such waves are commonly recorded hundreds of miles from their area of generation. Mariners have long known that large, long-period swells are often forewarning of an approaching storm.

Although the theoretical relationships between wave speed, length, and period—as well as water depth—are helpful in understanding general wave behavior, they are not of much use in actual forecasting. The time required for a wave system to travel a given distance is about twice that which would be expected from the speed of an individual wave. This is because an individual wave front gradually dissipates and transfers its energy to succeeding waves. This happens to each successive wave front at such a rate that the wave system advances at just half the speed of its individual waves. The speed at which the wave system advances is called its "group velocity."

The complex and irregular nature of actual ocean-wave systems cannot be entirely described in terms of theoretical wave-form relationships. This complexity is due, in part, to the coexistence of many independent wave systems and to the way they interfere with each other. In nature there seldom exists a single wave form, but rather a combination of waves of several sizes. This combination can best be described in terms of the frequency distribution of wave sizes known as the wave spectrum.

During the Second World War attention was di-

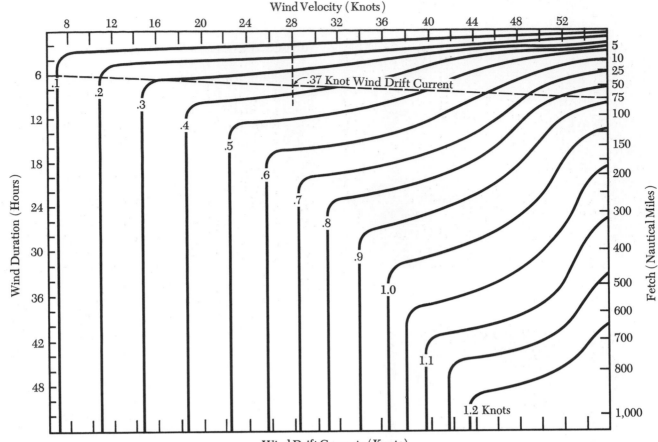

Wind drift current forecasting curves are based on duration, velocity, and fetch. For example, a wind velocity of 28 knots over a fetch of 75 miles for a duration of 6 hours will produce a wind drift current of 0.37 knots.

Waves, whether violent or symmetrical, consume much of the energy expended in propelling a ship. There is therefore a significant economic motivation behind wave study. (U.S. Navy)

rected to predicting the waves most likely to exist during amphibious landing operations. Harald Ulrik Sverdrup and Walter Heinrich Munk, who worked on this problem in its early stages, introduced the "significant wave" forecasting method in 1947. Their method is based on semiempirical relationships, as well as theoretical considerations. They introduced the concept of "significant wave height," which is the average height of the highest third of the waves in any wave train. The significant-wave period is simply an average of the periods of this same highest third. This average is important because it approximates the period of maximum wave energy.

In 1951 Charles L. Bretshneider revised the Sverdrup and Munk method of wave forecasting, and in 1955 another system was developed by W. F. Pierson, G. Neumann, and R. W. James. Combining theoretical and practical aspects, it resulted in improved description of the sea's surface and improved prediction of its reactions. The wave forecaster's aim is to predict a least-time track, a maximum cargo-safety track, or a maximum passenger-comfort track, depending upon the requirements of individual ships.

Experimental routing was initiated by the U.S.

Meteorological and sea-state observations are reported routinely from many ships at sea. Such information is used in wave forecasting. (Environmental Science Services Administration)

Submarine operations at depths in the vicinity of a sharp density-interface may be hampered by large internal waves. (U.S. Navy)

Naval Oceanographic Office in 1956, and by 1961 several commercial and governmental organizations were predicting optimum ship routes for the mariner.

Information from a network of meteorological and oceanographic stations and from ship reports is linked through computer installations for processing weather and oceanographic data. Wave and weather forecasts are then transmitted to ships at sea on a routine basis. Forecasts of sea and swell conditions are now broadcast twice a day by the U.S. Naval Weather Service to ships at sea.

Internal Waves

In addition to studying waves at the air-sea interface, twentieth-century oceanographers are turning their attention to the study of internal waves. An internal wave occurs well beneath the surface, along an abrupt interface separating water layers of different densities. Although poorly understood, the effect of internal waves has been observed by mariners since the advent of propeller-driven vessels. Their effect may be encountered in a region where river discharge spreads fresh water in a thin wedge

out over the denser seawater. A slow-moving ship may gradually stop, and seem "stuck," although the engine setting remains unchanged. This phenomenon occurs when the density interface is at approximately the depth of the vessel's propeller, and it is known to mariners as "dead water." What actually happens is that the normal propulsion energy of the ship is expended in generating internal waves along the boundary between the different water densities. This condition of "dead water" can be easily overcome by increasing the engine speed a few knots.

More important are the internal waves along density interfaces caused by temperature differences at depths of 300 to 1,000 feet. The full significance of these waves has not yet been determined, but it is known that they may cause submarines to rise and fall as a ship does at the surface, and they may also affect the transmission of sound in the sea, causing sonar aberrations.

Internal waves have been studied only during the last thirty or forty years. In 1935 Börje Kullenberg published studies made in the Kattegat between Denmark and Sweden. Since then, scientists such as C. S. Cox and E. C. La Fond at Scripps Institution of Oceanography, and O. S. Lee at the U.S. Naval Electronics Laboratory in California, have actively studied this phenomenon, which is such an important consideration in submarine operations. Because, in the open sea, major changes in the vertical density gradient are controlled by temperature, temperature measurement is the most common method of indirectly locating the density boundary along which internal waves occur. A vertical string of temperature-sensitive thermistor beads has been successfully used to record the change in depth of the density boundary during the passage of an internal wave.

Because of the relatively slight changes in the density gradient within the water column compared to the variation at the air-sea boundary, a comparatively small amount of energy is required to set up and maintain an internal wave. Internal waves are often much larger than surface waves: The maximum height of wind-waves on the surface is about 60 feet, but internal waves as high as 300 feet have been recorded. A submarine operating within an internal wave system will experience large changes in buoyancy as it encounters the successive crests and troughs of the internal waves.

Shoaling Waves

As waves impinge upon a shoreline, they no longer obey the laws that apply to deep-water waves. In deep water, the velocity of a wave is a function of its length. In shallow water, where water depth is less than half the wavelength, wave velocity is a function of the depth. As a wave shoals, the orbital wave path of a parcel of water is elliptically deformed by friction with the bottom, until it becomes essentially a back-and-forth translation of water along the bottom. This friction also slows the wave as it nears the shore. If a wave front approaches the shore from an angle, the shoaling effect can cause the front to be bent or refracted until it approaches as a line nearly parallel to the beach. The refraction of waves as they impinge on an irregular shoreline tends to concentrate wave energy on headlands and capes,

while reducing the wave energy expended in bays and coves. The direction of approach of refracted waves is not exactly normal (i.e., perpendicular) to the shoreline, but at some slight angle. Wave translation in the surf zone is resolved into a small component of movement in a direction parallel to the shoreline, which generates a longshore current. Such a current is capable of moving large volumes of sediment along the beach and the area of ocean bottom near the shore.

The effects of longshore currents in terms of beach erosion may be quite costly. The problem has been intensely investigated during the present century by many groups, such as the Army Corps of Engineers, which is charged with maintaining coastal navigational channels and harbor approaches. For the Navy, an understanding of shoaling waves is critical for amphibious operations and assessment of beach trafficability.

As a wave impinges on the shore, its speed and length are decreased, but its height and steepness tend to increase. When a critical ratio of wave height to wave length is reached, the wave becomes unstable and "breaks" forward. During periods of heavy weather, the effect of these breaking waves may be very damaging to life and property.

Tsunami Waves

Destructive ocean waves called seismic sea-waves or tsunamis (from the Japanese name), are generated by earthquakes and volcanic explosions. Throughout history the onslaught of seismic sea-waves has repeatedly devastated coastal settlements along the shores and islands of the Pacific.

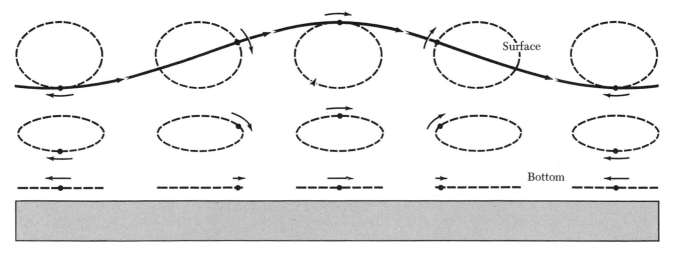

Waves "feel" the bottom when the water depth is less than half the wavelength. At that point, the water parcel's orbital path becomes elliptically deformed and the wave velocity slows.

88

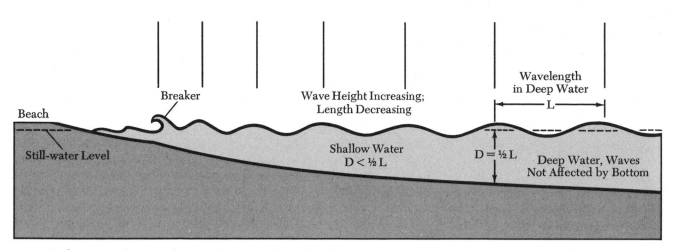

Beach

Breaker

Wave Height Increasing;
Length Decreasing

Wavelength
in Deep Water
L

Still-water Level

Shallow Water
D < ½ L

D = ½ L

Deep Water, Waves
Not Affected by Bottom

As the wave advances shoreward, its speed and length decrease, but its height and steepness tend to increase.

The Pacific Ocean's basin is ringed by a zone of major active fractures of the earth's crust and by an attendant ring of volcanic activity. Frequent earthquakes and volcanic events make this the most active seismic region on earth, and they spawn the great seismic sea-waves. These waves are generated by resonant vibration and displacement of the sea surface as vast quantities of energy are released during an earthquake or submarine volcanic eruption. The waves spread across the ocean as concentric rings of crests and troughs, like the ripples generated when a pebble is tossed into a pool.

Over the deep ocean the length of a tsunami from crest to crest may be several tens of miles, with a height of only a foot or two. Ships at sea are unaware of the passage of the wave form, which travels at speeds of up to 600 miles per hour. When a wave reaches coastal waters, where shoaling occurs, it is slowed down, and the water behind piles up to tremendously destructive heights. The arrival of a tsunami is often heralded by an initial gradual recession of the coastal waters. It is a warning to be heeded, for tsunami waves can crest to heights of more than a hundred feet and strike with devastating force.

Every island and coastal settlement in the Pacific Ocean area is vulnerable to the onslaught of seismic sea-waves. The waves of 1868 and 1877 devastated towns in northern Chile and caused death and destruction across the Pacific. A series of seismic sea-waves generated by the eruption and collapse of Krakatoa in 1883 killed more than 36,000 people in the East Indies. Japan lost 27,000 lives to the wave of 1896 and 1,000 more in 1933. In 1946 fifty-foot-high tsunami waves struck Hawaii, killing 159 people, injuring 163 others, and causing property damage estimated at $25 million. It was the thirty-sixth tsu-

Storm waves break with great force upon a sea wall along the coast of Britain. (Environmental Science Services Administration)

During periods of heavy weather, the effect of breaking waves may be damaging to life and property. This aerial view was taken over Galveston, Texas, after Hurricane Carla in the fall of 1961. (U.S. Navy)

89

In March 1964 seismic sea-waves washed these ships ashore and caused extensive damage to the coast at Kodiak, Alaska. (Environmental Science Services Administration)

nami recorded in Hawaii in 127 years. In March 1964 seismic sea-waves destroyed much of Valdez, Alaska, and brought extensive destruction and loss of life to Kodiak in that state. There have been hundreds of these savage waves; prior to 1946, they usually struck without warning.

The tsunami of 1946, which was the worst natural disaster in Hawaii's history, prompted scientists to attempt to establish a warning system. Within two years scientists of the U.S. Coast and Geodetic Survey at the Honolulu Observatory established a functioning Seismic Sea-Wave Warning System (SSWWS).

The Honolulu headquarters is the center of an extensive complex of seismograph stations, awaiting the first signal indicating an earthquake of sufficient magnitude to generate a tsunami. From the seismogram, scientists can determine the quake's magnitude and its epicenter, i.e., the position on the earth's surface directly over its place of origin. On the basis of seismographic evidence the Honolulu Observatory issues an advisory bulletin with the estimated time of the tsunami's arrival at various stations throughout the area. The first positive indication of the existence of a tsunami usually comes from tide-gauging stations nearest the disturbance. Tsunamis appear on the tide-gauge records as distinct abnormalities in the normal curve. When confirmation of a seismic sea-wave is received, the Honolulu Observatory issues a warning of the approach of a potentially destructive tsunami, with revised estimates of arrival times. Because of their great wavelength, sometimes in excess of a hundred miles, tsunamis behave essentially as shallow-water waves; their speed is determined solely by water depth, and this fixed relationship makes it possible to forecast their arrival times at distant locations with an error of less than about two minutes per hour of estimated travel time. Local evacuation and emergency procedures can then be initiated, often hours in advance of the waves.

Ocean Currents

Large-scale oceanic currents are the product of wind systems and density differences among water masses. Even today there is argument among oceanographers as to which of these two influences is of primary importance.

Reliable charts of prevailing winds and ocean currents have been available to the mariner since the work of Maury in the mid-1800's. The first evidence that slow-moving currents, transporting enormous volumes of water, could be generated by differences in the density of seawater masses was provided by density measurements made on the *Challenger* expedition. The general dynamic principles and theory of ocean circulation were not introduced, however, until the early twentieth century. The contributions of Norwegian physicist Vilhelm Bjerknes in the early 1900's established the theoretical principles of fluid dynamics that describe the current movements of the oceans. Bjerknes demonstrated that the very small forces resulting from pressure differences caused by nonuniform density of seawater could initiate and maintain fluid motion. The principle relating ocean currents to the density structure of the sea is generally known as the Bjerknes Circulation Theory.

The slope of an isobaric (uniform pressure) surface may be determined from analysis of water temperature and salinity. The deep-ocean currents generated are a function of the slope of the isobaric surface. Currents generated along the slope initially move directly "downhill," but once in motion, they are deflected by the Coriolis effect, which is an ap-

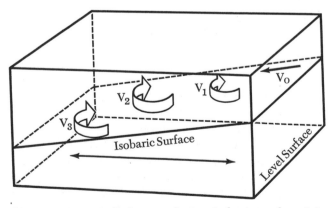

Currents generated along a sloping isobaric surface initially move in a downhill direction, but they are deflected by the Coriolis effect of the earth's rotation. Currents are deflected to the right in the Northern Hemisphere and to the left in the Southern Hemisphere.

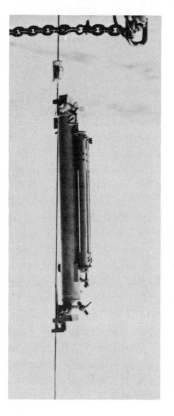

To collect seawater samples, a series of Nansen bottles is lowered along a wire cable. When each bottle has reached its desired depth, a small weight is slid down the wire, causing the top of the bottle to swing away from the cable, while the bottom remains attached. As the bottle reverses, a valve at each end closes, trapping a water sample. At the same time, a weight attached to the first bottle is released to trip the next bottle. (U.S. Naval Oceanographic Office)

Samples drawn from Nansen bottles are later analyzed to determine salinity, dissolved gases, and nutrient content. (U.S. Naval Oceanographic Office)

parent force "exerted" on a moving body by the earth's rotation. In the Northern Hemisphere a moving body is apparently deflected to the right from its direction of motion; in the Southern Hemisphere, deflection is to the left.

The movement of these deep, slow-moving ocean currents is usually measured indirectly by determining the vertical density structure of ocean waters from temperature and salinity profiles. A series of Nansen bottles (named after Fridtjof Nansen, the Norwegian explorer and naturalist who devised this sampling device) are used to collect seawater for salinity determinations, and reversing thermometers measure *in situ* temperature. Although instrument packages for continuous measurement of temperature, salinity, and pressure at an oceanographic station are in use, the bulk of the observations are still made by the classic Nansen cast.

Wind-Driven Currents

A mathematical treatment of ocean circulation was undertaken by V. Wolfrid Ekman at the beginning of this century. Working principally with currents caused by wind stress, he described the influence of winds and the earth's rotation on ocean currents. He found that fully developed surface currents are deflected 45 degrees to the right of the wind direction in the Northern Hemisphere and 45 degrees to the left in the Southern Hemisphere, due to the Coriolis effect. A current induced by wind in the ocean layer nearest the surface sets up current flow in the next deeper layer because of frictional forces; this deeper current, in turn, induces flow in a still deeper layer. Ekman theorized that in a homogeneous medium these deeper currents would gradually decrease in velocity, and would deviate more and more

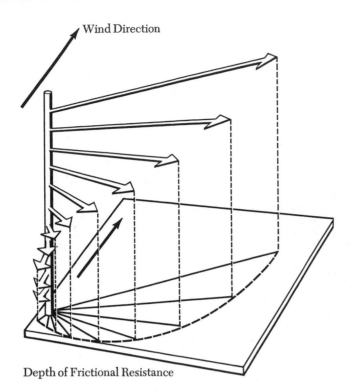

Wind Direction

Depth of Frictional Resistance

A current induced by wind in the surface layer sets up current flows in successive layers. The Ekman Spiral shows the effect of the earth's rotation in deflecting the lower layers in the direction of the Coriolis effect—to the right in the Northern Hemisphere and to the left in the Southern Hemisphere.

in the direction of the Coriolis effect. Plotted as vectors, these currents would eventually describe a spiral. At a depth known as the "depth of frictional resistance," Ekman calculated that the current would flow in a direction exactly opposite to that of the wind, but its velocity would be only about 4 percent of the velocity at the surface layer. The depth of frictional resistance usually falls between 300 to 500 feet.

A surface current in the Northern Hemisphere flows at an angle directed 45 degrees to the right of the wind direction, while net movement of the entire wind-induced current layer, to the depth of frictional resistance, is 90 degrees from that of the wind.

Diverging or converging wind systems at sea can cause a mixing of waters with different densities at the surface. In the open ocean, diverging surface currents are replaced by an upwelling of colder, denser water from below, while converging currents are responsible for a buildup of light surface water.

Along the edge of the continental shelf a similar relationship between wind and water transport manifests itself. In the Northern Hemisphere, when the

coast lies to the left of an observer facing downwind, warm surface water is moved by current transport away from the coast (to the right), and its volume is replaced by an upwelling of denser, cooler subsurface water. The deeper waters are commonly very rich in nutrients; the pronounced coastal upwellings of nutrient-rich water off the western United States and off the coast of Peru provide the nutritional base to support extensive commercial fisheries.

A new major subsurface current, the Cromwell Current in the Pacific, was discovered as recently as 1952. The Cromwell Current is a relatively thin countercurrent moving east at velocities of about one nautical mile per hour, and at a depth of about sixty-five feet, beneath the west-flowing equatorial current. Measurements by William G. Metcalf, Arthur D. Voorhis, and M. C. Stalcup (1962) of the Woods Hole Oceanographic Institution in March and April 1961 clearly demonstrated the existence of an equatorial undercurrent in the central Atlantic, which is analogous to the one in the Pacific. Both the Atlantic and Pacific equatorial undercurrents are characterized by a high-velocity current core near the depth region of the thermocline, which marks the rapid drop in temperature between warm surface waters and cooler, deeper waters.

Our understanding of the behavior of surface and subsurface currents is still incomplete. Movement of these surface water masses has far-ranging implications, not only as they affect the mariner's course, but also as they influence coastal climates, fisheries, and commerce, and the redistribution of such contaminations as atomic wastes. For these reasons, scientists at many institutions around the world, using a variety of tools, continue to investigate the behavior of wind-driven water circulation.

The design of a current meter that takes automatic readings has made possible more precise measurements both at the surface and at intermediate depths. The current at a given location can be monitored at different depths simultaneously by attaching meters at intervals to a deep-moored surface buoy, with a transmitter in the buoy relaying information to a remote receiving station. Currents at intermediate depths can also be investigated by deep-sea drifters—targets that will neither sink nor rise to the surface but remain neutrally buoyant in a water mass of a given density—which can be tracked by sonar from surface ships.

In spite of these more modern tools, surface-current studies still rely, as well, on drogues and dye dispersion, and on the age-old methods of examining navigators' reports and collecting drift bottles.

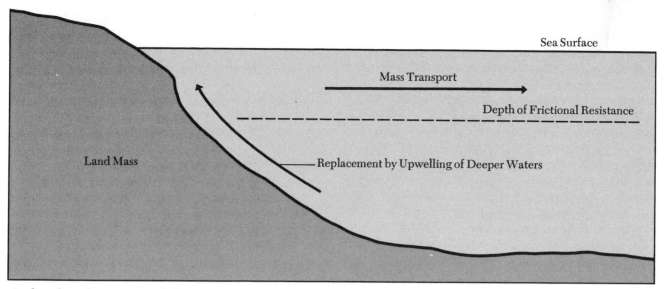

At the edge of a continental shelf the current carries the warm surface water away from the coast. Cooler subsurface water rises to replace it. Nutrient-rich upwellings such as the one in this diagram support extensive commercial fisheries off the coasts of the western United States and Peru.

The Dark Ocean

Because the ocean is a very poor transmitter of light, the sun warms and illumines only its surface waters. Even in the clearest ocean water the light intensity at depths of 600 feet is reduced to less than one percent of the surface illumination. The absorption is so complete that at depths little below 600 feet the ocean is cloaked in inky blackness.

The strong absorption of light and electromagnetic radiation has been most frustrating to man. Radio transmission beneath the sea is nonexistent. Oceanographers searching in depths greater than 600 feet for the nuclear device lost near Palomares, Spain, were severely limited by darkness and lack of radios. Even working from submersibles such as *Alvin* and *Aluminaut*, which carried powerful light sources, the search width of illumination was generally restricted to less than 100 feet. Furthermore, the exact position of a submersible and its search path cannot be determined aboard the submersible by conventional electronic navigational systems because of the absorption of electromagnetic radiation. Similarly, since conventional navigation systems are useless underwater, it has been necessary to develop an elaborate inertial navigation system and a system of acoustic positioning tied together by an electronic computer for nuclear submarines that remain submerged for weeks at a time.

The blue color of the open ocean has long been a puzzle to scientists and mariners alike. In 1895 the Swiss geographer, François Alphonse Forel, intro-duced a set of color standards for the observation of water color in Swiss lakes. His scale was made by mixing different proportions of end-member blue (copper sulfate) solution and a yellow (potassium chromate) solution and sealing the mixtures into tubes one centimeter in diameter. The colors in the tubes were then compared visually to the parent color of the lake water. This color scale, known as the Forel Scale, is still generally used by oceanographers to describe the color of the ocean. Numerous other scales have subsequently been introduced, but all are modifications of the Forel concept.

While methods of recording water color had been developed by the turn of the century, the theory to explain such color was still missing. In 1910 Lord Rayleigh (John William Strutt) published a study on the color of turbid media, and subsequently many attempts were made to explain the color of the sea entirely by the scattering of light from the fine suspended particles in the water. It is now recognized, however, that the relative absorption of the solar energy spectrums by the water molecules also plays a part in determining water color.

Visible light is composed of electromagnetic radiation ranging from the longer wavelengths of red-colored light through orange, yellow, green, to the shorter wavelengths of blue color. The various wavelengths are not absorbed in water at the same rate. Light of the colors with longer wavelengths, such as red and orange, is absorbed most rapidly.

For example, the red portion of the visible spectrum is reduced to 10 percent of its surface intensity at a depth of fifteen feet. The remaining wavelengths of visible illumination penetrate further. The blue colors of short wavelengths are the last to be extinguished. Ten percent of the surface intensity in the blue region of the spectrum remains at depths of 600 feet. This differential rate of absorption in part explains the blue color characteristic of the water of the open ocean, the color being due to the backscattering of the short wavelengths, which are the last to be absorbed.

It should be mentioned that although the color of the sea is often blue due to backscattering, this color changes constantly to reflect the color of the sky. A leaden overcast, an orange sunset, the passage of clouds across the face of the sun, all change the color and mood of the sea. In coastal regions green water is commonly reported, due to the pigments of plankton. Also, the discharge of suspended sediment from the outflow of large rivers often colors the water many miles offshore. The Yellow Sea is so named from the sediment-laden water discharged by several great rivers from the mainland of China. The muddy discharge of these rivers spreads seaward for nearly a hundred miles.

Sound in the Ocean

The ocean, while nearly opaque to light, is relatively transparent to sound. Since before the time of Columbus, mariners have known how to pick up the noises of distant ships by placing a tube in the water and listening at the other end. This primitive technique was used widely as recently as World War I for listening for submarines.

While the first measurement of the speed of sound in water was made in 1872 by Daniel Colladon and Charles Sturm in Lake Geneva, Switzerland, not much was known about the behavior of sound underwater until World War I provided the impetus for research in this field. By 1916 experimental underwater sonars had been developed. These were able to produce sound pulses and obtain reflected echoes at distances of up to about 600 feet. By 1918 echoes were being obtained from submarines as far as 4,650 feet away. The invention of these devices came too late in the war, however, to be effective against the German U-boat threat.

Today, following the many improvements in underwater listening and echo-ranging sonars that were inspired by World War II, oceanographers have at their disposal a wide variety of acoustic devices. These include acoustic depth-finders, acoustic de-

vices known as "pingers" for positioning an instrument a known distance above the bottom, and acoustic sub-bottom profilers for determining the configuration of rock units lying beneath the sea floor, as well as acoustic methods of underwater voice transmission. The mariner today, especially the submariner, depends heavily on the behavior of sound in the sea. Sound serves as the submariner's eyes; through the use of scanning sonar he can determine the location of underwater objects or targets.

From a source at one point in a homogeneous medium, sound spreads uniformly in all directions. With distance there is a gradual reduction in intensity. This is due to the spreading of the sound; to its absorption, as acoustic energy is converted to heat by friction resulting from water viscosity; and to the scattering of acoustic energy as the sound waves are randomly reflected by small suspended particles such as sediment, plankton, or bubbles. Reflection occurs whenever sound waves are obstructed by solid objects such as the sea floor or a ship's hull. This is, of course, the principle of sonar. An emitted acoustic pulse will be reflected from the target as an echo or "ping." The greater the range, or distance to the target, the smaller will be the amount of energy returned as an echo, due to the combined effects of spreading, absorption, and scattering. Nevertheless, the range of many sonar systems is quite great—up to several thousand yards, depending upon the initial intensity of the acoustic source. If sound velocity is known, the distance to the target can be determined by calculating the time between the emitted acoustic pulse and the returned echo. The calculation is a simple one if it is assumed that the sea is a homogeneous medium and that the speed of sound through it is therefore constant.

In actuality, however, sound speed in the sea varies from 4,700 to 5,300 feet per second, depending on the elasticity and density of the water. Elasticity is determined mainly by pressure or water depth. The density of seawater in the open ocean is more strongly influenced by temperature than by salinity. Since sound speed is inversely related to density, an increase in temperature will result in increased speed. From these relationships it is apparent that with regard to sound velocity the ocean is not homogeneous but stratified primarily in response to the combined influence of temperature and pressure.

Sound rays are bent or refracted due to changes in their velocity within the sea. Sound tends to be bent toward, and often confined to, zones within the ocean known as "sound channels." These low-velocity channels commonly occur in the dense cooler waters just beneath the permanent thermocline—the layer that separates the cold deeper waters from the less dense warm water of the surface. Sound traveling in a channel can be detected thousands of miles from its source. The discovery of permanent sound channels in the ocean has led to the development of many methods of underwater communications and many navigational aids.

"Sofar" (Sound Fixing and Ranging), which was developed during World War II, takes advantage of the sound channel as a method of determining the location of survivors at sea. By means of triangulation shore stations determine the approximate position of survivors by detecting sounds of exploding depth bombs preset for the approximate depth of the sound channel.

The mariner, particularly the submariner, has an urgent need for charts depicting expected sonar conditions based on the thermal structure of the oceans. For example, because of the refraction of sound rays downward through the thermocline submarine commanders during World War II would evade sonar by positioning their craft beneath the thermocline's depth. If they were detected, the refraction of the sound would prevent the sonar from indicating their true position.

The Tools of the Trade

The growth of oceanography in the past century has been remarkable. More than three hundred and

Self-recording bottom current meters detect the speed and direction of undersea "rivers." (U.S. Naval Oceanographic Office)

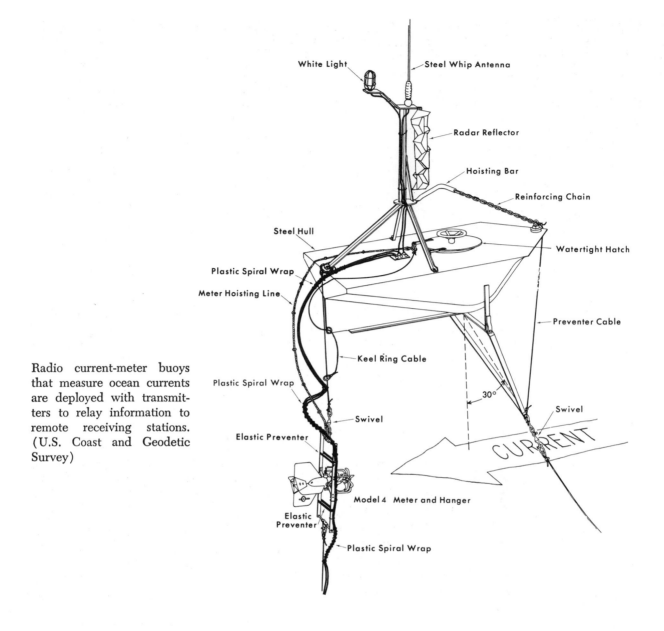

White Light
Steel Whip Antenna
Radar Reflector
Hoisting Bar
Reinforcing Chain
Steel Hull
Watertight Hatch
Plastic Spiral Wrap
Meter Hoisting Line
Preventer Cable
Keel Ring Cable
Plastic Spiral Wrap
30°
Swivel
Swivel
Elastic Preventer
CURRENT
Model 4 Meter and Hanger
Elastic Preventer
Plastic Spiral Wrap

Radio current-meter buoys that measure ocean currents are deployed with transmitters to relay information to remote receiving stations. (U.S. Coast and Geodetic Survey)

fifty oceanographic vessels from many nations now roam the seas. Numerous oceanographic laboratories and institutions have been founded, as well as several oceanographic data centers to facilitate exchange of the enormous quantities of information acquired in man's intensive study of the sea.

Increasing emphasis on the study of ocean currents and on wave forecasting has created a demand for improved oceanographic and meteorological data. Shipboard observations are unsatisfactory, not only because they are random but because they do not provide detailed information on such wave characteristics as spectral components, period of maximum

energy, and directional behavior. To meet the demands for more and better data, oceanographers have developed an impressive array of new instruments.

Airborne wave meters, utilizing electromagnetic radiation, are now used to record actual profiles of the ocean's wave structure. Although still in the experimental stage, satellite sensors may be used in the near future to collect routine oceanographic data.

Additional wave and weather observations are obtained from sensors installed on oceanographic research towers such as Argus Island near Bermuda. The use of such towers has permitted study of the

Argus Island, a deep-ocean laboratory thirty-two miles south of Bermuda, is run by the Office of Naval Research to study storm-wave heights and meteorological conditions. (U.S. Naval Oceanographic Office)

heights of storm waves and of meteorological conditions during periods when it would have been imprudent to make similar observations from ships at sea.

Buoy-mounted systems have been developed as well; the Naval Oceanographic-Meteorological Automatic Device (NOMAD) buoy, for example, is able to remain unmanned on station for months at a time, relaying data to shore facilities. The eleven-ton, aluminum-hulled buoy is equipped with a complex of electronic instruments in its floating framework.

While the tools of physical oceanography have been greatly improved, and major technical barriers in electronics and computerization breached, much of the modern instrumentation is still patterned after hardware assembled for the *Challenger* expedition nearly a century ago. The mariner's log, the Nansen bottle, the reversing thermometer, current drogues, and drift bottles are still widely used, and they serve to remind us that oceanography is still struggling through its infancy and adolescence. At hand lie airborne oceanography, spacecraft photography, satellite interrogation of buoys, and the use of submersibles to explore the mysteries of the physical behavior of the sea.

5 CHEMISTRY OF THE

by Neil R. Andersen

Most of the investigations of modern oceanographers are related in one way or another to the chemistry of the oceans. Although the chemical nature of the ocean is extremely complex, the problems of marine geochemistry can be simply stated: Why is the sea full of seawater? How did the salts get there? Why do they stay there? Men have pondered these problems since long before the beginning of the Christian era.

As far back as 400 B.C. Greek philosophers constructed theories to explain the dissolving power of water and the saltiness of the sea. Aristotle indicated that several individuals during the fourth century B.C. hypothesized that the ocean was salty for the same reason that fresh water becomes salty when passed through ashes. Therefore, they reasoned, the sea must be saline because it contains material similar to ashes. Another ancient idea was that heat from the sun caused the earth to perspire, forming the oceans, which, like sweat, were naturally salty.

Seneca, writing in the first century A.D., probably was one of the first thinkers to hint at the more plausible reasons—at least to us at this time—for the salinity of the oceans. He noticed that rivers and streams significantly eroded land areas and rocks, transporting the eroded material great distances. From this realization, it was but a step to the concept of the oceans as a vast "bathtub" or "giant sewer," as some people classify it today, receiving both solid and dissolved constituents from the land, either by water runoff or some other physical process.

Men have continually been interested in the sea, even before the time when the great thinkers of the day were hypothesizing on the chemical makeup of the oceans and the processes that make seawater what it is. With few exceptions, however, man's interests seem to have been stirred more by practical reasons than by a pure scientific curiosity about the fundamental questions of nature. The maritime interests of the Phoenicians and the Greeks, as far back as 800 or 900 B.C., stirred them to seek answers to questions about the oceans, but even with the increasing activities of man on the sea in the following hundreds of years—up to and including the expansion of the Arabian Empire—few inroads were made on the study of the sea aside from marine exploration and geography. In this respect, the field of chemical oceanography is quite similar to that of biological oceanography. Had the interests and ideas of Aristotle, himself an expert marine biologist, been pursued after his death by other scientists, rather than lying dormant for about two thousand years, marine biology would be considerably more advanced than it is today. The same can be said about chemical oceanography, but to a much greater extent. Aristotle, who lived several hundred years before Christ, is considered by many as the father of marine biology. His counterpart in marine chemistry, the English scientist Robert Boyle, lived in the seventeenth century—from 1627 to 1691.

Chemical oceanography was born when Robert Boyle published his chemical investigations of seawater in *Observations and Experiments on the Saltiness of the Sea* in 1670. Boyle's interests were far-ranging, including, in addition to his analysis of seawater, his more notable work on gases, in which he found a constant relationship between the pressure and volume of a gas; scriptural studies; and biblical translations. To measure the chloride content of his seawater samples, Boyle introduced the silver nitrate test, which is still used today. His analyses convinced him that all natural waters contained some degree of dissolved solids from the land areas, or lithosphere—a conclusion that Seneca had nearly reached at the opening of the Christian era.

Other investigators during the seventeenth century added to Boyle's studies, but it wasn't until the latter part of the eighteenth century, when the first attempts were made to determine the composition of seawater, that the development of chemical oceanog-

SEA

raphy took another significant step forward. The French chemist Antoine Lavoisier (1743–94), the father of modern chemistry and discoverer of oxygen, analyzed seawater for its possible use as a medicinal remedy. At approximately the same time, but independently of Lavoisier, the Swedish analytical chemist and physicist Torbern Bergman (1735–84), who was on the teaching staff at the University of Uppsala, Sweden, and was the first experimentalist to isolate nickel in its pure state, undertook similar studies. All their studies, however, were confined to surface waters.

The next major advance did not occur until the mid-nineteenth century, when Johann Georg Forchhammer (1794–1865), a Danish mineralogist, chemist, and geologist who was professor of mineralogy at the University of Copenhagen and director of the Polytechnic Institute in Copenhagen, completed the analyses of several hundred samples of seawater from all parts of the world. He reported his results in "On the Composition of Sea Water in Different Parts of the Oceans" in 1865 in the *Philosophical Transactions*. His analyses are not considered today as being highly accurate in describing the world's oceans from a chemical point of view, inasmuch as only surface waters were involved and some of the most abundant elements were not determined.

Although more detailed analyses were made a generation after those of Forchhammer by Konrad Natterer (1860–1901), the true beginnings of modern chemical oceanography, in common with most other branches of oceanography, date from the *Challenger* expedition of 1872–76. During the *Challenger's* world-circling cruise of 68,890 nautical miles, the scientific staff obtained information and numerous types of environmental samples at 354 locations. Many of the samples were analyzed on board by the expedition's twenty-nine-year-old chief chemist, John Buchanan. His measurements, however, were limited primarily to density and temperature. More com-

plicated analyses had to await *Challenger's* return home, when the samples could be studied with the superior equipment of land-based laboratories.

For chemical oceanography, the most important feature of the *Challenger* expedition was the collection of seventy-seven samples of seawater that was analyzed after the voyage by Wilhelm Dittmar (1833–94), a German-born scientist then teaching at Andersonian University in Glasgow, Scotland. Dittmar very carefully analyzed the samples (which had been obtained at different locations and at various depths down to about 5,000 feet) for chloride, bromide, sulfate, sodium, calcium, magnesium, and potassium. Although criticisms have been made of his analyses, several with some validity, the high degree of reliability of his results testifies to the excellent work conducted under his supervision, especially when one considers the relatively crude methods of the time. His measurements of the variations in the chemical constituents have been verified very recently, although to a lesser degree. His work probably represents the major building block in the foundation of modern chemical oceanography.

The field of chemical oceanography has enjoyed continual growth since the time of Dittmar, with rapid advances occurring since World War II. This growth is the result not only of the greater emphasis being placed on oceanography generally, but also in part a consequence of the technological advances that have direct applications in marine chemistry. We are at a point of development now where studies of the chemistry of the oceans have become so complex that the investigator in marine chemistry must be not only informed about oceanography but must be initially a trained chemist. Depending on the nature of the problem, he may find it necessary to be even more specialized—for example, to be an analytical, organic, or physical chemist. In a similar manner, research vessels are continually becoming more sophisticated to enable the increasingly com-

Studying the chemistry of seawater in the seventeenth century, Britain's famous pioneer scientist, Robert Boyle, concluded that all natural waters contained solids dissolved from land areas.

Seawater samples collected on the *Challenger* were analyzed in the ship's chemical laboratory. (*Challenger Reports*)

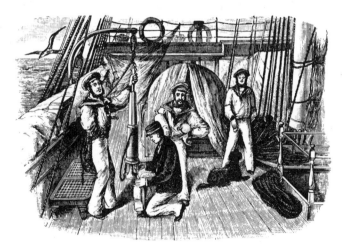

The *Challenger's* most important contribution to chemical oceanography was the collection of seventy-seven samples of seawater, obtained from varying depths and locations. The crew is shown here emptying a water bottle on the deck. (*Challenger Reports*)

This sketch of the hydrographic laboratory on board the *Discovery* was made by Sir Alister Hardy for his book, *Great Waters*. (William Collins Sons & Co.)

plex studies to be accomplished. The time when an oceanographer can be defined as "a sailor who uses big words" has long since passed.

Chemical Properties of the Sea

The amount of dissolved material in seawater is relatively large, 3.5 percent by weight, on the average, enabling the oceanographer to distinguish sea-

water from other natural waters and, by determining the precise percentage of dissolved material in a sample, to label numerous masses of seawater from different parts of the ocean. The amount of dissolved material in a sample is expressed in terms of its salinity; and salinity, in turn, is one of the principal variables (along with temperature) used to determine density. As a result, salinity must almost always be considered whenever one discusses any phenomenon of the oceans; it is probably the most common measurement that the chemical oceanographer is called upon to make.

Salinity is very difficult to measure directly, as indicated by the discrepancies in the various nine-

Commissioned in the early 1930's, the United States ketch *Atlantis I* was outfitted specifically for oceanographic research. (Woods Hole Oceanographic Institution)

Modern research vessels, such as the U.S.N.S. *Lynch*, conduct increasingly sophisticated and specialized chemical oceanographic studies. (U.S. Navy)

teenth-century studies of the composition of seawater. To reconcile these conflicts, an International Commission was established at a conference held in 1899 at Stockholm, shortly after the publication of Dittmar's work. This scientific body was under the supervision of Martin Knudsen (1871–1949), a Dane who had ascended to the position of professor of physics at the University of Copenhagen, where he was previously enrolled as a student.

The International Commission presented its findings and recommendations at the Second International Conference on the Exploration of the Sea in 1901. One of the most significant outgrowths of this meeting was the establishment of the Standard Seawater Service of the Hydrographic Laboratories in Copenhagen. This organization prepares standard seawater (*Eau de Mer Normale*), sometimes called "normal water," by dissolving in very pure water the appropriate kinds and amounts of salts to approximate an average sample of seawater. The preparation is rigorously controlled and the resulting liquid

is distributed throughout the world for use in standardizing methods for determining the salinity of seawater. Martin Knudsen directed this service until his death in 1949.

Because of the experimental difficulties that resulted from measuring salinity directly, the International Commission also considered alternative ways in which to determine the salinity of a sample of seawater. The commission measured the salinity, density, and chloride content of different seawater samples and studied the relationships of their findings. The investigators—Carl Forch, Martin Knudsen, and Søren Sørensen, a Dane who is best remembered in chemistry for his suggestion of how to describe quantitatively the acidity of a solution, and who used the symbol pH for this purpose—found that there was a reasonably constant linear relationship between salinity and chloride content.

The relative simplicity of making a chlorinity measurement (using a method based on Boyle's silver nitrate test) so greatly overshadowed the salinity de-

101

An instrument package for continuously sensing temperature, salinity, and pressure during immersion is lowered from the oceanographic vessel U.S.N.S. *Silas Bent*. (U.S. Naval Oceanographic Office)

A scientist attaches a Nansen bottle to the cable. The safety rope is used as a precautionary measure until the bottle is securely fastened. (U.S. Naval Oceanographic Office)

termination, as developed by Sørensen, that the commission recommended the calculation of salinity from chlorinity data. Actually, standard seawater is being prepared today using the proper proportions of salts to give a solution having an accurate and precise value of chlorinity rather than salinity. Chlorinities are always determined, and often reported, in describing the salt content of seawater throughout the world. When salinity is reported, it has been converted from the measured chlorinity.

In the mid-1930's a difficulty arose in the definition of chlorinity as set forth some three decades earlier. Through the improvements in technology, it was possible to measure the atomic weights of the elements with a much higher degree of accuracy and precision than was formerly the case, with the result that some of the previously accepted values were changed. To compensate for this advance in science, the definition of the chlorinity of seawater was changed in 1940 by Martin Knudsen and by another Danish scientist, Jacob Jacobsen.

The salinity of the world's oceans is expressed in terms of parts per thousand, the units being designated by the symbol $^0/_{00}$. The average value is 35 $^0/_{00}$. The different values of the salinity of ocean waters that are used to obtain the average figure range from 33 to 37 $^0/_{00}$, this range representing some 97 percent of the marine environment. This average thus does not take into account the 3 percent of the environment having extreme salinity values.

The variation in salinity results from a number of physical processes that control the salt content of seawater and are constantly at work in affecting the salinity distributions of the world's oceans. In warm areas where there is not an excessive amount of annual rainfall, the resulting evaporation causes high values of salinity. When seawater freezes, the result is similar; pure water is effectively removed, leaving the salt behind. Conversely, the addition of pure water through melting ice, precipitation, and land runoff tends to diminish salinity values. The chemical makeup of a mass of seawater also may be altered when one volume of water, with its characteristic salinity, mixes with another volume of water with a different salinity. The net result is a mass of seawater whose chemical characteristics are intermediate between those of the two volumes of water that went into making up the mixture.

Melting and freezing, which occur only in the higher latitudes, are negligible in causing salinity variations when one considers the world ocean as a whole. Similarly, land runoff is only important in coastal areas. Therefore, evaporation, precipitation, and water-mass mixing remain as the three most important processes affecting changes in surface salinity, with the latter also being able to affect subsurface salinity distributions. The accompanying diagram illustrates the surface salinity distribution of the North Atlantic Ocean.

102

A seawater sample is poured into a salinometer for salinity determination. (U.S. Navy)

A seawater sample is drawn for chemical analysis from one of a battery of Nansen bottles. (U.S. Navy)

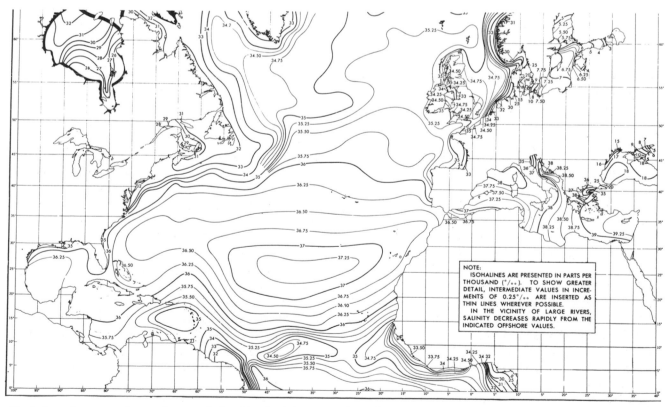

NOTE:
ISOHALINES ARE PRESENTED IN PARTS PER THOUSAND (°/₀₀). TO SHOW GREATER DETAIL, INTERMEDIATE VALUES IN INCREMENTS OF 0.25°/₀₀ ARE INSERTED AS THIN LINES WHEREVER POSSIBLE.
IN THE VICINITY OF LARGE RIVERS, SALINITY DECREASES RAPIDLY FROM THE INDICATED OFFSHORE VALUES.

Variations in sea-surface salinity of the North Atlantic Ocean are shown on this chart by "isohalines," or lines of equal salinity. The world average ocean salinity is about thirty-five parts per thousand. (U.S. Naval Oceanographic Office)

Two other important chemical properties of seawater are its alkalinity and acidity (pH). These two measurements are indispensable for thorough studies of the air-sea carbon dioxide system.

It is not difficult for anyone who has ever been to sea during a period of stormy weather to realize that the turbulent surface of the ocean must interact intimately with the atmosphere. In fact, the water at the surface tends toward a state of equilibrium with the carbon dioxide and other gases of the atmosphere, the gases dissolving into the seawater in amounts determined by the chemical makeup of the water and the prevailing physical conditions. A portion of the carbon dioxide thus dissolved reacts with the water to form carbonic acid, bicarbonate, and carbonate ions. The alkalinity of a sample of seawater is considered as an indicator of the bicarbonate and carbonate concentration in the sample.

The pH value of seawater generally ranges between 7.5 and 8.4; a neutral solution is 7.0. The higher values, 8.0 to 8.4, are usually found in the wind-driven surface layers, where atmospheric carbon dioxide is in equilibrium with the water. Minimum values of about 7.5 are found at a depth of about 2,500 feet, with a gradual increase at deeper depths. In unusual areas, where seawater is diluted, or in closed basins where there is virtually no oxygen, hydrogen sulfide is produced and the pH can actually drop into the acid range—that is, below 7.0. The fact that it was not until after 1910 that a satisfactory measurement of pH could be made again underscores the youth of the field of chemical oceanography.

Major Constituents of the Sea

Wilhelm Dittmar's "complete analyses" of the seventy-seven seawater samples collected by the *Challenger* expedition stood until just recently as the most extensive study of its type. He performed tests for only seven constituents: sodium, calcium, magnesium, potassium, chloride, bromide, and sulfate. During the past two decades, however, great strides in technology have made it possible to detect and analyze many other elements that are present in extremely minute quantities in the sea. It now appears that the chemist is justified in considering that the ocean contains all the natural elements present on the planet. Their detection depends only on the "state of the art" of the technology required to measure them.

The chemical content of seawater, as accepted at this time, is represented in Table I. Taking the elements that are positively charged, it can be seen that the overwhelming majority of the material dissolved in seawater comes from the presence of sodium, magnesium, calcium, and potassium—the latter having an abundance of 380 milligrams per liter (about one ounce per twenty gallons). The next most abundant positive ion is strontium, with an abundance of eight milligrams per liter, two orders of magnitude less abundant than potassium. According to the classifications of some investigators, the major components of seawater are arbitrarily chosen as those with abundances greater than one milligram per liter. The break between potassium and strontium, however, appears to represent a more natural line for dividing the major and minor constituents. On this basis, the major chemical constituents of the sea are oxygen, hydrogen, sodium, magnesium, calcium, potassium, chlorine, sulfur, and bromine—the latter three elements being in the form of the negative ions, chloride, sulfate, and bromide.

Another indication of the youth of marine chemistry is that as recently as 1962 theoretical considerations were being set forth as to the forms in which the major positive ions (i.e., cations) existed in the ocean. As a result of these investigations, it appears that about 9 percent of the calcium, 13 percent of the magnesium, and 1 percent of the sodium and potassium found in seawater are combined with negative ions (i.e., anions). Additionally, investigations of negative ions in seawater have indicated that 50 percent of the sulfate, 90 percent of the bicarbonate, and virtually none of the chloride are combined with positive ions. Comparable studies of the trace constituents are now being made.

As a result, the chemist has a highly complex solution with which to work. His difficulties are especially great when the goal is to measure ingredients of seawater samples whose concentrations are many orders of magnitude less than that of the salt predominantly making up the solution (i.e., sodium chloride). An additional problem facing the analyst is that elements whose concentrations are quite different may have similar chemical properties. For example, magnesium and calcium, calcium and strontium, and potassium and cesium are just a few such groups of elements having similar chemical properties. While such situations cause many analytical problems, it is just this fact, the tremendously complex chemical nature of seawater, that intrigues and drives the marine chemist. On a practical level, it is the presence of the large amount of dissolved material in seawater that makes it so difficult to obtain fresh water through desalination processes at a reasonable cost. At the same time, however, it is the

TABLE I

Chemical Content of Seawater[*]

Element	Abundance, Mg./Liter	Principal Species (g) = gas; (s) = solid
Hydrogen	108,000	H_2O
Helium	0.000005	He (g)
Lithium	0.17	Li^+
Beryllium	0.0000006	. . . .
Boron	4.6	$B(OH)_3$; $B(OH)_2O^-$
Carbon	28	HCO_3^-; H_2CO_3; CO_3^-; organic compounds
Nitrogen	0.5	NO_3^-; NO_2^-; NH_4^+; N_2 (g); organic compounds
Oxygen	857,000	H_2O; O_2 (g); SO_4^{-2} and other anions
Fluorine	1.3	F^-
Neon	0.0001	Ne (g)
Sodium	10,500	Na^+
Magnesium	1,350	Mg^{+2}; $MgSO_4$
Aluminum	0.01	. . . .
Silicon	3	$Si(OH)_4$; $Si(OH)_3O^-$
Phosphorus	0.07	HPO_4^{-2}; $H_2PO_4^-$; PO_4^{-3}; H_3PO_4
Sulfur	885	SO_4^{-2}
Chlorine	19,000	Cl^-
Argon	0.6	A (g)
Potassium	380	K^+
Calcium	400	Ca^{+2}; $CaSO_4$
Scandium	0.00004	. . . .
Titanium	0.001	
Vanadium	0.002	$VO_2(OH)_3^{-2}$
Chromium	0.00005	. . . .
Manganese	0.002	Mn^{+2}; $MnSO_4$
Iron	0.01	$Fe(OH)_3$ (s)
Cobalt	0.0005	Co^{+2}; $CoSO_4$
Nickel	0.002	Ni^{+2}; $NiSO_4$
Copper	0.003	Cu^{+2}; $CuSO_4$
Zinc	0.01	Zn^{+2}; $ZnSO_4$
Gallium	0.00003	. . . .
Germanium	0.00007	$Ge(OH)_4$; $Ge(OH)_3O^-$
Arsenic	0.003	$HAsO_4^{-2}$; $H_2AsO_4^-$; H_3AsO_4; H_3AsO_3
Selenium	0.004	SeO_4
Bromine	65	Br^-
Krypton	0.0003	Kr (g)
Rubidium	0.12	Rb^+
Strontium	8	Sr^{+2}; $SrSO_4$
Yttrium	0.0003	. . . .
Zirconium	. . . .	. . . .
Niobium	0.00001	. . . .
Molybdenum	0.01	MoO_4^{-2}
Technetium	. . . .	. . . .
Ruthenium	. . . .	. . . .
Rhodium	. . . .	. . . .
Palladium	. . . .	. . . .
Silver	0.0003	$AgCl_2^-$; $AgCl_3^{-2}$
Cadmium	0.00011	Cd^{+2}; $CdSO_4$
Indium	0.02	. . . .

[*] Adapted from *The Sea*, Vol. II, ed. M. N. Hill. New York: John Wiley & Sons, Inc., 1963.

TABLE I (continued)

Element	Abundance, Mg./Liter	Principal Species
Tin	0.003	....
Antimony	0.0005	....
Tellurium	....	....
Iodine	0.06	IO_3^-; I^-
Xenon	0.0001	Xe (g)
Cesium	0.0005	Cs^+
Barium	0.03	Ba^{+2}; $BaSO_4$
Lanthanum	0.0003	....
Cerium	0.0004	....
Praseodymium	....	....
Neodymium	....	....
Promethium	....	....
Samarium	....	....
Europium	....	....
Gadolinium	....	....
Terbium	....	....
Dysprosium	....	....
Holmium	....	....
Erbium	....	....
Thulium	....	....
Ytterbium	....	....
Lutecium	....	....
Hafnium	....	....
Tantalum	....	....
Tungsten	0.0001	WO_4^{-2}
Rhenium	....	....
Osmium	....	....
Iridium	....	....
Platinum	....	....
Gold	0.000004	$AuCl_4^-$
Mercury	0.00003	$HgCl_3^-$; $HgCl_4^{-2}$
Thallium	0.00001	Tl^+
Lead	0.00003	Pb^{+2}; $PbSO_4$
Bismuth	0.00002	....
Polonium	....	....
Astatine	....	....
Radon	0.6×10^{-15}	Rn (g)
Francium	....	....
Radium	1.0×10^{-10}	Ra^{+2}; $RaSO_4$
Actinium	....	....
Thorium	0.00005	....
Protoactinium	2.0×10^{-9}	....
Uranium	0.003	$UO_2(CO_3)_3^{-4}$

presence of these salts that may make the process commercially feasible, if the removed elements can be marketed as byproduct material.

Surprising as it may seem, no highly accurate and precise methods yet exist for the separation and determination of sodium, the most abundant cation of seawater (excluding, of course, hydrogen). The situation for many other elements is not much better, although it is improving rapidly at this time because of the increasing emphasis being placed on the marine sciences. Once again, the youth of marine chemistry is evident.

Minor Constituents of the Sea

According to the definition of the major and minor constituents just set forth, the minor elements in seawater are defined as strontium and all other elements whose concentrations are approximately the same or less than that of strontium. When attempting to separate and isolate these tremendously dilute elements, the first problem one faces is to remove a major proportion of the dissolved constituents of the system so that the necessary chemical manipulations can be carried out to determine the minor elements. As if the initial task were not difficult enough, there are additional problems. The chemistries of certain minor constituents, for example, may be very similar to those of a component of the system whose abundance is much greater. Moreover, as was pointed out in the case of the major constituents, the minor elements also do not exist just as simple uncombined elements, but rather as numerous complicated chemical entities, either dissolved or particulate. Additionally, some of the elements are combined with suspended particulate matter, which is not the case in a pure chemical system. It is largely because of these many variables that marine chemistry is so fertile an area of research today, and one in which chemical investigations must be conducted along multidisciplinary lines in order to be successful.

Thus far, only the minor inorganic chemicals in seawater have been considered. There is, of course, another suite of chemicals that complement the inorganics in seawater; these are the dissolved organic compounds, about which comparatively little is known.

Konrad Natterer, in the late nineteenth century, was one of the first to investigate the dissolved organic makeup of seawater. His major contribution was not the fact that he found dissolved organics in seawater (since it had already been observed that fresh water contained these substances), but that the amount was so large relative to that which was found in suspended particles. Some research activity continued on this aspect of chemical oceanography—primarily from the biologist's viewpoint—during the years following Konrad Natterer; Schack August Steenberg Krogh (1874–1949), professor of zoology at Copenhagen University and winner of the Nobel Prize in physiology and medicine in 1920 for his discovery of the regulation of the motor mechanism of capillaries, being especially prolific during the 1930's. However, not until the middle 1950's, following development of instruments and techniques capable of analyzing seawater for its dissolved organic compounds, were any significant advances made in understanding this area of marine chemistry.

Prior to the 1960's it was generally thought that most of the dissolved organic material in the oceans was the result of excretions from living cells. Since the beginning of this decade, however, another school of thought has emerged, primarily as a result of the work of Egbert Duursma, formerly at the Netherlands Instituiit voor Onderzoek der Zee, Den Helder, Holland (formerly known as Zoölogische Station, Den Helder) and now associated with the Oceanographic Museum at Monaco. The new hypothesis is that the majority of dissolved organic material derives from dead phytoplankton and detritus. More information is needed on the chemical composition of not only living organisms but also dead particulate matter, before a better understanding of the problem is possible. There has been a recent increase in research activities dealing with this phase of chemical oceanography, but it is still in an embryonic stage.

Dissolved Gases of the Sea

The various chemicals discussed so far are all either solids or liquids. This leaves yet another group of chemicals to be considered: the dissolved gases.

Through the air-sea interaction processes (another aspect of oceanography in which intense research is just beginning), all the components of air can be expected to find their way into the ocean and play significant roles in the unceasing chemical reactions of the sea. Additionally, there are other sources and mechanisms producing gases within the ocean that supplement those supplied from the atmosphere.

The dissolved gases that exist in seawater have been classified into four general groups. The first group contains the inert gases: nitrogen, argon, helium, neon, xenon, and krypton. These gases enter the oceans through the air-sea interface or through the introduction of aerated water by land runoff. The second group contains but a single member, oxygen. The sources of this gas are the same as those of the previous group, plus another: photosynthesis by the myriads of plants that exist in the ocean. The third group also contains only one member, carbon dioxide. This gas is introduced into the sea through the large chemical equilibrium system—the carbonate system—already discussed. Specific sources of carbon dioxide include the atmosphere, land runoff, and the ocean floor. The fourth group is simply the collection of all the remaining gaseous ingredients found in seawater, and its sources are air pollution, usually from industry, and chemical reactions other than photosynthesis. Hydrogen sulfide resulting from the reduction of sulfate in the absence of oxygen is one member of this fourth group.

Carbon dioxide and oxygen are the two most important dissolved gases present in the ocean. Logically enough, they are the two that have historically received the greatest amount of attention from oceanographers.

Consideration was given to the oxygen content of seawater even prior to the *Challenger* expedition, on the cruise of the H.M.S. *Porcupine*. During the *Challenger* voyage, dissolved oxygen was one of the properties of seawater that was investigated. However, it was not until 1888, when the German scientist Ludwig Winkler reported a new method for the determination of dissolved oxygen in seawater, that oceanographers could begin to study its distribution as an oceanographic variable with any accuracy whatsoever. The Winkler method for determining dissolved oxygen did not solve all the chemist's problems, however. While more reliable than any previous test, Winkler's procedure was far from perfect. As recently as the early 1960's, comparative oxygen analyses conducted at sea by different chemists demonstrated the errors inherent in this method. Modifications have been made in the Winkler method, and today very few oceanographers use the method in its original form to determine the dissolved oxygen concentrations of seawater samples. At the present time, some sixty years after the first extensive utilization of the Winkler method on oceanographic cruises, this method is being superseded, even in its modified form, by instrumental methods of analysis using a gas chromatograph—an instrument that isolates and measures the gas content of complex mixtures by passing the sample through long absorbing columns or tubes.

Nitrogen, the most abundant gas in the atmosphere, and until recently one of the more difficult elements to measure because of its inert chemical nature, is of exceptional interest to the marine chemist. It exists in seawater in several oxidation states and it is an important nutrient for marine plants. However, not until the early 1960's—when U.S. Navy scientists John Swinnerton and Victor Linnenbom of the Naval Research Laboratory in Washington, and James Sullivan of the Naval Oceanographic Office, also in Washington, developed instrumental techniques (gas chromatography) for the separation and determination of nitrogen (and oxygen) in seawater—was it possible to undertake extensive studies of nitrogen distributions in the marine environment with a very high degree of confidence. Such studies are now being conducted by a number of scientists.

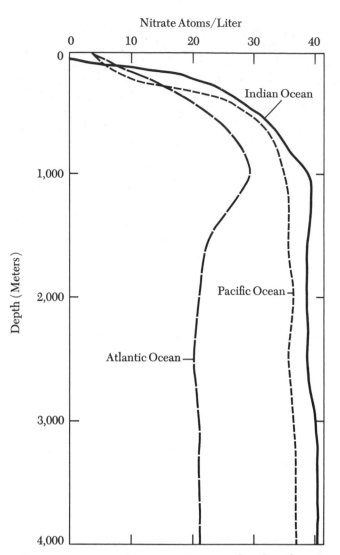

The concentration of nitrates, the most abundant form in which combined nitrogen exists, varies from ocean to ocean. The nitrogen supply, vital to the existence of life, is depleted at shallower depths by biological utilization. (Redrawn with permission, Prentice-Hall, Inc.)

Hydrogen sulfide is yet one more gas found in seawater that deserves special mention. This gas is not usually found in the atmosphere, and when oxygen is present in seawater in amounts greater than about 0.5 milliliters per liter, no significant amounts of hydrogen sulfide will be present, for chemical reasons. However, there are several unique ocean areas where the water is not able to freely circulate, and as a result anaerobic or stagnant conditions exist (i.e., virtually no oxygen is present). In these anoxic waters the hydrogen sulfide content can reach concentrations as high as thirty milligrams per liter. With the exception of certain bacteria no marine life can exist in this type of environment. Whenever water is

Water samples are chemically analyzed in a shore-based laboratory. (R. Housell)

Seawater is continuously pumped aboard to collect suspended particles for chemical and mineralogical analyses. (U.S. Navy)

collected from such areas, the fact that the sample contains hydrogen sulfide becomes very apparent within an extremely short time. As soon as the water sample is exposed to the environment, the smell of "rotten eggs" immediately permeates the air.

Although the importance of dissolved gases in the marine environment has been recognized for many years, aside from studies of carbon dioxide and oxygen there has been surprisingly little activity in investigating these chemical oceanographic variables. This lack of activity is due to the degree of technological sophistication required for the measurement of such gases and the fact that the utilization of such variables to understand marine environmental processes has only recently been recognized. At the present time, the oceanographic distribution and mechanisms controlling other dissolved gases, such as methane, carbon monoxide, and ethane, are not well understood. It has only been within the last few years that analytical techniques, once again instrumental, have been developed to the point that studies in this area could be undertaken.

Radioactive Elements

The final major group of seawater constituents is made up of the radioactive isotopes. Stable representatives of these elements appear in each one of the aforementioned classifications. Only fourteen radioisotopes occur naturally in the marine environment. They come from various terrestrial and extraterrestrial sources. The total number of natural radioiso-

topes, however, is much larger, due to the normal decay processes.

The naturally occurring radioisotopes undergo nuclear decay, giving rise to other isotopes, their respective "daughters," which in turn may also be radioactive. This disintegration process continues until a stable isotope is formed—the end product. A typical example is the decay of uranium 238 through sixteen successive intermediate daughters to the end product, stable lead 206. This combination of nuclear transformations is known as the uranium series. There are two additional analogous series—thorium and actinium—with the parents and stable end products being thorium 232 and lead 208, and uranium 235 and lead 207, respectively. The thorium series has eleven intermediate radioisotopes, and the actinium series has thirteen. In the stratosphere radioactive carbon 14 is produced. This isotope is deposited on the earth's surface along with the nonradioactive carbon. Several other isotopes are also deposited on the earth's surface, hydrogen 3 (tritium), beryllium 10, and silicon 32 being examples.

Radioisotopes are new and useful tools in solving oceanographic problems. Investigators of the pre-Atomic Age, however, were by no means ignorant of nuclear considerations in the environment. One of the earliest recorded determinations of radioactive isotopes for application to a problem in the earth sciences—the heat flow through the crust of the earth —was carried out as far back as 1906 by the English

Large bottles are used to collect seawater for radioactivity determinations. (U.S. Navy)

A sensor for measuring gamma radioactivity is lowered into the water. (U.S. Navy)

physicist and Nobel Prize winner John William Strutt (1842–1919). Just two years later, John Joly (1857–1933), an Irish physicist, discovered the high radium 226 content of certain types of sedimentary red clays. It has been only within the last three decades, however, that any degree of emphasis has been placed on utilizing radioisotopes. Present interest in radioisotopes stems primarily from their geochronological (i.e., dating) applications, with two Americans, Charles Piggot and William Urry, laying the groundwork in the late thirties and early forties with the research they conducted at the Geophysical Laboratory of the Carnegie Institution of Washington, D.C.

In 1946 a new era of nuclear applications to environmental problems was ushered in by the research of Willard Libby, now at the University of California at Los Angeles—research he conducted at the Institute for Nuclear Studies at the University of Chicago, for which he received the Nobel Prize in chemistry in 1960. He predicted that the isotopes carbon 14 and hydrogen 3 would occur in nature as the result of nuclear reactions between small atomic particles, called neutrons, and nitrogen. Since these isotopes decay at a fixed rate over an extremely long period of time, the rates of various oceanic processes can be measured by accurately determining the quantity of isotopes remaining in a sample. In addition, carbon 14 can be used to date deep-water masses and to measure sedimentation rates on the ocean floor. This, then, provided an additional "atomic clock" for solving geochronological problems. The utiliza-

tion of radioisotopes in understanding oceanographic processes, such as circulation and diffusion, received widespread attention and is presently an active area of research in the marine sciences. Samples of seawater obtained from great depths in the ocean are often preliminarily processed on board the research vessels for removal of the carbon 14 for later counting in the shore-based laboratory.

Prior to the mid-1940's, the only radioactive isotopes available to the marine scientist were those that were naturally present. However, with the birth of the Nuclear Age, another suite of isotopes was introduced into the marine environment and became available as tools for use in studying oceanographic phenomena. This new group of isotopes is comprised of the artificial radioisotopes resulting from nuclear-weapons explosions and other utilizations of nuclear technology, such as nuclear reactors.

There is, however, a major problem in carrying out radiochemical determinations of the artificial isotopes present in seawater. Just as in the case of the stable trace constituents, the activities that are sought are usually a very small fraction of the total activity naturally present, due to dilution. As a result, they must be separated from the environmental sample and rigorously purified, before counting. Only within the last ten to fifteen years has a significant amount of attention been given to developing techniques for chemically separating and for measuring these isotopes. Another factor that prevents the marine scientist from making use of all the many available radioisotopes present in the marine environment is his incomplete understanding of the marine chemistry of some of the elements that have isotopes available for usage. It is of very little value to utilize the distributions of a particular isotope in trying to understand a particular oceanographic process if the basic marine chemistry of the element itself is not understood. Such is the situation with several of the possibly useful radioisotopes present in the ocean. Until there is a complete, unequivocal understanding of the marine geochemistry of such isotopes, their use as tracers will have definite limitations.

The Relative Constancy of Proportions

If a marine scientist had been asked fifteen years ago about the composition of seawater, he probably would have replied without hesitation: "There are approximately 35 parts per thousand of dissolved constituents in seawater, and their relative abundance is constant." By relative consistency he would have meant that if one had a sample of seawater whose chlorinity was, for example, 20 $^o/_{oo}$, and if there was associated with this an abundance of 12 milligrams per kilogram of a particular element, then in another sample of seawater that had a different chlorinity value, for example, 15 $^o/_{oo}$, it would follow that the abundance of the associated element would be 9 milligrams per kilogram. In other words, the ratio of the element to the chlorinity of the sample is constant. In the above case, $12/20 = 3/5 = 9/15$, disregarding the units. This is a fairly all-encompassing concept, which recently, due to improvements in the chemist's ability to make accurate and precise determinations of the constituents of seawater, has suffered a number of serious setbacks.

The idea that the relative proportions of the ocean's dissolved constituents are constant is quite venerable, dating back to 1819, when Alexander Marcet (1770–1822) analyzed seawater from the Arctic and Atlantic Oceans and the Mediterranean, Black, Baltic, White, and China Seas. He was of the opinion that seawater contained the same ingredients all over the world and that these ingredients had nearly the same proportions to one another. The concept received some reinforcement in 1865 when Johann Forchhammer, completing what is considered to be the first major analysis of seawater, wrote that "ionic ratios showed only very slight variations if the results for certain restricted seas were neglected." His results came from several hundred surface-seawater samples from all over the world. Wilhelm Dittmar, however, found what appeared to be flaws in the concept. His analyses of the now-famous seventy-seven *Challenger* samples, the cornerstone of chemical oceanography, prompted him

to report: "When we compare the percentages of the several components with the respective means, we frequently meet with differences *which lie decidedly beyond the probable limits of all the analytical errors* [italics added], hence, the variations must be owing partly to natural causes."

Dittmar's research showed, in effect, that seawater has major constituent/chlorinity ratios that are *nearly* constant. After his time, however, the word "nearly," or "approximately," used by some other investigators, somehow was misplaced in the literature, with the result that the concept of constancy became dominant. Recent research, however, appears to be reversing this ingrained concept in chemical oceanography. In 1962 British oceanographer Roland Cox, and his co-workers from the National Institute of Oceanography in England, reported highly accurate and precise electrometric measurements of seawater in which they observed certain ratio variations. Interestingly, they found statistically significant calcium/chlorinity ratio variations that were similar to, although smaller than, Dittmar's. This not only substantiated the earlier work to some degree but underscored its exceptionally high quality, especially when one considers the relatively crude methods of analysis available to Dittmar. As recently as 1965 this author, and David Hume of the Massachusetts Institute of Technology, then associated with the Woods Hole Oceanographic Institution and M.I.T., developed a highly precise and accurate method for the separation and determination of strontium and barium from seawater. Our results showed statistically significant variations in the strontium/chlorinity ratio of approximately three percent. This type of information, together with knowledge of the mechanisms causing such trace-element distributions, is indispensable if one is to understand the distributions of the radioactive counterparts of the stable trace constituents, and is yet another example of a fertile research area in chemical oceanography. As the analytical techniques for determining both the major and minor trace constituents of seawater become more precise and accurate, undoubtedly more variations, which previously were thought to be the results of experimental errors, will be recognized.

It is apparent that the marine chemist's concept of the relative composition of seawater is now similar to what Dittmar's was in 1884. The emphasis again is on the word "nearly." It is important to recognize the fact that seawater has approximately a constant relative composition. However, it is much more important to realize that this is an approximation and that variations, although small in some cases, do in

Understanding the Oceans

The origins of oceanography are as old as man's curiosity about the sea. As a science, however, it is a comparative late-bloomer. After the time of Aristotle, whose careful observations of 180 marine animals mark the beginnings of marine biology, oceanographic studies underwent a long hiatus. In the first century A.D.—400 years after Aristotle—Pliny the Elder could account for only 176 marine species.

Knowledge of the physical dimensions of the world and its oceans also languished. Eratosthenes, in the third century B.C., believed the earth to be spherical and calculated its circumference to be 25,000 miles. Five centuries later Ptolemy erroneously reduced this figure to 18,000 miles, a mistake that was perpetuated for nearly 1,500 years.

In the field of chemistry, too, the development of man's understanding was painfully slow. The first careful investigation of the elements present in the oceans was not made until the seventeenth century, when Robert Boyle, Aristotle's counterpart in the field of marine chemistry, published his *Observations and Experiments on the Saltiness of the Sea.*

Modern oceanography began to flower in the eighteenth century. Near the beginning of the century, Carolus Linnaeus, a Swedish botanist, set the biological world in order with his system for classifying plants and animals. Benjamin Franklin mapped the Gulf Stream in an effort to shorten the sailing times of mail packets across the Atlantic. At the very close of the century, young Alexander von Humboldt left on an expedition to study the great ocean current off the west coast of South America that now bears his name.

The momentum of oceanography increased rapidly in the nineteenth century. Traveling as an unpaid naturalist on the *Beagle* during 1831–36, Charles Darwin developed his theory of the formation of coral reefs, which has held up in its major aspects to this day. Darwin was also one of the first scientists to study plankton —the microscopic plants and animals that drift passively in the ocean and provide, directly or indirectly, nearly all the food for the creatures of the sea.

As late as 1840, Edward Forbes, a marine biologist from the University of Edinburgh, speculated that all life ceases at depths below 300 fathoms. His error was soon refuted by British ships that dredged up living creatures from far deeper water. Some of the most exciting finds were living "fossils"—creatures formerly thought to be extinct for millions of years—which were brought in alive by the *Challenger* and other research ships.

While the British, led by Forbes, were pioneering in marine biology, important strides in revealing the physical geography of the sea were being made by Matthew Fontaine Maury, who headed the U.S. Navy's Depot of Charts and Instruments. By systematically compiling observations from the logs of ships, he produced a series of wind and current charts that significantly reduced the sailing times on many voyages.

Oceanographers today have sophisticated techniques and devices for detecting the current systems Maury so laboriously correlated. Surface currents are made visible from the air by using dyes. Swallow buoys, which can be made to float at predetermined depths, are used to track subsurface currents. In 1957 these buoys helped oceanographers find a large current that drifts southward beneath the north-drifting Gulf Stream.

The most striking feature of the world ocean floor is the Mid-Oceanic Ridge, the greatest mountain range on earth. A continuous, broad swell, it extends through all five oceans for a length of 40,000 miles—or nearly 20,000 leagues under the sea. Running down the center of the ridge is a remarkable rift, ranging from fifteen to thirty miles wide and more than a mile deep. Earth scientists now believe that the continents, once joined in one or two supercontinents, have divided along the lines of the Mid-Oceanic Ridge and are continuing to drift apart. The tensions along the rift line cause it to be a zone of earthquakes and volcanoes.

The active forces of the interior of the earth are also evident in the submarine canyons that plunge in winding axes down the slopes of the continental shelves. Their origins are still not fully understood, but turbidity currents—raging torrents of mud—appear to be a likely cause. Undersea cables have disappeared at their points of intersection with submarine canyons, and piers placed at canyon heads have mysteriously been swallowed up into the deep.

In spite of advanced technology and firsthand observation, the deep-sea floor remains an unknown and shadowy region. But the information explosion applies as much to oceanographic data as to anything else. Today more oceanographers and oceanographic vessels ply the surface and depths of the oceans than ever before, probing the secrets of the earth's last frontier.

112

In a famous five-year voyage around the world, Charles Darwin studied marine life and coral reefs, and gathered data for *The Origin of Species*. His ship, the H.M.S. *Beagle*, is shown in Sydney in 1841 in this watercolor by Owen Stanley. (National Maritime Museum, London)

Really not a flower at all, the sea lily is an animal related to the starfish. This particular species attracted great attention when it was first caught in Norwegian waters in 1864, since zoologists had previously known it only as a fossil dating from 160 million years ago. (Matthew Kalmenoff)

Living proof that life exists well below 300 fathoms, this abyssal crab was dredged up from 2,250 feet of water off the Philippine Islands by the H.M.S. *Challenger*. (Matthew Kalmenoff)

Wyville Thomson, who later led the *Challenger* expedition, described the large scarlet sea urchin discovered by the *Porcupine* as "a specimen of extraordinary beauty and interest." Formerly known only as a fossil, the creature was observed to "pant" when it rolled out of the dredge. (Matthew Kalmenoff)

Built up from the skeletons of tiny colonial animals, colorful coral reefs abound in the Pacific and West Indies. This photograph of delicate coral growth with a small butterfly fish was taken off the island of Wake. (Ron Church)

Myriad microscopic plants float passively in the sea, soaking up the abundant energy of sunlight with which they manufacture carbohydrates. These organisms, part of the plankton, are gathered in finely meshed nets for study of the complex food chain of the oceans. (R. Morris)

Delving into the past with increasingly sophisticated techniques, marine archaeologists stake out precise measuring grids to survey a site. In this photograph Dr. George Bass of the University of Pennsylvania explores a Byzantine shipwreck off Yassi Ada, Turkey. (© National Geographic Society)

Jacques-Yves Cousteau's diving saucer maneuvers skillfully within the towering walls of the La Jolla Submarine Canyon off the California coast. Capable of diving to 1,000 feet, the saucer permits visual checking of data difficult to obtain from instruments lowered from the surface. (U.S. Navy Electronics Laboratory)

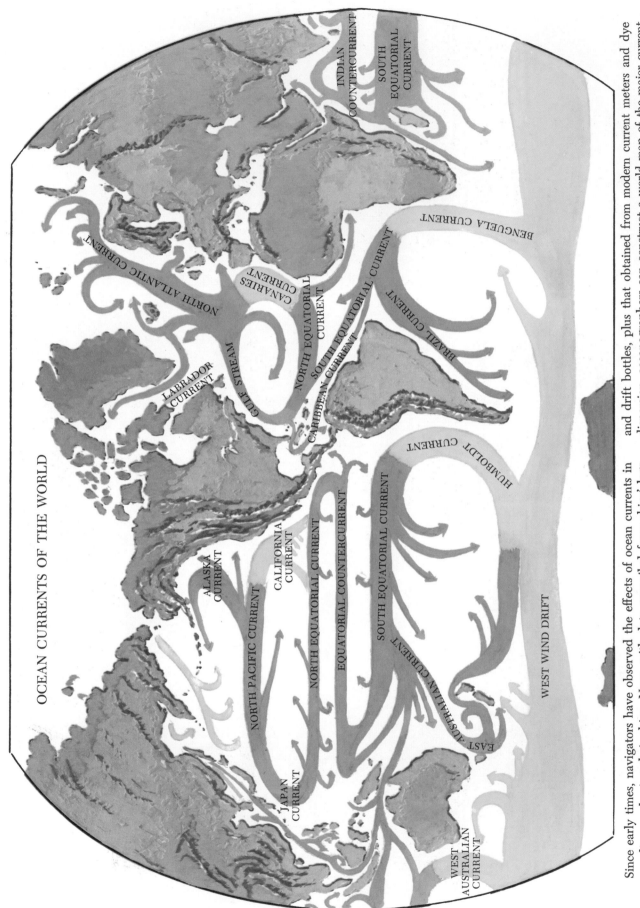

OCEAN CURRENTS OF THE WORLD

Since early times, navigators have observed the effects of ocean currents in speeding or slowing their ships. Now, with data compiled from ships' logs and drift bottles, plus that obtained from modern current meters and dye dispersion, oceanographers can construct a world map of the major current systems. (Matthew Kalmenoff)

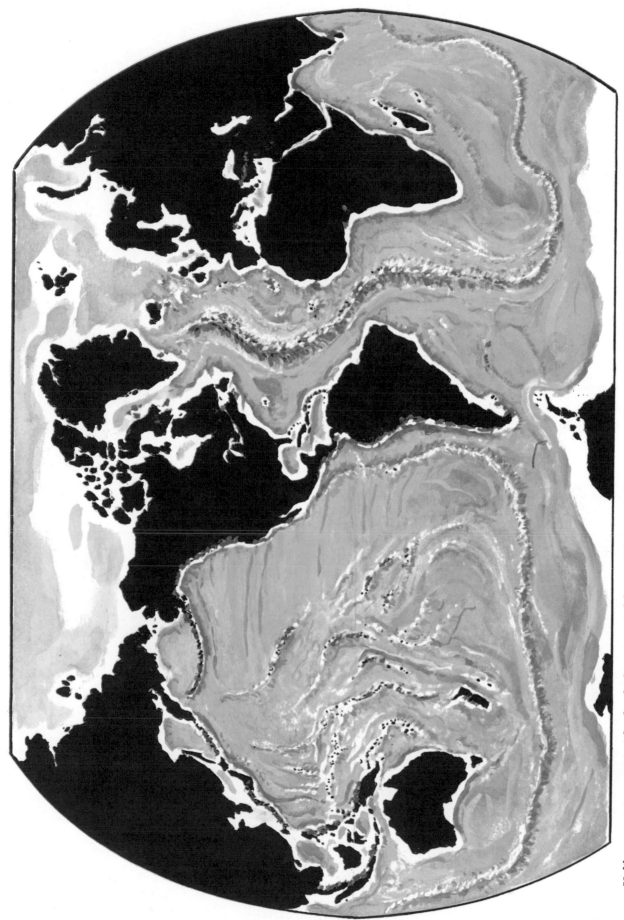

Unlike mountain ranges on land, which are created by crustal compression, the Mid-Oceanic Ridge may have been created by crustal stretching as the earth's continents drifted apart. On this map, the world ocean ridge is traced with a solid gray line; the earthquake zones are shown in red dots. (Matthew Kalmenoff)

The tensions of the earth's interior create continuing geologic forces along the rift zone of the Mid-Oceanic Ridge. A volcanic eruption near Iceland, on the Mid-Atlantic Ridge, created the new island of Surtsey in 1964 and 1965. (Solarfilma, Reykjavik; courtesy, Consulate General of Iceland)

The port area of Seward, Alaska, was a scene of debris and destruction following the onslaught of a tidal wave during the earthquake of March 1964. The tidal wave was caused by an earthquake that occurred in the Pacific rift zone. (Environmental Science Services Administration)

fact exist and have been reported. The evidence for this nonconstancy is too overwhelming to even begin to entertain the thought of the absolute relative constancy of the chemical constituents of seawater.

Chemical Distributions

There are a number of causes for the variations that exist in the chemical content of seawater from one geographical location to the next, as well as with respect to depth. The gross salt content of the oceans, that is to say the salinity, is altered by a number of physical processes, which have previously been mentioned. Inasmuch as the density of seawater is determined not only by its salt content, but by its temperature, the salinity profile of a column of water may either increase or decrease with depth, depending on the thermal structure. However, the combination of temperature and salinity, represented graphically, can and does effectively distinguish different water masses.

The one chemical characteristic that is probably used more than any other to categorize a mass of seawater, aside from salinity, is the dissolved oxygen content. The chief processes that alter the distribution of dissolved oxygen are photosynthesis, utilization by animals, and decomposition of organic matter. The oxygen distribution with depth usually follows a general pattern, latitudinally as well as vertically. The accompanying drawing shows a typical vertical profile of the dissolved oxygen distribution in the South Atlantic. The vertical distribution of dissolved oxygen in the ocean is affected by several factors. In the upper layer there is a uniform concentration of oxygen due to the mixing action of the wind. In certain localities, in those periods of the year when plankton blooms occur, a subsurface maximum will also be observed, associated with photosynthesis. Below these surface depths, the oxygen concentration will decrease as the depth increases, with oxygen reaching a minimum at depths of about 2,200 to 3,100 feet. Below this minimum, the oxygen concentration usually increases to a value that then remains virtually constant to the bottom of the water column. This constancy is basically a result of the deep oceans continually being renewed as a consequence of circulatory processes. In general, the concentration of dissolved oxygen in seawater has a value of from four to six parts per thousand. The floor is established essentially by the circulatory processes replenishing oxygen. Where there is no new intrusion of oxygenated water, such as in stagnant basins, there will be essentially no dissolved oxygen in the seawater. The

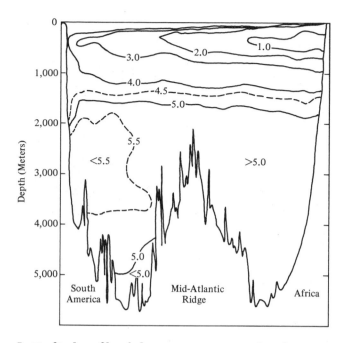

Latitudinal profile of the oxygen concentration (c.c. per liter) of the South Atlantic Ocean at 9°S. (Redrawn with permission of John Wiley & Sons, Inc.)

upper limit of the dissolved oxygen content of seawater is controlled by the physical variables affecting the oxygen solubility at the air-sea interface. The seawater content of dissolved oxygen between these two limits is controlled by the *in situ* processes already mentioned above, and additionally, by the horizontal shifting of water masses and the diffusion phenomena characteristic of a particular area.

The distributional patterns of the world's oceans have been established for only a limited number of the trace constituents of seawater—a significant deficiency in chemical oceanography. With increasing activity in marine chemistry, however, more trace-constituent distributional patterns, both horizontal and vertical, will be learned. The variations in the distributions of the elements in seawater and their relationships to different water masses, are a result of some of the processes that cause elemental alteration. The evaporation of water and the melting of ice, for example, alter the overall salt content of seawater. In addition, the oceans are fed continually by rivers and streams, whose own characteristic distributions of elements will affect the distributions of these same elements in the oceans. Such land runoff also introduces particulate matter into the oceans; this can affect the elemental distribution of seawater

TABLE II

Accumulation Factors for Elements in Marine Organisms*

Element	Algae	Invertebrates		Fish		Plankton
		Soft Parts	Skeletal	Soft Parts	Skeletal	
Beryllium	15000	....	....	....	....	....
Sodium	1	0.5	0	0.07	1	....
Phosphorus	10000	10000	10000	40000	2000000	....
Sulfur	10	1–5	1–5	2	....	....
Potassium	3–50	1–15	0	5–20	20	....
Calcium	0.5–10	110	1000	0.5–3	50–250	10
Scandium	1500–2600	....	....	....	....	....
Titanium	1000	1000–1500	....	40	....	....
Vanadium	1000	100–5000	....	20	....	5000
Chromium	300–100000	70–1000	....	2000	....	....
Manganese	6500	10000	....	1000	2000	750
Iron	20000–35000	10000	100000	300–1000	5000	2000–140000
Cobalt	450	500 —— 3000 (whole)		160	1500	....
Nickel	500	200	200	100	....	....
Copper	100	5000	5000	200–1000	1000	....
Zinc	100–13000	5000–40000	1000	300–3000	30000	1000
Germanium	15–200	10	16	....	....	....
Arsenic	200–6000	20000 (whole)		....	....	....
Rubidium	5–50	3–50	30	3–30	40	....
Strontium	1–40	1–25	180–1000	0.1–2	100–200	9
Yttrium	100–1000	20–100	300	5	....	....
Zirconium	350–1000	....	....	....	....	1500–3000
Niobium	450–1000	20000 (whole)		....	....	....
Molybdenum	10–100	100	....	20	....	....
Ruthenium	15–2000	2000	....	3	....	600–3000
Antimony	....	100(?)	....	100(?)	....	50
Iodine	3000–10000	50–100	50	7–10	....	....
Cesium	1–100	10–100	0	10–100	....	0–5
Cerium	300–900	300–2000	45	12	....	7500
Bismuth	....	500	....	50	....	....
Radon	....	....	....	....	....	....
Radium	100	50	650	15–200	....	2750
Thorium	10	50–300	....	100	....	....
Protoactinium	....	....	....	....	....	....
Uranium	10	....	400	20	....	....
Plutonium	....	....	....	13	....	....

* Mauchline, J., and Templeton, W. L., in *Oceanography and Marine Biology Annual Revue*, Vol. II., ed. H. Barnes. London: George Allen & Unwin, Ltd., 1964.

by absorption processes. Reactions of this type also may occur because of the settling of organic particulate matter. Moreover, biological activity sometimes alters the elemental composition of seawater by extreme amounts. Plants and animals may collect certain elements to concentrations 100,000 times their normal amounts in seawater. A tabulation of biological accumulation factors for several elements is given in Table II.

All the processes just mentioned for altering the distributional patterns of the constituents of seawater can be collectively considered, as in Table III. When,

TABLE III

Processes Affecting Chemical Distributions in the Sea*

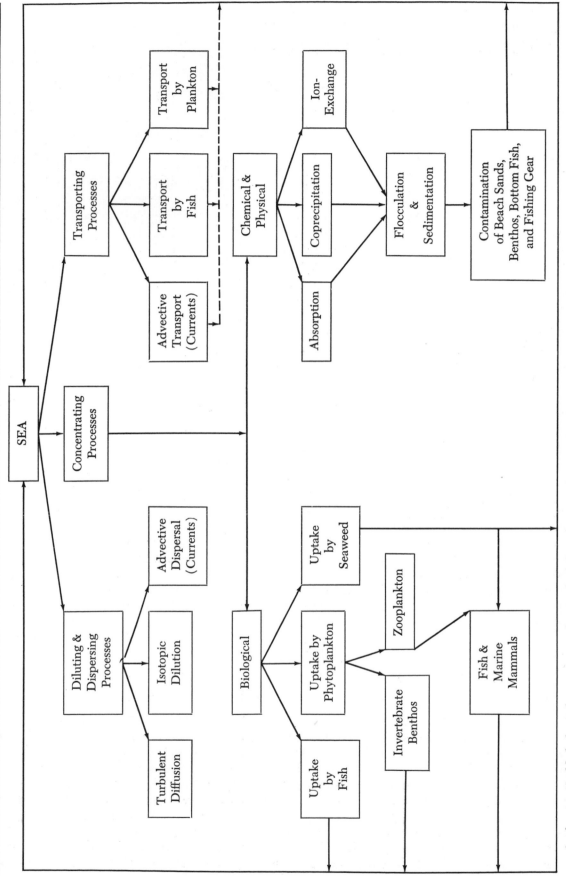

* After M. Waldichuk, Fisheries Research Board of Canada, Circular No. 59 (1961).

in addition, one takes into account all the possible complex chemical and biochemical reactions that can take place in the marine environment, together with the fact that most of the constituents of seawater do not exist as completely free chemicals, but are usually tied up as some chemical complex, it is no small wonder that a complete understanding of the system requires efforts not only from specialists within the various fields but also different disciplines, such as geology and biology.

If a chemical is altered at a particular location, and this process of alteration can be identified and understood, it may be possible to take advantage of the situation in tracing or locating a water mass. A large biological population may deplete the waters of particular elements, for instance, producing distinctive ratios between the amounts of those elements and the chlorinity. As long as there is no significant mixing of the water, different water masses can effectively be identified by their chemical composition in this way. For example, at about 50 to 60 degrees south latitude, water which is at the surface sinks to about 2,500 feet and then flows northward, being clearly identifiable as far north as about 10 to 15 degrees north latitude. This water mass is known as Antarctic Intermediate Water. Appreciably more data is needed on the distributions of specific elements, however, before this tracing technique can be generally utilized.

The chemical distributions of the world's oceans can be discussed generally, as long as one keeps in mind that, just like everything else in nature, there are very significant exceptions to most general statements. One of the major exceptions in the oceans is the stagnant (i.e., anoxic) areas, in which circulatory processes are not replenishing the waters. There are many such areas, the Black Sea being a notable example. Many of the fjords and inlets in Norway also tend toward stagnancy. The Cariaco Trench off the coast of Venezuela in the Caribbean Sea is another such area. From a chemical standpoint, all of these oceanic locations are similar in that they have virtually no dissolved oxygen and, as a result, are devoid of the normal biological populations.

As recently as the early 1960's, it was observed that other chemical characteristics of the anoxic areas differed from those of the open oceans. American oceanographer Francis Richards, now at the University of Washington, found that the dissolved copper concentration in the Cariaco Trench, for example, is so low that the usual methods of analysis are not sensitive enough to detect the presence of this element. In the oxygen-containing waters above the sill of the trench, however, the copper distribu-

A corer is assembled for obtaining samples of sediment from the ocean bottom. (U.S. Navy)

tion is similar to that of the open oceans. The iron concentration, on the other hand, has a thousandfold greater concentration in the anoxic trench waters.

There are other areas of the world's oceans that also differ dramatically from the open sea. The "hot spots" recently found in the Red Sea by oceanographers on the American R.V. *Atlantis II* and the English *Discovery* should prove extremely interesting from a marine geochemical point of view. The elevated temperatures (over 120° F.) of these areas should produce a correspondingly significant alteration in the chemical equilibrium. This has been observed already in the salinity; values of over 280 $^o/_{oo}$ have been measured. Chemical investigations in these areas should prove rewarding, especially in connection with the trace element distribution. These areas may even be exploited for their mineral resources.

Nutrient Chemicals

The growth of plants in the sea, as is the case everywhere else on our planet, is dependent on the nutrients available for photosynthesis. Table IV lists several physiologically important elements. The elemental content of 100 grams of biota is designated by N, and the abundance of that particular element in seawater available for utilization by A. The smaller the ratio of supply to demand (A/N) for a certain element, the more likely this element is to be a limiting factor in the metabolic processes. As shown

TABLE IV

Distribution of Physiologically Important Elements*

Elements	Content per 100 Grams Organic Material (Dry) N	Content per M.^{3} Seawater (35 °/oo Salinity) A	A/N	Elements	Content per 100 Grams Organic Material (Dry) N	Content per M.^{3} Seawater (35 °/oo Salinity) A	A/N
Hydrogen	7 g.	....	......	Copper	5 mg.	10 mg.	2
Sodium	3 g.	10.75 kg.	3,600	Zinc	20 mg.	5 mg.	4
Potassium	1 g.	390 g.	390	Boron	2 mg.	5 g.	2,500
Magnesium	0.4 g.	1.3 kg.	3,300	Vanadium	3 mg.	0.3 mg.	0.1
Calcium	0.5 g.	416 g.	830	Arsenic	0.1 mg.	15 mg.	150
Carbon	30 g.	28 g.	1	Manganese	2 mg.	5 mg.	2.5
Silicon	10 g.	500 mg.	0.05	Fluorine	1 mg.	1.4 g.	1,400
Nitrogen	5 g.	300 mg.	0.06	Bromine	2.5 mg.	66 g.	26,000
Phosphorus	0.6 g.	30 mg.	0.05	Iron	1 g.	50 mg.	0.05
Oxygen, as O$_2$ and CO$_2$	47 g.	90 g.	2	Cobalt	0.05 mg.	0.1 mg.	2
				Aluminum	1 mg.	120 mg.	120
Sulfur	1 g.	900 g.	900	Titanium	100 mg.	....	....
Chlorine	4 g.	19.3 kg.	4,800	Radium	4×10^{-12}g.	10^{-10}g.	25

* Adapted from *General Oceanography, an Introduction,* Günter Dietrich. New York: John Wiley & Sons, Inc., 1963.

in the table, aside from silicon, nitrogen, and phosphorus, the more important essential elements are carbon, oxygen, cobalt, iron, manganese, vanadium, zinc, and copper.

Should the ratio A/N reach some critical minimum value (such a value being dependent upon the specific organism and element), further metabolic processes may cease, although other nutrients necessary for continued life of the organism are abundant. This is essentially the oceanic application of the "minimum law" established by the German chemist Baron Justus von Liebig (1803–73), who conducted his research at Giessen and Munich in the last century in connection with agricultural problems.

Accumulation factors of elements for certain biota are listed in Table II. The tremendous concentrations of certain elements by the biosphere reflect both nutritional and non-nutritional needs. The concentrations probably are achieved through several different mechanisms by tissues having great affinities for certain elements. The human thyroid gland, which has an affinity for iodine, is an example of such a tissue. In such a case, the element is then incorporated into an organic chemical. On several occasions in the past, when analytical techniques were less sensitive than they are today, elements were discovered in the remains of marine organisms that had not been found previously in seawater. Vanadium, cobalt, nickel, copper, iodine, and lead were first found in the sea in this manner. Additionally, there are other dissolved chemicals, such as vitamins, that are needed for life processes and as such can be classified as nutrients.

While many different chemicals are required to sustain life, some are more essential than others. In his soil studies, Baron von Liebig found that phosphorus, nitrogen, and potassium had to be present for life to be successful. In the ocean, the corresponding minimum substances necessary for sustaining life are nitrogen, phosphorus, and silicon. These elements are present in the marine environment in the forms of nitrates, nitrites, ammonia, and organically bound nitrogen, as well as phosphates and silicates both organically and inorganically bound.

Studies of the importance of the chemical composition of seawater in sustaining marine life are not new. In the nineteenth century Johann Forchhammer recognized the need for nitrogen in the production of phytoplankton. By 1900, knowledge had advanced to a point where it was being shown by the German zoologist Karl Brandt (1855–1931), professor of zoology at the University of Kiel and director of the Marine Biological Laboratory of the German Scientific Commission for International Sea Research in that city, that nitrogen and phosphorus were assimilated by marine organisms and incorporated into organic compounds. It was at about this time, also, that the importance of silicon to marine life, specifically in relation to the needs of diatoms and silicous algae, was being recognized. Unfortunately, the methods of analysis were so crude by present-day standards that nutrient distributions were known with only poor accuracy, which necessarily limited the degree of marine biological interpretation. As a result, a rigorous quantitative study of the biological relationships could not be undertaken successfully. In the mid-1920's, however, a new era was introduced in the studies of the nutrients in the oceans, when English scientists William Atkins (1884–1959) and Hildebrand Harvey, from the laboratory of the Marine Biological Association in Plymouth, introduced rapid procedures for the determination of the nitrate, phosphate, and silicate content of seawater. They followed up their initial breakthrough with intensive investigations of nutrients, making significant contributions to the topic of the fertility of seawater.

An equally significant consequence of the technical developments was the fact that the new methods made it possible for the first time to analyze samples for their nutrient content while on board research vessels at sea. The German oceanographer Hermann Wattenberg (1901–44), who later became the director of the Institut für Meereskunde at Kiel and was killed during a World War II air-raid, was the first researcher to take advantage of this technological advance, during the *Meteor* cruise in the mid-1920's. Today the laboratory facilities of most research vessels engaged in chemical or biological investigations are sufficient for analyzing seawater for its nitrate, phosphate, and silicate content. In fact, present research is making it possible to carry out aboard ship the automatic and simultaneous analysis of samples for their nutrient content.

Although nitrogen, phosphorus, and silica are each indispensable for marine life, nitrogen is perhaps a shade more important than the other two, because it is a component of the amino acids, which, in turn, are necessary for protein synthesis. The first investigations of the nitrate content of seawater were conducted in 1923 by the Finnish chemist Kurt Buch, head of the chemistry department at the Institute of Marine Research in Helsinki. Not until 1926, however, when Harvey (with whom Buch later collaborated) developed the rapid method of nitrate, was the door opened for making extensive quantitative studies of nitrate in seawater. Inorganic nitrogen exists chiefly as nitrates, and these serve as the principal reservoir for combined nitrogen in the sea. Nitrates are depleted in the shallower depths of the euphotic zone by the photosynthetic action of plants, and this depletion may be carried to a point of exhaustion in some arctic and temperate regions.

Nitrate, however, is not the only form in which nitrogen is utilized by the biosphere. It was recognized in the mid-1930's that ammonia and nitrite may also be utilized by plants, with the former being used preferentially when all three forms of nitrogen are available. Studies in the early 1940's by Theodor von Brand and Norris Rakestraw, presently at the Scripps Institution of Oceanography, provided a considerable amount of information on the biological oxidation processes of ammonia. Much has yet to be learned about the quantitative distributional features of ammonia. With recent breakthroughs in methods for the analytical determination of ammonia in seawater, however, it is expected that more precise and accurate information on the oceanic distribution of this micronutrient will become available.

Several oxidation-reduction reactions involving nitrogen deserve special mention. Certain bacteria in the marine environment can reduce nitrate or nitrite to either nitrous oxide or free nitrogen: the process of denitrification. The reverse process—the oxidation of ammonia to nitrite or nitrate—is classified as nitrification. Unlike denitrification, the exact biological and chemical mechanisms involved in nitrification still are uncertain, in spite of the time that has passed since the initial consideration of this phenomenon more than twenty-five years ago.

The second of the three basic micronutrients is phosphorus, whose concentration in the oceans varies from almost undetectable amounts to about 124 parts per million. The extremely low levels of phosphorus, usually found in approximately the upper three hundred feet of the ocean, result from the concentration of this element by plankton, in some cases at more than eight orders of magnitude (i.e., 10^8) greater than the phosphorus level in the associated seawater. After the plankton die, either naturally, or by being consumed by other ocean creatures, the element is returned to the ocean, either through the death of the predator or excretion of the nonassimilated nutrient. Such a chain of events produces the higher concentrations of phosphorus often found at depths greater than three hundred feet.

Phosphorus exists in seawater in a number of forms: inorganic, organic, dissolved, and particulate. The dissolved organic phosphorus—not unexpectedly, in view of the general lack of knowledge

A corer is about to be lowered from the deck of an oceanographic vessel. (U.S. Navy)

about dissolved organics in seawater—is not very well known. The dissolved inorganic counterparts, however, are somewhat better understood.

Considerations of the phosphorus concentrations in seawater date back to the latter part of the nineteenth century, when Karl Brandt studied the marine cycle of this element along with that of nitrogen. The need for phosphate in supporting the nutritional requirements of the biological community also was realized by other investigators. Additionally, at that time, this nutrient, as well as nitrogen, was recognized as being associated with waters of high productivity. As was the case with nitrogen, however, intensive investigations were not possible until rapid and accurate methods of analysis were developed. This technological breakthrough occurred in 1923, when William Atkins developed a rapid colorimetric method for phosphorus determinations of seawater. With this development he began the study of the phosphorus cycle in the marine environment, and his first reports appeared in 1932. This work initiated studies on the horizontal and vertical distributions of phosphorus in the oceans that are still being vigorously pursued. Since Atkins' time, however, methods of analysis have undergone continuous improvements, especially in shipboard applications. The latter are very important, not only for nutrient analysis, but for all measurements made

at sea, since on-the-spot analysis provides real time data and eliminates the need to wait up to several months for results of analyses from a shore-based laboratory. The tendency now in nutrient chemistry is to conduct such analyses automatically and simultaneously at sea.

Early investigators, primarily Hildebrand Harvey in 1926 and the American physiologist Alfred Redfield in 1934, observed a very interesting phenomenon with respect to nitrogen and phosphorus. They found that phytoplankton appeared to utilize the two micronutrients in a constant ratio of 16:1, with the same ratio also being found in the water. Later investigators, however, have demonstrated that exceptions do exist in certain ocean areas with respect to depth and the season of the year. As a result, although this ratio can be considered as a convenient "rule of thumb," it is important to recognize that it is an approximation and not an absolute fact. The danger, of course, is that the rule of thumb will assume the status of a universal law, as did the concept about the constancy of relative proportions of the ocean's constituents.

Studies of the silicon content of the marine environment and requirements of this element by the biomass developed in much the same manner as did studies of the two micronutrients, nitrogen and phosphorus, previously discussed. As early as 1907 the need for silicon in the growth of diatoms was recognized. It is now known that silicon is essential in building the solid structures not only of diatoms but also of other ocean creatures, such as sponges. It was also realized early in the twentieth century that the growth of such forms of marine life can result in the complete removal of silicate in the water in a very short period of time.

Although some consideration was given to the determination of silicon in seawater prior to the turn of the century, it was not until Atkins applied his technique for nutrient analysis to the analysis of seawater in 1923 that analytical methods became precise enough for quantitative studies of the silicate cycle in the marine environment to commence. Studies that followed have demonstrated that the silicon in the oceans can have concentrations ranging from undetectable amounts to about one part per thousand.

It is not uncommon to find that the inorganic matter of certain organisms includes silicon dioxide in quantities amounting to almost two-thirds of the total weight, attesting to the extreme importance of this nutrient in the oceans. The distribution of the silicate content of the oceans varies, as is the case with the other nutrients, according to depth, the biological community, and physical circulatory processes. In oceanic areas where the usual water-mass movements are not occurring, the silicate distribution can be affected markedly. Such is the case in the Caribbean Sea. In the mid-1950's, Francis Richards observed significantly higher silicate values for the deep waters in the Caribbean relative to the open ocean. Such findings have been attributed to the fact that the Caribbean's depths are not being continually renewed. When silicon-concentrating organisms die, their bodies subsequently settle and redissolve, resulting in abnormally high silicate concentrations at depths.

Future Needs of Chemical Oceanography

Knowledge of the chemistry of the sea has made tremendous strides since its conception as a field of study in its own right. Nevertheless, we are still far removed from completely understanding the system, irrespective of the present absolute amount of knowledge possessed. It has been suggested recently that we know less about the oceans than about outer space. Although such a statement may be disputable, the fact that it is a point of discussion emphasizes the relatively slow progress that has been made, not only in chemical oceanography, but in the overall study of a natural system covering approximately 72 percent of the earth's surface. Whether or not a "wet NASA" is necessary to enhance "inner space" studies, as has been proposed in certain oceanographic circles, is a subject for debate, but it is quite obvious that an increase in research activities, specifically in the field of chemical oceanography, is needed. To make use of the ocean, and to preserve it against pollution, the system's chemical properties must be understood. Some oceanographers believe that physical oceanography is at a point now in its development where continued physical measurements will not completely suffice for solutions to all the problems. Other data are needed, and specifically chemical oceanographic data, such as dating with carbon 14, for example. In the spring of 1968 a group of leading American oceanographers met at the Maury Center for Ocean Sciences located at the Naval Research Laboratory in Washington, D.C., to formulate an organized program for the U.S. Navy to follow in chemical oceanography. That such a meeting was held so recently underscores the administratively underdeveloped nature of the field.

In spite of the tremendous advances which have been made in instruments and research vessels since Wilhelm Dittmar worked on the *Challenger* water samples, investigators for the most part are still forced to conduct environmental chemical studies by collecting samples at the ends of mile-long cables hung over the sides of ships. The samplers attached to these cables for the purpose of water collections have become more and more sophisticated with the passage of time, taking on various shapes and sizes, and being made of various types of material (plastics being the more popular today) to combat the problem of samples being chemically altered. The sampler typically is lowered in an open position for flushing purposes, and the water is collected by sending a weight down the cable, mechanically triggering a closing mechanism on the sampler upon impact. In some cases, the sampler is closed electronically, the sampler being lowered at the end of a conductor cable. Recently a sampler has been constructed that is capable of the aseptic collection of samples. But despite the sophistication of such samplers, collections are still made for the most part as they were fifty to sixty years ago, namely at the end of a long cable. In fact, the water sampler designed by the Norwegian arctic explorer, zoologist, and statesman, Fridtjof Nansen (1861–1930), at the turn of the century, and called the Nansen bottle, still is extensively used, virtually in its original design.

At the present time a sound-producing instrument is often placed on the end of the cable to determine how far from the bottom an instrument package or water sampler is, as indicated by a receiving unit located on the research vessel. More often than not, however, the oceanographer does not know the exact location of the intermediate segment of the cable. This information is critical if water-sampling equipment or instruments are on that portion of the cable. Another difficulty arises when samples are brought from considerable depths, because they are subjected to tremendous pressure and temperature changes that may cause profound changes in their chemical characteristics. As a result, the sample may have no resemblance to its original chemical character when it is ultimately analyzed. What is needed, naturally, are either collecting techniques that circumvent this problem or, better yet, instruments that will actually measure the variable of interest in the water at the desired location (i.e., *in situ* measurement). The needs and problems associated with obtaining such a capability are not unlike those present in making measurements in outer space. There have been some advances made recently, but many improvements still are needed.

Yet one more problem, common to the whole field of oceanography rather than just to chemical oceanography, is that of obtaining continuous measurements. For the most part, the properties of an

Daily salinity readings are taken on the *Ambrose Light-ship*, at the entrance to New York Harbor. (U.S. Coast Guard)

oceanic location—for example, vertical profiles of a chemical constituent—are derived from data from a number of discrete points. By the application of *in situ* measurement systems, continuous measurements will be possible. The detection of such phenomena as narrow lenses of water having anomalous properties will then not be dependent on the fortuitous placement of water samplers on a cable or the paying out of the cable to a particular depth. That such variations do go undetected, as a result of drawing smooth curves through data obtained from widely spaced samplers, was demonstrated in 1965 by John Strickland and his co-workers at the Scripps Institution of Oceanography.

Researchers are now active in developing instruments that will provide automatic and simultaneous analyses of seawater. Aside from the problems relating to chemical alteration of seawater samples, indicated previously, there is another problem: Whenever samples are not analyzed immediately on board

the research vessel, but are stored for later analyses, serious errors can be introduced into the resulting data because of sample aging. As a result, whenever possible, chemical analyses are carried out on the research vessels. The usual rigors of shipboard life, however, together with the fact that marine scientists are as prone as any other group of individuals to seasickness, often combine to make shipboard analyses less accurate and precise than would be the case under other circumstances. The automation of analyses will circumvent these problems and permit analyses to be made at a much higher rate. The increase in speed, in turn, will not only present the investigators with "real time" data, which can be used to alter a ship's course to reinvestigate waters giving anomalous values (an opportunity obviously lost when analyses are performed long after the sample is taken), but also will result in significant monetary savings.

Although automated techniques are applicable to any seawater analyses utilizing wet chemical methods, present efforts are being directed toward nutrient analyses. Manual analyses of the nitrate, phosphate, and silicate content of seawater still utilize chemical manipulations. Using basically the same chemistry, the instrumental procedure is automated to a point where the analyst places the sample in a cup at the beginning of the analysis train and reads the concentration of the sample from a recorder only several moments later.

A significant aspect of the new automated technique is, of course, the realization of simultaneous analyses. By accomplishing analyses in this way for several components of a water sample, the possibility of incorporating errors by division of the sample and by different handling of its various portions is completely removed. In addition, the results of the analyses will be completely compatible, inasmuch as the analyses are accomplished at the same time, eliminating the aging factor. The ultimate utilization of such a concept of measurement, of course, will be to accomplish the analyses in the oceans rather than on their surfaces.

Research activity in the field of chemical oceanography is not only beginning to enjoy increasing attention as a field in its own right but is also reaping the rewards that naturally follow from capable men of science becoming interested in the field. Although chemical oceanography is still in its infancy, a continued expansion of efforts on the part of research organizations and capable scientists will begin to answer the question, "Why is the ocean full of seawater, and what are the controlling factors making it do what it does?"

6 FOOD FROM THE

by C. P. Idyll, Hiroshi Kasahara

The sea has provided food for man as long as he has lived on its borders. The archaeological records of every seaside culture contain evidence of this in the form of fishhooks and other artifacts. Bones of marine fish discarded in 40,000 B.C. by Late Stone Age men have been found in the kitchen middens from caves along the Dordogne River in France, while Neolithic cave dwellers used harpoons, hooks, and nets. Drawings on ancient temple walls show nets and netting needles. Long before Christ, probably beginning about 1250 B.C., dried fish was a trade item in the Persian Gulf. The ancient Greeks devised complicated traps for catching the fish of their seas, and the Phoenicians and Carthaginians founded many cities as fishing settlements—Sidon ("the fisher's town"), Malaga ("the salting place"), Cadiz, Sinope, and even Byzantium (now Istanbul). From about 400 B.C. to A.D. 450 Rome's colonies in the eastern Mediterranean supplied the Imperial City with preserved seafood of many varieties—tuna, swordfish, sturgeon, mackerel, eels, oysters, and sea urchins.

Millions of years after man started to fish he is still taking most of his catch from coastal waters. This is partly because the nearshore areas are easiest to reach. In addition, despite the common impression that the sea is monotonously the same over its whole vast expanse, there are important differences in productivity among parts of the ocean. Depending upon whether the water is deep or shallow, cold or warm, salty or brackish, the quantities and kinds of animals to be found in it vary greatly. The result is that some areas of the oceans are deserts while others are extremely rich in life. The richest are the shallow, well-lit waters along the coasts.

The sea contains an extraordinary variety of creatures, comprising millions of species of plants and animals. There are, for example, more than twenty thousand species of fishes. A great many marine animals and plants have been used by man-

kind for food, but a surprisingly small number of them are consumed in large quantities. Only about a dozen kinds form three-quarters of the amount used. More than one-third of the world's catch consists of herrings, anchovies, and sardines; more than a sixth is cod-like fishes; the rest, in rapidly declining proportions, consists of mackerel and jack mackerel, tuna, flatfish, capelin, salmon, and shark, followed by hundreds of other varieties.

The overwhelming proportion of food from the sea—approaching 90 percent—consists of fish. Shellfish supplies 7 or 8 percent and whales about 1 percent. Plants from the sea provide much less than 1 percent of man's food, compared to a figure of 80 percent for plants from the land in some parts of the world.

The scarcest and most valuable of all human food materials is animal protein, needed to build muscle and other body tissue. Plant protein does a good job, but some of the amino acids required by the human body are in short supply in many plant proteins. Fish is one of the best of all sources of these, and one of the cheapest. The productivity of ocean water is determined in much the same way as the richness of earth; the greater the amount of fertilizing minerals —chiefly nitrates and phosphates—in the water, the greater its productivity. And just as the productivity of land is increased by returning minerals to the surface layers through plowing, in the sea a kind of plowing serves this purpose, too.

The similarity between production of living material on land and in the sea is not mere coincidence; it is a consequence of the fact that in both regions the creation of edible animals depends on the growth of plants. These are eaten by animals; man either consumes the plants directly or he eats the animals that have eaten plants or other animals.

In the sea the plants are mostly very different from those on land, being microscopic in size so that

SEA

they are rarely seen (unless they bloom in prodigious numbers in phenomena called "red tides") and often being encased in skeletons of hard calcium carbonate or silica. Their small size makes them hard to collect and their skeletons render them unpalatable to man. The relatively scarce large marine plants we call seaweed are eaten to a minor extent, but generally speaking, the quantity of marine plants providing food for man is insignificant.

But the role of the plants in the economy of the sea is enormous, and in a particular region the abundance of the sea creatures consumed by man—the fishes, the crustaceans (crabs, shrimps, lobsters), the mollusks (oysters, clams, scallops)—is first determined by the ability of that region to grow plants. The nutrient minerals are carried to the depths of the sea in the dead bodies of animals and plants. Once they sink below the level where light is strong enough to carry on photosynthesis, plant production ceases. This depth varies from about one hundred to three hundred feet according to the clarity of the water. The rich areas of the ocean are those where some mechanical process overturns the water layers and brings mineral-laden water to the sunlit surface again. In the temperate zones this takes place when the surface cools in the autumn, increasing the density so that the water sinks, which forces deep layers to the top. Productivity is also renewed when prevailing offshore winds skim off the surface waters; these are replaced by water welling up from the depths, bringing with it fertilizing mineral salts.

The Fishing Industry

The first fishermen engaged in "subsistence" fishing—that is, they caught fish and shellfish only for themselves and their families. This pattern of fishing persists in many parts of the world, and it sometimes accounts for a high proportion of the food

TABLE I

World Catch by Groups of Species, 1970*

Species	Catch in Thousands of Metric Tons**
Herring, sardines, anchovies, etc	21,150
Cod, pollack, hake, haddock, etc.	10,320
Redfish, bass, etc.	3,640
Mackerel, cutlassfish, etc.	3,040
Jack, mullet, saury, etc.	2,310
Salmon, trout, smelt, etc.	2,090
Tuna, bonito, billfish, etc.	1,680
Flounder, halibut, sole, etc.	1,240
Sharks, rays, etc.	520
Crustaceans (marine)	1,565
Mollusks (marine)	3,239
Miscellaneous marine fishes	9,580

* SOURCE: Food and Agriculture Organization *Yearbook of Fishery Statistics*, Vol. 30.
** One metric ton = 2,204 pounds.

consumed in shore communities. But early in the history of man fishing became an industry, as certain bold and skillful individuals undertook to supply other members of the community with seafood. Since the gathering of some kinds of shallow-water animals is easier than hunting on land, fishing may be the oldest industry in the world, predating even agriculture and herding.

In the Middle Ages fishing became an important industry in northern Europe, centering on the rich North Sea. By the twelfth century such English cities as Hull, Grimsby, and Barking were already famous as fishing ports, and in that same era the Hanseatic League was organized by North German merchants to exploit the sea, including its fisheries.

As the centuries passed fishermen became bolder and their ships bigger, and a few tough and heroic

Herrings, sardines, and anchovies account for about 35 percent of the world's catch. These empty barrels for herring are stacked along the docks at Vegesack, West Germany, alongside a North Sea herring boat. (C. P. Idyll)

By the mid-1500's the Portuguese were exploiting the teeming cod schools of Newfoundland's Grand Banks. In this picture a modern Portuguese dory fishes for cod, using traditional methods and equipment. (F. Krügler)

men began leaving their own shores for distant and deeper fishing grounds. As early as the mid-1500's the Portuguese were exploiting the teeming cod schools of Newfoundland's Grand Banks. They fished with hooks and preserved the catch on board with salt. More than four hundred years later they are still carrying on this rugged operation, in some cases with sailing ships. But long after the Portuguese ventured forth, the majority of fishermen of most nations were still tied to shallow bays and nearshore areas. They moved out gradually, as vessels became bigger and gear more efficient.

These improvements made additional productive grounds accessible, and more time available for fishing. Sail trawlers created a new era of fishing in the quarter of a century from 1825 to 1850. The rich Dogger Bank and other offshore areas of the North Sea were exploited by the 1840's. A trawl is a conical net dragged over the bottom of the sea, scooping up fish as it passes. It is an efficient piece of equipment that catches more fish more quickly than hooks and lines or stationary nets. The earliest trawls were held open at the mouth by a beam supported on runners. By 1893 the rigid and clumsy beam had been replaced by otter doors—kitelike devices that kept the mouth of the trawl net open as the gear was dragged forward in the water. The otter trawl extended the depth of water in which dragging could be done, and another part of the ocean was opened to exploitation. Machines for knitting fishing nets were developed by

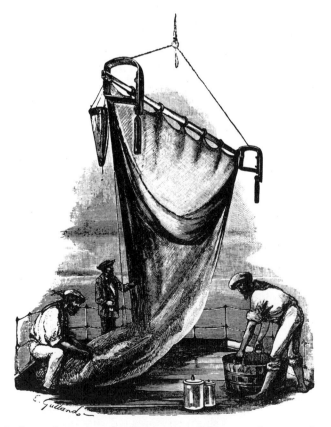

Sail trawlers opened a new era in fishing in the second quarter of the nineteenth century. This drawing of crewmen examining the contents of a scientist's beam trawl appeared in the *Challenger Reports*.

132

A later improvement in trawling was the device of otter doors, which forced the mouth of the trawl net open as it dragged forward in the water. The steamer *Brazilian*, shown above, was one of the first vessels to work the otter trawl. (D. H. Cushing, *The Arctic Cod*)

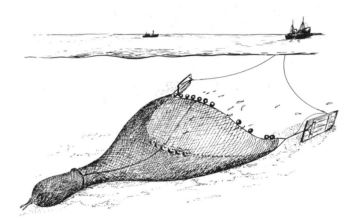

The diagram of a North Atlantic otter trawl shows the kite-like doors at the mouth, floats to buoy the top of the mouth, and weights to keep the bottom of the net on the ocean floor. Most of the fish are funneled into the smaller "cod end" in the after part of the net. (U.S. Fish and Wildlife Service, Woods Hole, Mass.)

1840, making large quantities of cheap netting available for the first time.

Starting in 1854, when sails were replaced by steam, another great surge forward took place. The vagaries of wind no longer frustrated the trawlerman, and the power and speed of fishing vessels could be greatly increased, with corresponding increases in the size and efficiency of gear. By the 1890's trawlers had ventured to Iceland, and in the early 1900's they were fishing a type of flounder called plaice from the

Gill nets are used to catch fish during the high tides at Nova Scotia. The fish are removed when the tide is out. (Nova Scotia Film Bureau)

Barents Sea near Murmansk in Russia. The introduction of gasoline engines in 1910 and diesels in 1921 provided still more dependability, power, and efficiency. By the latter date, cod were being trawled off Spitsbergen, Cape Kanin, and Nova Zembla, in the far North Atlantic.

Early in the twentieth century fishing was advanced still further when the purse seine was developed. This is a great wall of netting (nowadays as much as three-quarters of a mile long and 500 feet deep) that is dragged around a school of fish. When the ends are brought together a "drawstring" cable, strung through rings in the lower edge of the net, closes the bottom like an old-fashioned purse, trapping the fish in a big bowl.

While these advances were being made, human population was rising and the Industrial Revolution was building railroads that carried fresh fish to new and greatly expanded markets. All over Europe fleets were increased as fishermen reacted to this demand. In 1855 there were only twenty-four fishing boats in Grimsby, England; twenty-two years later the number had risen to 600, and the boats were bigger; by 1887 there were 830. From 1897 to 1903 the number of fishing vessels in Aberdeen, Scotland, increased 258 percent. In America, Japan, and elsewhere similar gains were taking place.

Thus in the space of several decades fishing increased enormously, and for the first time attention began to be given to the possibility that harm might be done to the stocks. Until late in the eighteenth century it was almost universally believed that the sea was so enormous and its fish swarmed so abun-

The whaling industry took a great step forward with the invention of the harpoon by a Norwegian, Svend Foyn, in 1870. Here a gunner takes aim with a modern version of this weapon. (Gustav Hansson, Black Star Publishing Company)

In many parts of the world, fish are caught in simple gear, like this beach seine net in Ceylon. An FAO expert, part of a team assisting local fishermen to improve their efficiency, watches the operation. (United Nations)

dantly that nothing man did could affect their numbers. In 1883 the famous and highly respected English scientist Professor T. H. Huxley declared in his presidential address to the Royal Society, "I believe . . . that all the great sea fisheries are inexhaustible; that is to say that nothing we do seriously affects the number of fish." Even as Dr. Huxley made this statement, however, there were signs that man's harvest of certain popular species was heavy enough in some places to reduce their numbers seriously. Trawlers doubled their catches in the 1880's alone, depleting nearby fishing grounds. Fishermen had to go farther and farther afield or accept smaller catches for the same amount of effort.

Fishery Management

In 1883, the same year that Professor Huxley made his pronouncement about the inexhaustibility of fish resources, an International Fisheries Exhibition was held in London, at which it was urged that research be started on the sea fisheries. This resulted in the founding of the Marine Biological Association in 1884 and the building of a laboratory in Plymouth four years later. Here, in 1892, a remarkable new kind of scientist began his work on the fisheries of the North Sea. He was E. W. L. Holt, and his fresh approach was to get out of the laboratory and into the fishery—on the boats and on the docks. By calculating the average catch per unit of fishing effort, Holt was able to demonstrate that there had been a real reduction in stocks. He pioneered in many phases of fishery biology, striving to find how many fish were breeding within a given area, the smallest size at which they reproduced, the numbers of fish of various sizes. Holt's researches swept away many uncertainties about the state of British fishing, and he proposed some of the earliest regulations—including a minimum-size limit of thirteen inches for plaice. His suggestions were largely ignored at the time, but his influence on the later course of fishery science was great.

Meanwhile a Dane, C. G. J. Petersen, developed concepts about the rational development of a fishery that are still sound today. He showed that the amount of fish in a particular stock is the result of the interaction of additions by reproduction and growth, and losses by death. More than that, he believed that the rate of the latter is significantly intensified by fishing, so that exploitation sometimes needs to be controlled.

Petersen also showed that it would be profitable to transplant small plaice from overcrowded areas to rich feeding grounds. Neither of Petersen's ideas bore

Soviet fishermen bring cod on the deck of the stern trawler *Nekrasov*. (D. Kozlov)

Anchovies off the coast of Peru trapped within a purse seine, a wall of netting that is dragged around the school of fish. A drawstring on the bottom cuts off the escape of the fish. (*Pesca*)

135

fruit, largely because of the difficulties of international agreement as to the relative responsibilities and profits of the various countries necessarily involved. To this day no fully satisfactory legal system has been devised to control fishing in international waters so that each fishing nation gets its share without overexploitation taking place. Yet the necessity to work out such a system provides one of the most urgent and baffling problems facing mankind.

Before the turn of the century the United States began to be concerned about marine resources. The U.S. Commission of Fish and Fisheries was created in 1871, and Spencer F. Baird was appointed the first commissioner. Convinced of an alarming decrease in the abundance of commercially important fishes, Baird initiated studies of their biology. These included research on the distribution of marine organisms, their reproduction and movements, the ecology of sea-bottom communities, the parasites and diseases of fish. Thus he laid the foundation for the new science of fishery biology in America.

Baird was anxious to find some positive measures to create bigger catches, rather than negative rules that served only to reduce production, so he became an enthusiastic supporter of hatcheries. In 1885 the vessel *Fish Hawk* was put into service to make surveys of the fishing grounds, and in particular to conduct fish-culture operations. Millions of shad fry, and millions of other species of fishes and lobsters, were distributed. For many years it was believed (with little evidence, or indeed, with the evidence to the contrary) that hatching operations were of great benefit. The American effort in this field was applauded in other parts of the world, and American fish-culture exhibits won awards in the Berlin Fisheries Exhibition of 1881 and the London Exhibition of 1883. Slowly evidence accumulated that hatching fish to release them back into the sea was having no measurable effect on catches, and many hatcheries were closed. But, the preoccupation with this procedure seriously slowed the development of fishery research.

As fishing increased in intensity, plaice, salmon, and other stocks declined. Scientific evidence began accumulating to support the theory that fishing could deplete oceanic species. In 1918 a Russian biologist, F. I. Baranov, developed a theory of how fishing affected the size of populations; in the next two or three decades more support gathered for this idea. The English biologists E. S. Russell and Michael Graham were among the earliest to show what effect fishing had had on the stocks in Europe. The dramatic recoveries of North Sea fish following relaxation of fishing during the two world wars revealed clearly

Spencer F. Baird, the first American fish commissioner, encouraged sound biological studies, as well as hatcheries, to replenish diminishing fish populations. (Remington Kellogg)

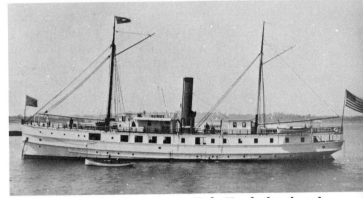

Beginning in 1885 the steamer *Fish Hawk* distributed shad, lobsters, and other fish along the Atlantic coast for the U.S. Commission of Fish and Fisheries. (Gareth W. Coffin, U.S. Bureau of Commercial Fisheries)

that fishermen had a profound effect on the stocks. In America William F. Thompson, using an effective combination of biology and records of landings, established convincing proof that fishing had reduced the level of Pacific halibut populations and that reduction in fishing effort to rational levels could restore these stocks. As a consequence great emphasis was put on conserving fish stocks, and the potential of the seas as a source of food for mankind was sometimes underrated.

And international politics complicated matters. As nearby grounds became crowded and overfished, boats moved farther offshore. Larger, faster, and better-equipped ships made possible the exploitation of rich distant grounds, and problems arose when ships of various nations began to compete for the same fish. For example, unbridled competition

By sound biological research and international management, the North Pacific halibut fishery has increased its yield by 50 percent since 1930. (International Pacific Halibut Commission)

Streamer tags, such as the one shown here in the back of a California scorpion fish, help biologists estimate the size and movements of marine populations. (Charles H. Turner, California Department of Fish and Game)

A marine biologist tags a tiny kelp plant, of which monthly measurements will be taken, on an overturned car off the California coast. (Charles H. Turner, California Department of Fish and Game)

among Great Britain, Germany, Holland, France, and other countries threatened to drive North Sea stocks to unprofitable levels, and international co-operation became essential.

In 1899 most of the countries having a stake in the North Sea fisheries met in Sweden, and in 1902 they formed the International Council for the Exploration of the Sea. They agreed to study the biology of fishes and to try to find ways of conserving them. With the economic spur, interest in ocean science increased, and many able men turned their attention not only to the exploited species but to the other organisms that influenced the size of their populations. And they studied the sea itself to discover how changes in the watery environment affected the fishes for good or ill.

Despite the obvious need for international co-operation in fisheries, it has been hard to obtain. The International Council for the Exploration of the Sea has stimulated brilliant research but has achieved little in the way of management on a regional basis, since the member countries have been unable to agree on how to divide the catch.

In North America results have been somewhat better, but progress has been slow. In 1905 serious talks began between Canada and the United States about joint investigation and control of the dwindling sockeye-salmon fishery of the Fraser River, shared by

both countries. But it was not until 1937 that a treaty was signed creating the International Pacific Salmon Fisheries Commission. Under this agreement, the two countries share equally in the catch and the costs of scientific investigation and comply with conservation regulations. These regulations prohibit fishing at times when depleted races of salmon are approaching the river, so that runs damaged in the past by overexploitation have a chance to reestablish themselves. The clearing of blockades to free passage in the river has also been effective, and these measures have resulted in bigger catches.

The long negotiations that preceded the salmon treaty had constructive effects even before the treaty was signed. In 1923 the International Fisheries Commission was created to study and control the halibut fishery of the North Pacific Ocean. As a result of the soundly based biological research designed by William F. Thompson this body was able to restore the badly damaged fishery so that, with a greatly reduced amount of fishing, it yields a steady 65 to 70 million pounds a year, compared to only 44 million pounds in 1930. The United States is now a member of several international fishery bodies, which deal with the conservation of whales (not very successful), tuna, salmon, fur seals and other species of the North Pacific, ground fishes of the northwest Atlantic, and other marine creatures.

The undoubted need to control overzealous fishing of certain popular species has prompted many governments to put a brake on exploitation not only of these species but of others that did not need to be protected by restrictive regulations. This had the unfortunate effect of discouraging what could have been rapid and profitable development of many fisheries that were still unused, or underused to various degrees. The pendulum of protection had swung too far.

Then, after World War II, came the second massive wave of fishery development. This was nearly worldwide in its effect, although most United States fishermen have not shared in the bigger catches.

The entry of more and more countries into open-ocean fishing has progressively increased international problems. In attempts to solve these, nations have joined together in various fishery agreements. And the United Nations has provided the base from which the Department of Fisheries of the Food and Agriculture Organization (FAO) has sprung. This agency encourages development of fisheries over the world by lending expert assistance to those needing it and by encouraging exchange of information. FAO is in charge of implementing massive projects for fishery development financed by the United Nations Development Programme. In one UNDP fishery project, for example, FAO has assisted the Peruvian Sea Institute in making the fullest use of Peru's enormous anchovy resource. Other fishery projects financed by the UNDP are being carried out in countries of Africa, Latin America, and Asia.

The World Catch

Statistics of world fish production are incomplete even today, and their accuracy diminishes as one goes backward in time. The reports of landings of a century or more ago are to be used with cau-

---

TABLE II

World Catch of Marine Fish and Other Products*

Year	Catch in Millions of Metric Tons**	Year	Catch in Millions of Metric Tons**
1850	1.50–2.25	1958	28.38
1900	4.50	1959	31.21
1930	10.00	1960	34.00
1938	18.44	1961	37.25
1948	17.54	1962	40.77
1950	18.43	1963	41.88
1951	20.49	1964	46.70
1952	21.91	1965	47.35
1953	22.50	1966	51.10
1954	23.35	1967	54.51
1955	25.09	1968	57.64
1956	26.44	1969	56.08
1957	27.18	1970	62.35

---

* SOURCE: Food and Agriculture Organization *Yearbook of Fishery Statistics,* Vol. 30.

** One metric ton = 2,204 pounds.

tious reservation since they are not comparable to those of modern times. Nonetheless it is obvious even from these imperfect records that ocean catches have risen rapidly, especially in very recent years. In 1850 the world catch of food from the sea may have been somewhere between 1.5 and 2.25 million metric tons. The bustle of activity in the next half century doubled this to about 4.5 million metric tons annually. Catches rose moderately in the first decades of the present century and reached 18.44 million tons by 1938. Then, after the setback of the war, catches increased at a much higher rate; in fact they went up more quickly than the rate of human population and substantially more quickly than the overall rate of increase in food production. By 1950 production from the sea totaled 18.43 million tons; by 1960, 34.00 million; and by 1970, 62.35 million. Although total food supplies are still increasing more slowly than the human population, the increases in supplies of fish represent an average annual rise of 6.0 percent, compared to a rise in population of 2.1 percent.

Some of the principal fishing nations of the present are new in the field, and have overshadowed some of the older ones—notably the United States. While this nation used to rank second only to Japan in production, it is now sixth. Ahead of it are some recent newcomers, Peru and the U.S.S.R., and some old hands, Japan, China, and Norway. Crowding not too far behind are many others, some of which rarely sent boats offshore a generation or so ago—Chile, Thailand, India, the Philippines, and South Africa.

Anchovies from Peru

Without doubt the most extraordinary fisheries story is that of Peru. She is now the leading fishing nation in terms of total production, and, remarkably, her catch consists almost entirely of a single species— the anchovy, *Engraulis ringens*, a small silvery fish beloved by sea birds but not often eaten by man. Its value lies in its high nutritive content, its cheapness, and the convenience to users of the dried fish meal that can be made from it. The fish is extremely abundant and easy to catch in quantity. Because of these advantages, Peru's anchovy fishery developed rapidly to meet a soaring demand for fish meal as a protein supplement for poultry and other animals. As recently as 1957 Peru was only the twenty-sixth nation in total fish production; by 1962 she was first. In 1970 she landed the enormous quantity of 12,277,000 tons of anchovies.

The demand for fish meal since World War II has been a major factor in rapidly increased catches not only in Peru but in many other parts of the world.

TABLE III

Catch of Fish and Other Aquatic products by Country, 1970*

Rank	Country	Catch in Thousands of Metric Tons**
1	Peru	12,613
2	Japan	9,308
3	U.S.S.R.	7,252
4	China (mainland)	5,800[1]
5	Norway	2,980
6	United States	2,714
7	India	1,745
8	Thailand	1,595
9	South Africa	1,519
10	Spain	1,497
11	Canada	1,377
12	Indonesia	1,249
13	Denmark	1,226
14	Chile	1,161
15	United Kingdom	1,099
16	Philippines	990
17	Republic of Korea	934
18	France	775
19	Iceland	734
20	China (Taiwan)	613
21	Federal Republic of Germany	613
22	Republic of Viet-Nam	577
23	Brazil	493[2]
24	Poland	469
25	Portugal	457[2]

* SOURCE: Food and Agriculture Organization *Yearbook of Fishery Statistics*, Vol. 30.

** One metric ton = 2,204 pounds.

[1] Estimated (1960).

[2] 1969 data.

Fishes of almost any kind, or a mixture of different species, can be used as long as they are caught in large quantities (a modern plant requires hundreds of tons a day) at low cost. Thus the industry can utilize resources that would otherwise remain unexploited due to lack of markets. From 1938 to the mid-1960's the increase in catches of herrings, sardines, and anchovies—used largely to make meal— was far greater than that of the high-priced salmon, cod, tuna, and halibut.

Fish meal—cooked fish from which most of the oil and nearly all the water have been removed— contains up to 80 percent protein. With its content of minerals and other nutrients, this makes it one of the most valuable of all food materials. It is nutritious for both human beings and farm animals. The common idea that it is used as fertilizer is long outdated— it is much too valuable for that. Fish meal is not consumed by people directly, since it has a strong and not especially palatable flavor, but indirectly it figures largely in the diet of the people of the United States, Germany, and other nations, where it is used as a food supplement for chickens, pigs, and other livestock. The use of fish meal has increased especially because of its role in modern chicken farming, which has grown enormously and, in its turn, substantially changed the nature of the fishing industry. To date, most fish meal has been used in the industrialized countries, but its consumption has begun to rise sharply in developing nations. Many of them already are importing substantial quantities.

The Far-Ranging Japanese

Fishing is a very important factor in helping Japan maintain a population on the order of 100 million, with a very small amount of arable land on four none-too-big mountainous islands. For a great many years—until the rising demand for fish meal spurred Peru to make use of her swarming anchovies—Japanese fishermen landed more fish than any other nation, and they did this by harvesting the waters of virtually every part of the world ocean. More than half their animal protein comes from the sea, compared to only about 3 percent for Americans.

Japanese began this worldwide activity long ago. Their salmon fishery on the Far-Eastern coast of Russia can be traced back to the 1880's, and they were operating crab-canning ships in Kamchatka waters in the early 1920's. In the 1930's refrigerated Japanese trawlers were fishing in the South China Sea, the Bering Sea, and the shelf areas of Australia, Mexico, and Argentina. Their antarctic whaling started in 1934. Before World War II the presence of Japanese fishing vessels in the Bering Sea and in Aleutian waters caused great anxiety among American and Canadian fishermen. After the setbacks of the war, when most of Japan's offshore fishing fleet was destroyed, its industry recovered swiftly, and by 1950 it was establishing new records. Now Japanese boats are encountered in the remotest oceans of the world. They share with the Russians enormous catches in the Bering Sea—an area adjacent to the United States coast that is almost ignored by American fishermen. The Japanese land tunas from wide offshore areas of the North and South Pacific, the Indian Ocean, and the Atlantic. Japan also pioneered on the high seas off the west coast of Africa, an area with the same kinds of oceanic advantages as Peru and Chile that has become one of the world's most prolific fishing regions.

Frozen tuna line the docks at the Tokyo fish market. Japan still leads the world in the value of fish landed, although she now ranks second to Peru in tonnage. (C. P. Idyll)

The Japanese are especially active in fishery research and exploration. The research boat *Shunyo Maru* performs investigations on tuna. (Japan Ministry of Agriculture and Forestry)

The Japanese success in world fishing results from the willingness of her people to work hard under difficult conditions, from skill in devising efficient gear and fishing methods, and from the development of markets for everything that can be caught. With the rise of living standards in Japan the availability of cheap labor for her boats is dwindling, but she still has a wide advantage in this respect over the United States and some other countries.

Many other Asian nations have turned increasingly to the sea as a source of protein. South Korea, the Philippines, China (Taiwan), Thailand, Malaysia, Ceylon, and Vietnam have more than doubled their catches during the last decade. They have accomplished this mainly through increasing mechanization of their fishing vessels and the introduction of modern fishing techniques.

The Rise of the U.S.S.R.

Drawing much less attention than the firing of *Sputnik* in 1957, the rapid rise of the U.S.S.R. as one of the leading fishing nations may be more significant in the end. At the time of the Russian Revolution in 1917 the fish consumed in that nation came almost entirely from fresh waters, and this changed only slowly in subsequent years. Few Soviet boats engaged in long-distance fishing before World War II. But another revolution took place after the war. A greatly intensified fishing effort vaulted the U.S.S.R. into third place among all nations—well ahead of the United States.

The Soviets decided to increase fishery production because their economists concluded that it was cheaper to wrest animal protein from the sea than from the land. This led to the appearance of Russian vessels off scores of foreign coasts. Large numbers of Soviet trawlers now operate in such international fishing grounds as the Grand Banks of Newfoundland, the Bering Sea, the Gulf of Alaska, and the waters off the west coast of Africa. They fish the high

Rather than sending individual boats to a fishing ground, the Russians send whole armadas. This Soviet refrigerator ship, which will accompany a fleet, is under construction in the Leningrad shipyard. When completed it will hold 10,000 tons of fish. (Novosti Press Agency)

seas off British Columbia, Washington, and Oregon; their boats have reached waters off Mexico; sizeable fleets of Russian trawlers operate along the Atlantic coast of the United States. In addition, the Soviet Union has helped construct a multimillion-dollar port for fishing vessels in Havana and her ships fish the Caribbean.

Soviet methods are unique—and highly effective. Fishing operations are planned and mounted as elaborately as military campaigns. Instead of sending individual, competing boats or even fleets to a fishing ground, the Russians send whole armadas. These are dispatched after intensive oceanographic research

Soviet ships, such as the fish factory *Eugene Niskishin,* have facilities to use everything caught, by canning and freezing high-value fish and reducing the rest to meal and oil. (Novosti Press Agency)

and exploratory fishing have indicated the potential of the area. An armada may include huge stern trawlers up to 275 feet long, smaller trawlers up to 170 feet (like the biggest of the present-day New England trawlers), seiners, and other specialized kinds of fishing craft. Enormous mother factory ships like the *Serverodonetsk,* 571 feet long, may go along, too. The group is supported by fuel tankers, salvage tugs, and repair ships (one of these, the *Neva,* has eleven different kinds of workshops). A commodore aboard a command ship controls the entire operation.

Virtually everything that can be caught is used by the Russians. Their ships are equipped to process high-quality fish into frozen or canned products, while cheaper varieties—including those often spurned by American fishermen—are frozen in bulk or reduced aboard the vessels into meal and oil.

While the U.S.S.R., Peru, and a host of other nations were increasing their harvest of the sea, the fishing industry of the United States was stagnating. Second only to Japan in production before World War II, this country had dropped to sixth place by

1967. National policy, which puts fishing behind other interests, is partly responsible for this. The rising cost of vessels, gear, and labor in relation to nearly stable prices for fish is the other major factor. The increased demand for fish in the United States that has accompanied the population rise has been supplied largely by imports, which now exceed domestic production.

The Threat of Overexploitation

The vigorous fishing done by the Soviets and some other countries has alarmed other nations, who regard it as tough competition and poor conservation. There is real cause for concern if this fishing is so intense that overexploitation takes place, but it is important to realize that conservation should be concerned not with *how* fish are caught, but with *how many* in relation to the capacity of the population to maintain itself. Hence, rapid and efficient capture methods and the fullest possible use of whatever is caught should be encouraged. Vigorous biological research must be conducted at the same time to determine if the fleets are overfishing. Certainly, some species—salmon, sardines, whales, some types of groundfish—have been seriously damaged by over-exploitation.

Because some heavily fished stocks require control it has been assumed that all fisheries should be restricted. As a consequence, unnecessary legal and economic barriers have slowed the development of some fisheries and have prevented much food being made available. This trend has been strongly apparent in the United States, and this is one reason the industry in this country has fallen behind that of some other nations. What is required, of course, is a balanced approach, giving vigorous encouragement to the expansion of fisheries like those for Argentine hake, California anchovy, or Alaska herring, whose stocks can withstand more pressure, and maintaining flexible control over those that have reached the limit of rational exploitation. Some fisheries, like that for the tuna of the eastern Pacific, have both kinds of situations simultaneously: The yellowfin stocks are now probably fished at the limit of their capacity, while skipjack, caught by the same gear, are still underexploited.

Meanwhile, the search goes on for new stocks that will allow mankind to continue the rapid increase in the ocean harvest that has characterized recent years. There is, however, a good deal of doubt that this trend can continue at the same high rate unless more efficient harvesting methods are developed.

The fishing industries of many countries are underdeveloped. Primitive methods still prevail in this community in British Honduras. (C. P. Idyll)

While it is certain that the food resources of the sea are not being fully used, no one knows exactly how large these resources are. In spite of recent progress in oceanography and fisheries research, forecasting of future production is still a matter of educated guesswork. Much of the necessary scientific data is missing, and a number of unknown technological and economic factors are involved. The recent development of the world fishing industry has clearly shown that most of the estimates made by scientists in the 1940's and 1950's were much too low. The question is being reexamined in the light of more recent information.

There are different methods of estimating the potential yield of food from the sea. One way is to start by calculating the amount of plant material produced in the whole ocean during a year. This difficult job must then be followed by estimations, based on uncertain data, of how much of this plant material is passed along to the herbivorous animals, how much of the substance and energy locked up in these is passed along to the first level of carnivores, and so on up through several more levels of the food pyramid. The results have been estimates of greatly different magnitude, ranging from twice to ten times

(or more) the present harvest. Such estimates have little practical significance, for they do not indicate under what circumstances, from what areas, or in what kinds of animals the estimated potential yields might be realized.

The second method of estimating the potential yield is to extrapolate from known production of existing fisheries and from presumed possible exploitation of unused or under-utilized stocks. For example, it is believed that the Arabian Sea (the northwestern part of the Indian Ocean) contains virtually untapped fishery resources of large magnitude; fully exploited, these fish stocks might support several million metric tons of fish landings annually.

Only recently has the food potential of the rich waters off the west coast of Africa begun to be appreciated. Some twenty countries bordering the sea there, together with many foreign nations—including the Soviet Union, Japan, Poland, Spain, Portugal, France, Italy, Germany, Greece, Israel, Korea, China (Taiwan), the United States, and others—have been devoting increasing energy to fishing for pilchards, sardines, anchovies, mackerel, and tuna, as well as various bottom-living forms, such as porgies, hake, and octopus. The combined catch from west African

waters exceeded five million tons in 1970, but further expansion appears to be feasible, particularly for mid-water species.

On the other side of the Atlantic, off South America, the waters are less rich, but there seems to be room for considerable expansion. The immense continental shelf off the Guianas has been fished for shrimp since 1959, but only recently have trawls begun to catch the great quantities of fish that exist there. Brazil can produce much more, and the Patagonian shelf of southern Argentina has hardly been touched. Argentine waters may not be as productive as comparable areas in the northern temperate zone, but perhaps two or three million more tons can be taken annually. Many other areas of the world ocean, including various parts of the Indian Ocean, also appear to be greatly underexploited.

Even some parts of the sea already under heavy exploitation may be capable of greatly increased fish production. California was for many years the leading producer of fish in the United States, and even after the collapse of her big sardine fishery in the mid-1950's—caused by altered oceanographic conditions and overfishing—she still ranks high, with about 250,000 tons a year. But despite the vigorous exploitation of California waters, additional millions of tons of fish could be caught. These are hake, anchovy, jack mackerel, and small deep-sea fish—all species not familiar to, or sought by, buyers in the United States, and this, of course, is why they are underexploited. But they are species that can make fish meal and fish-protein concentrate.

Indeed, according to fishery scientists, the waters adjacent to the United States (including the Pacific, Atlantic, and Gulf of Mexico) are potentially capable of supporting annual yields of fish and shellfish totaling some 20 million tons, compared to 2.5 million tons caught by United States flag vessels and about an equal amount taken by foreign fishermen in 1969.

This does not by any means exhaust the list of underexploited fish stocks of the world, and there is no doubt that enterprising fishermen and energetic government research ships will uncover resources whose potentials are now only dimly guessed or are quite unknown. But there is a finite limit to these, and the day will come when there will be few new stocks to be harvested by conventional methods. Unless spectacular new ways of fishing are invented, it seem likely that production of the kinds of fish now being used will about double in the next fifteen to twenty years. The rapid rate of increase of the last two decades will probably slow down as readily exploitable new resources become more and more difficult to find. At that point the hope of substantial increases in food from the sea will depend on man's ability to find uses for the truly enormous variety of marine organisms now wholly or largely unused, and to develop methods for harvesting them profitably.

Stocks for the Future

The greatest of these untapped resources is plant plankton, of which perhaps some 400 billion tons in wet weight are produced in the sea in a year. Because plant plankton is unpalatable and very expensive to collect, it seems unlikely that we can use it soon, if ever. Animal plankton also is produced in such large quantities that every person in the world could have several tons every year if it could be caught, processed, and delivered to him—and if he would eat it. Unfortunately, despite its total bulk, animal plankton is scattered thinly throughout the enormous mass of the ocean—some 330 million cubic miles—and most attempts to harvest it have failed to pass the test of economics. Furthermore, much of this material is unpalatable or inedible (for example, stinging jellyfishes).

Nonetheless there are exceptions, and in some areas of the world there exist concentrations of animal plankton large enough to make economic harvest seem possible. The most likely of such plankton is the krill of the Antarctic, *Euphausia superba*. Krill are small crustaceans, relatives of the shrimp. Those found in the Antarctic are about one and a half or two inches long, and they occur in enormous numbers up to the edge of the ice pack. Observations by British whale biologists and more recently by Soviet fishery scientists put estimates of their production at somewhere between 40 million and a fantastic 1.3 billion tons a year. It is probable that a conservative 50 million tons a year could be caught without endangering the stocks—nearly as much as the present total world seafood production.

It seems hopeful that animal meal (for poultry and pigs) can be made from the krill without too much more experimentation. If this is so it should also be possible to produce protein concentrate for direct human consumption, because the same raw material can serve both purposes. The difference in the two processes is largely a matter of removing a greater proportion of the oil from the protein concentrate, since its degradation is responsible for most of the problems of flavor and odor that make fish meal unpalatable.

Even the first of these steps—the making of meal of acceptable quality from the krill—is not at the point of being solved. The Russians, with their eagerness to develop new sources of animal protein for their people, have been the most vigorous in this research, and they are developing methods of capture and of processing. These are promising, if not yet fully satisfactory. The krill meal has a strange orange color derived from the bodies of the little shrimplike animals. It has a high protein concentration—more than sixty percent. The high proportion of unusable chitin (the shell of the krill) and the very rapid spoilage rate (even in the cold southern ocean) are problems that so far have prevented an industry from developing. But technology has solved problems more difficult than these, and it seems likely that krill, in northern seas as well as in the Antarctic, will be an important crop in the future.

The "red crab" (*Pleuroncodes planipes*), in its young stages, may be another animal plankton creature that will eventually supply food for man. This animal exists in incredibly large numbers off the coasts of California and northern Mexico. One scientist describes how his vessel "crunched through" immense swarms of red crabs for miles. Hundreds of thousands of tons may be available for harvest after efficient methods of catching them and converting them into some saleable product have been devised.

Once the proper techniques have been developed, many animals may be successfully harvested. Herein lies the greatest hope for truly substantial increases in the sea harvest, since the abundant marine herbivores would be tapped—the level of the food pyramid in the sea equivalent to the cows, pigs, and sheep of the land. At present most of the catch consists of carnivores—salmon, cod, tuna—which are the equivalent of tigers, foxes, and weasels. The significance of the comparison is not so much the unattractiveness of tiger meat compared to beef (maybe the public would prefer tiger if they had been raised on it) but the far smaller total quantity of the carnivores compared to the animals that feed on grass. The enormous differences in size and in other characteristics between land and ocean herbivores have prevented wide use of the latter in the past; perhaps this can be changed, and if so the world will be less hungry.

Squid are not herbivores, but they are another potential food resource of enormous size. They are distributed over most oceans of the world, and may be one of the most abundant of the sea's larger animals. They are not nearly as well known as the fishes, largely because little use is made of them. Many types of squid are nocturnal and some forms live most of their lives in deeper parts of the ocean. They are curious in appearance, and to most people they are not attractive as food. Yet they are palatable and nutritious, ranking not far behind fish in the composition of their protein. Only the peoples of East and Southeast Asia, notably the Japanese, make any substantial use of them as food, although they are eaten to a small extent in southern Europe and in other parts of the world. In proportion to their abundance they are an enormously under-utilized resource. If more people accept them and if efficient methods of harvesting them can be developed, another great source of human food will be available.

Possibly the commonest fish in the ocean, the three-inch bristlemouth is one of the untapped resources of the sea. If techniques were developed to catch bristlemouths at low cost, they could become an important source of fish meal and fish-protein concentrate. (G. B. Goode and T. H. Bean, *Oceanic Ichthyology*)

It is also known that huge numbers of small, odd-looking bristlemouths and lantern fishes (so called because of their luminous organs) occur in deep layers of vast oceanic areas. They can be considered an important potential resource for fish meal. So far no techniques have been developed to catch them in large quantities at low cost.

Realizing the Sea's Potential

If mankind is to realize what is clearly a great potential for food from the ocean, efforts should be made in three directions. First, in order to maintain catches at the highest possible level, conservation measures based on scientific research will continue to be important. As increased pressure is put on more stocks by more nations, protection from overexploitation will be increasingly necessary. But this will produce *more* fish only in the relatively few stocks now severely overexploited, which can be restored to greater abundance by relaxing fishing pressure; these are a minority of the sea's population. Perhaps only a few million tons of fish (a small amount in terms of the total needs of mankind) can be saved in this way, but they are highly valuable components of our food supplies.

The second requirement is a greatly increased level of search ·for under-utilized stocks throughout the oceans of the world. Even after years of active investigation there still remain large areas that are apparently productive, judging from oceanographic data, but for which knowledge of potential resources

Ocean perch have spiny projections that help to entrap them in the meshes of trawl nets. The right-sized netting is important to hold in the big ones and to permit easy escape of those too small for the market. (Albert C. Jensen)

Power blocks mounted on two small boats—devices to wind in the net mechanically—retrieve a purse seine filled with menhaden, greatly increasing the efficiency of this kind of fishing. Two larger carrier vessels stand by in the background. (Bureau of Commercial Fisheries, Beaufort, N.C.)

is limited. These include the Arabian Sea, the Andaman Sea (in the northeastern part of the Indian Ocean), waters around the Malay Archipelago, Australia and New Zealand, various parts of the South American coast, and other areas. The stocks available in the mid-oceanic areas are largely unknown. Although in recent years major fishing companies of several nations have searched for new resources, there is a limit to such efforts by private enterprise, and the time has come when much greater public funds and scientific manpower must be put into the survey and evaluation of food resources in the ocean. The Soviet Union is ahead of other countries in this respect, but it is unlikely that a worldwide evaluation of oceanic food resources can be made through the efforts of one nation.

Thirdly, if full use is to be made of the ocean's protein material, there is a need for research to develop new fishing techniques. The present methods are not basically different from those employed many years ago. The trawl net, introduced in the late nineteenth century, is still one of the most important types of fishing gear. The purse seine and most of the other important methods have been employed for at least sixty years, and some for centuries. Factory ships are not new either, since Japanese crab-canning mother ships were in operation in the early 1920's and the first Norwegian factory ship for whaling went into the Antarctic Ocean during the 1923–24 season. Although improvements such as synthetic netting and the power block (which relieves the fisherman of much of the task of retrieving the net) have increased the rate of catch of purse seines, there have been no revolutionary changes in the basic technology of sea harvesting. In order for food production from the ocean to keep growing at a high rate, spectacular technological developments will have to take place to utilize its enormous potential resources, which cannot be harvested economically by present methods.

Research into the behavior of fish should be encouraged to provide scientific information on which to base new fishing techniques. In this way we can learn to know the fish more intimately: how they are attracted to their food, how they signal their mates or their fellows in a school, how they detect danger and react to it. Experiments are going on now to determine the reaction of fish to such stimuli as light, sound, and electricity.

The behavior of many animals, both terrestrial and marine, is influenced by light. The reactions are complex, the creatures being attracted by some colors or intensities of light and repelled by others. A considerable variety of gear already exists that depends on these reactions—ranging from a spear and a pitch flare at the bow of a canoe to a highly sophisticated combination of a powerful underwater electric light and a pump to sweep fish into the hold of a vessel. The latter method is used by the Russians in the Caspian Sea to catch a fish called hilsa, a herringlike species. Most of the purse-seine fisheries in Japan also employ lights. A much greater use of lights in combination with various kinds of gear is likely to prove profitable.

Herring are pumped aboard a carrier boat to be taken to a cannery. (William Underwood Company)

Many fishes and other marine creatures make sounds of bewildering variety—beeps, grunts, honks, moans, whistles, clucks. These are for a variety of purposes, also: to keep companions informed of their position, to give warning of the presence of enemies, to announce sexual maturity, to protest injury. Some animals produce noise by swimming and eating. There is evidence that fish are attracted by playbacks of recorded sounds, either because they think a meal or mate is available, or because they are merely curious. It seems fully possible that they can be accumulated for easier capture by more sophisticated development of this phenomenon.

Fish react curiously to electricity. An electrical field in the water frightens them at low levels of power; at a higher level it lines them up in the electrical field with their heads toward the positive pole. As the power is gradually increased many kinds of fish are forced to swim toward the positive pole, and eventually they may be killed. With the proper intensity of field and with suitable characteristics of the current, they can be attracted and brought aboard the vessels much more efficiently than with standard types of gear. In the sea this is more difficult than in fresh water, since more power is required to produce the same effect on the fish. As a consequence, development of electrical fishing gear has been delayed and has not come up to the early exciting promises of the period right after World War II. In the United States electricity has been used effectively to concentrate menhaden (a herringlike fish) in purse seines. After they are caught, a pump is put among them in the net. The end of the pump is made the positive pole and a current is set up with the negative pole on the hull of the vessel. Electricity forces the fish to swim to the pump head and they are swept aboard the boat with far less effort than was once necessary. But this device is only for emptying nets faster and is not a means of *catching* fish;

In fresh and brackish water electricity is being used to force fish to swim to a pump intake, where they are swept aboard, as in this drawing of a Soviet device. In seawater the method is less successful, since much more power is required. But in the United States, menhaden, already caught by the purse seine, are concentrated by electricity around a pumphead, increasing efficiency of operation. (G. H. Davis, *Illustrated London News*)

The *Kompas*, a Soviet refrigerated fish carrier, has equipment for trawling as well as for processing the catches from smaller fishing boats. This vessel is also used as a seagoing school. (Burmeister & Wain)

moreover, it has not been perfected to the extent that it is used by all menhaden fishermen.

It has also been found that shrimp can be affected by certain kinds of electrical impulses, which force them to leap out of the bottom mud and sand. Many shrimp are nocturnal, and the catch is poor during the day when they are buried in the mud. High hopes are held that by attaching electrical devices to shrimp trawls, fishing might be possible around the clock and at a high rate of efficiency. Other aids, some of them unguessed now, may very well promote great increases in production.

Man has by no means reached the limit of the harvest of the sea. In the years ahead he will continue to increase the catch of familiar kinds of seafood as his needs expand and his technology improves. He will learn to use new and unfamiliar kinds of animals from the sea—squid, krill, lantern fish. He will get some additional fish and shrimp from the new shallow-water farms that are now being established.

But the importance and potential of food production from the ocean must be assessed realistically. There is no evidence to support the view that the problem of feeding the growing world population can be solved by exploiting sea resources. An overwhelmingly large proportion of the food requirement has been and will continue to be met from agricultural products, and the main contribution from the ocean will be to increase the supply of animal protein. The amount of animal protein available may well increase greatly in the future if man is wise enough to make full use of the sea's rich resources.

7 FARMING THE

by C. F. Hickling

The art and industry of farming, including the farming of seafood, signifies a benevolent interference with nature, whereby the production of desired crops of plants and animals is enhanced by artificial management measures, which include the accumulation of fertilizer, the protection and propagation of valuable species, and the elimination of unwanted predators and competitors.

The effect of such measures can be very considerable, as may be seen if we compare the crops of fish produced naturally with those produced when there is some degree of cultivation.

Different parts of the seas and oceans may have very different rates of fish production, much as the land has its fertile and barren areas. The seas and oceans have an area of about ninety billion acres, and from these men took, in 1969, some sixty-three million tons—which amounts to a crude rate of extraction of only about 1.4 pounds per acre annually. But most of the ocean is not fished at all, and certain parts are fished fairly heavily; for example, in the North Sea and the Baltic, the rate of fish harvesting in 1964 was respectively thirty-three pounds and eight pounds per acre.

Estuaries are much richer, and have been described as "nutrient traps." Chesapeake Bay, for example, crops at about eighty pounds per acre annually, and the Sea of Azov at not much less.

But by more or less elaborate farming measures, far larger fish crops can be obtained. A fishpond in Indonesian estuarine waters is considered poor if it produces only 150 pounds per acre, while good ponds there will produce 560 pounds and more. In Taiwan, crops in the marine fishponds are now on the order of 1,700 pounds per acre annually.

Fish farming in seawater or estuarine water may have developed in more than one way. It undoubtedly developed from saltings—artificial shallow lagoons in which seawater is evaporated to dryness to make salt. The laborers must have soon noticed that the fish and prawns unintentionally admitted with the water flourished and grew well in the ponds, where conditions were obviously to their liking. Both in Biscay and in Java, fish farming in seawater is believed to have arisen in this way in the Middle Ages.

The farming of fish and mollusks in cages floating in the sea, which is practiced in Japan and beginning in Britain, arose from the long-established practice of keeping bait-fish, such as sardines, alive in wickerwork cages trailing in the water.

Fish farming in fresh water arose in China about 1000 B.C., or even earlier. No doubt experience gained over the millennia led to the extension of the practice to brackish water and seawater. Certainly, there were marine fish farms in Java in the late Middle Ages.

Fish farming spread because it was profitable and because it guaranteed a supply of fresh fish. The same two considerations lie behind the renewed interest in fish farming in marine waters today. The decline in the abundance of the most valuable sizes and species of fish, such as flatfishes and salmon, makes their raising under controlled conditions commercially attractive at this time, and much pilot-scale experimental work is being done.

The rate of organic production in the sea depends on the amount of sunlight and of dissolved carbon dioxide and nutrient salts in the water. The deeper parts of the oceans, below the zone to which the sunlight penetrates, contain vast reserves of nutrient salts such as nitrates, ammonium, phosphates, sulfates, and trace elements. In favored areas of the sea this nutrient-rich water comes up to the surface, where there is plentiful sunlight for the growth of plant life, and where other requirements are at an optimum. The result is a tremendous outburst of plant life, leading to prodigious fish production. Such rich areas occur where the deeper water is dragged up to the surface by the divergence of two ocean currents, or where offshore winds drive away the

SEA

surface layers of water, so that deeper water upwells to take its place. This last process is especially common off desert shores; and we have the paradox that the offshore winds, which blow the life-giving moisture out to sea and so create desert conditions on land, cause the sea to become very fertile. A recently exploited case is that of the coastal waters of Peru, where the Humboldt Current brings up to the surface nutrient-rich deep water. In 1964 some nine million tons of fish were landed from this narrow strip—a quantity equal to one-sixth of that year's entire fish production throughout the world. A somewhat similar fishery takes place off the desert coast of southwest Africa, due in this case to the Agulhas Current.

These accessions of nutrient-rich water, an obvious source of wealth and prosperity, are a present of nature to the lucky countries and fisheries that enjoy them. Suggestions have been made that nutrient-rich deep water might be pumped artificially to the surface, and though this would be unprofitable in the open ocean, enclosed areas such as atolls could be made to produce immense crops of seafood. Pilot trials have already begun.

Another important source of nutrient salts is land drainage, and especially the drainage containing the waste products of human populations. It has long been recognized as a paradox that farmers are obliged to spend large sums of money buying fertilizer, while at the same time thousands of tons of valuable fertilizer are returned to the sea as domestic sewage. In many countries sewage is treated so that some of the fertilizer value is returned to the land as fermented sludge; but even in this case the clarified effluent carries most of the plant-nutrients in the sewage down the drains to the sea.

However, it is not all wasted there. Land drainage does have a significant fertilizing effect in the sea, though this cannot always be demonstrated. A favorable case is the treated sewage of the London conurbation of ten million people. Hydrographic conditions in the southern North Sea are such that this nutrient-rich water is surrounded by water masses derived from the Atlantic, so that it remains in the southern North Sea long enough for its fertilizing effect to become manifest. There is, in the first instance, an increase throughout the year in the rate of organic production, in the form of plant plankton, as compared with neighboring unfertilized areas of sea. Phosphorus is the key fertilizer, and about 3,000 tons a year are supplied by London's sewage.

The total fish crop harvested from the southern North Sea in 1948 was about 300,000 tons, at a rate of forty-six pounds per acre. This rate is about twice that of adjacent unfertilized areas of sea, and about 100,000 tons can be considered as due to the nutrients from London's sewage. The sewage of New York has a similar effect, though it has not been quantified in terms of fish.

Such spectacular results of natural and domestic fertilization have suggested that the artificial addition of nutrient salts to the sea might give notable and perhaps paying additions to the fish crop. It is a fact that, on a small scale in freshwater fishponds, the production of fish can be increased tenfold or more by the addition of fertilizer, chiefly phosphate.

During World War II some fairly large-scale trials were made in two Scottish sea-lochs by the late Dr. Fabius Gross and his associates. These narrow and rather deep arms of the sea were given considerable quantities of fertilizer in the form of nitrates, ammonium salts, and phosphates. These fertilizers, as could have been expected, caused great increases in the production, firstly, of the plant plankton, and then of the animals living on the sea bottom. The lattter, in fact, increased sevenfold. But too few fish were naturally present in these lochs to use this increased food supply, and the trials failed to show that this kind of fish farming would pay. Fish had to be introduced from elsewhere, or specially bred for the purpose, in the same way that a farmer might

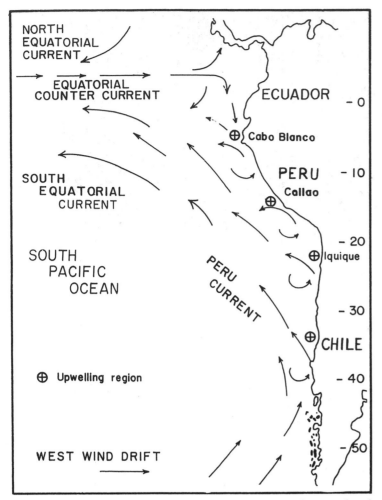

As on land, food production in the sea depends on the supply of chemical nutrients. Off the west coast of South America a natural "plowing" occurs when surface waters are swept away by offshore winds and deeper, nutrient-rich waters are brought to the surface. This supports the enormous commercial fisheries of the region. A simplified version of the ocean current system in the area is shown here.

purchase cattle to use his crop of grass. But the fish which survived grew much faster than they would have in the sea under unfertilized conditions, partly because there were now enough food organisms to keep them growing well even during the winter months.

But competing "weed" fish of no value, such as gobies and sticklebacks, also flourished in these favorable conditions, and in fact ate eighteen times as much of the food produced as the population of commercially valuable fish. Among the plants, too, the brown seaweeds, which have no direct value as food, took a large part of the added fertilizer. The weeding-out of such useless, competing plants and animals would not be economic.

One of the two sea-lochs used, Loch Craiglin, is eighteen acres in area and is connected with the sea by a narrow channel that was dammed to reduce the loss of fertilizer. But this barrier to the free movement of the water soon caused a layer of warmer, fresher, and therefore lighter water to overlie a deeper layer of cooler and more salty water. The oxygen in this deeper layer was soon used up, creating conditions unsuitable for fish life.

The other, larger loch, Kyle Scotnish, has an area of 180 acres and is open to the sea. Consequently there was no deoxygenation of the deeper water, but since there was no fencing to confine the rapidly growing fish, they soon quit the loch and were lost to the experiment.

These trials suffered from the stresses and shortages of wartime conditions. It has been suggested that a sea-loch closed by a dam should have two sluice gates—one at the bottom, to admit nutrient-rich deeper water on the rising tide, and one at the

The fertility of marine ponds can be increased naturally by using dams with two sluice gates—one at the bottom to admit the richer, deeper water, and one at the top to skim off the poorer surface layer. This sluice gate in Java controls the movement of water between the fish ponds at the right and the seawater access canal on the left. (C. F. Hickling)

top, to discharge surface water, from which the nutrients will have been used up, on the falling tide. In this way, fertility would build up and stratification be prevented. Even better results might be obtained than by the addition of fertilizer. To admit new water is one of the ways of maintaining the fertility of marine fishponds, and it is well known that water discharged is poorer in nutrients than the new water admitted, showing that a proportion of the nutrients has remained to enrich the pond. Indeed, in the years ahead, man is more likely to extract fertilizer from the oceans than he is to add it to them.

In some of the marine bays and channels on the coast of Yugoslavia, some interesting fertilizer trials have been made. Whereas the Scottish trials gave no opportunity of testing different combinations of fertilizers, the Yugoslav trials were partly based on the known ability of certain algae and bacteria to build up protein directly from nitrogen dissolved in seawater from the air. Therefore very little expensive nitrogen fertilizer was used; the chief ingredient was superphosphate at a rate of 102 pounds per acre, together with trace elements, humic acids, and salts of tungsten and molybdenum in soil extract.

In a bay of 454 acres, with an average depth of seventy-five feet, connected with the open sea by a narrow channel, the growth of marine plants was increased sixfold, and there was a thirtyfold increase in the phytoplankton and phytobenthos (the blue-green algae and benthic diatoms). There was no increase in the zooplankton, but the growth of oysters was accelerated fivefold. The oysters and mussels showed so much benefit from these fertilizations that the commercial oyster and mussel farmers paid for some of the trials.

As in the case of Loch Craiglin, there was some deoxygenation of the deeper water, but clearly this did not affect the shellfish beds.

Enclosure of Marine Fish Farms

Marine fish farms must be surrounded by fish-proof fences or other barriers to prevent the escape of the livestock being farmed. Control of livestock for management and cropping, and exclusion of predatory or competitive species, are thereby secured.

Such enclosure may be done in three ways: (a) by embankment, (b) by nets or fencing staked to the sea bottom, or (c) by floating net-rafts.

In the case of embankment, obviously the amount of fill needed, and therefore the capital cost of the work, will increase very sharply with each additional foot of height. Thus fish farms enclosed in this way are unlikely to be made in water much deeper than at present, namely, about five feet. Stratification is unlikely in such shallow ponds, and since they can be drained dry at very low tides, the whole crop of fish and prawns can be taken. There can even be a

Research is conducted on the vitamin C content of seaweeds and the nutritive value of fish and oysters at the Yugoslav Institute of Oceanography and Fisheries. (United Nations)

tilling of the soil, including the working-in of fertilizer, thus stimulating the production of algae on which the marine crop depends.

The enclosure of farms by fencing is simple in the case of seaweed and most shellfish culture; since these organisms cannot move, a fence may chiefly function as demarcating each farmer's claim. As some sites are more favorable than others, the government may have to referee claims and maintain order.

But with shellfish such as abalones, which can move about, and with fish and prawns, which are highly mobile, fences must confine the livestock even if they are stocked at small sizes, without preventing free movement of the water. The fencing may be of wood, fabric, or metal. In deeper water, and where there are strong currents, the fence will have to be supported by posts driven into the seabed; the deeper the water, the more substantial the posts, which may also have to be stayed to concrete anchor blocks. Once again, capital costs will increase sharply with increase of depth, and the repayment of interest and amortization on loans is always one of the heaviest charges against income in farming. Therefore a depth of about twenty feet seems likely to be the economic limit for fish farms enclosed by fencing or netting. In the Seto Inland Sea of Japan, where this kind of fish farm has been developed, capital costs are reduced, where possible, by using the configuration of the land and adjacent islands.

Marine fish farms enclosed by open fences allow free movement of the water and so prevent stratification; but where the rise and fall of the tide is great, low spring tides may leave the cultured stock with dangerously little living space. Fertilizer cannot be

151

economically used because of the movements of the water.

As to net-rafts, these may be used for the cultivation of oysters and mussels, which are grown on ropes or wires suspended from them. But in Japan they are used for the cultivation of fish and prawns; here they are made of plastic netting, supported in shape by a framework of bamboo and floated by steel drums. These floating cages may be as small as a few square yards, or as large as 100 square yards. Floating cages make fish farming largely independent of the sea bottom, but even so, practical considerations would decide that shallow and sheltered water be used.

This likely limitation of marine fish farming to shallower coastal waters is reinforced by the concept of international waters. A fish farmer staking out a claim outside his country's territorial waters would have no enforceable title. An illustration of the limiting effect of territorial waters is the Dogger Bank transplantation scheme. For half a century it has been known that the Dogger Bank, in the North Sea, grows a heavy crop of the small shellfish on which the plaice feeds; and it has also been known that too few plaice are naturally present on the Bank to make full use of this bountiful fodder. The English naturalist Walter Garstang was a leading advocate in suggesting how sensible it would be to transport millions of baby plaice from their overcrowded nursery grounds of the eastern North Sea, and release them

Hanging collectors for "spat" (young oysters) in Hong Kong allow the growing oysters to tap the entire water column for food. Thus suspended, the oysters are also kept clear of starfish and other ocean-bottom enemies. (C. F. Hickling)

on the Dogger Bank, where many tagging experiments have shown that they would grow to marketable size in quick time. But, over the decades, plans to take this simple and potentially profitable step have foundered on the question of who pays for the cost of transplantation to an area where all nations are free to fish without paying a cent. Garstang foresaw this difficulty from the start, and the situation has not changed in sixty years.

Transplantations

Transplantation of fish and shellfish from overcrowded conditions to situations where food is plentiful but fish scarce (as in the case of the Dogger Bank plaice mentioned above) is a measure of husbandry practiced on a large scale in some countries. Two examples will be given here.

The first was the result of the work of the great Danish biologist Johannes Schmidt, whose classic series of experiments, much interrupted by World War I, ended triumphantly in 1922 with the revelation that the common European eel does not breed in this or that local pond or marlpit, but makes a fantastic migration across the Atlantic to breed. This knowledge inspired the German government to begin transporting live young eels, or elvers, from the estuary of the English River Severn, where migrating elvers appear in great abundance, to eastern Germany, where there are vast bodies of water but few of the valuable eels. The fact that these transplantations were continued for several decades suggests that the German government found the modest expenditure worthwhile.

For many years plaice have been successfully transplanted from the overcrowded nursery grounds of the eastern North Sea to the inner waters of the Danish Liim Fjord. Here the peninsula of Jutland is cut across by a system of sheltered creeks and lagoons, wholly inside Danish territorial waters, where there are good feeding conditions for plaice but where plaice were naturally scarce. The transplanted fish grew fast, so that the value of the plaice harvested was much greater than the cost of transplantation. Fishermen's associations, in fact, paid most of the cost.

In the farming of shellfish such as oysters, clams, and cockles, the spat (or settling, young stages) are collected on the settling grounds where they may be naturally much overcrowded, or they may be induced to settle on special artificial spat collectors. The settled spat are then transplanted and relaid at the most favorable density per square meter on grounds where feeding conditions are good.

Transplantation invites new predators and parasites, such as this oyster-boring snail, which is abundant in the waters of southern England. The snail preys on young oysters by drilling through their shells and eating the flesh. (P. J. Warren)

Nori, or laver, an alga with high vitamin but relatively low nutrient value, is cultivated for human consumption in Japan. Until recently, farmers collected spores of the seaweed on brush bundles that they stuck into the mud bottom, as shown in this old drawing. The bundles were then removed and replanted where the *nori* would grow fastest. Modern *nori* farmers collect spores on netting stretched parallel to the surface of the water and held in place by bamboo poles. (U.S. Bureau of Commercial Fisheries)

Two-year-old oysters are inspected in the winter holding grounds of New South Wales, Australia. (Australian News and Information Bureau)

Farming Seaweeds

Several species of marine algae are grown for human consumption, chiefly in Japan, but formerly in other parts of the Pacific also. This is a genuine cultivation. The best known of these cultivated algae is the *Porphyra* plant, grown to make the popular seafood item in Japan that is called *nori*. The dried fronds are used to wrap and add flavor to seafoods such as shrimp and fish. *Nori* has an excellent protein and vitamin analysis. Similar algae, known as laver or dulse, are gathered on strands at low tides in some countries of western Europe, but no cultivation is done. In Japan the growing of *nori* is a substantial business, occupying some 57 percent of the total cultured sea-areas of the country and earning some nine billion yen ($21 million) annually. The spores of *Porphyra* grow into filaments, which inhabit empty mollusk shells and there produce sporangia. These sporangia in turn produce spores, which are distributed by the tides and which settle on suitable substrates to grow into the edible fronds.

In earlier days, sticks or bamboos were staked in rows on the sea bottom to serve as collectors of the *Porphyra* spores. When the collectors were well covered with sporelings, they were removed and replanted where the fronds could grow fastest. At a length of three to five inches, the fronds were plucked off, spread on mats, and dried.

Growth of the alga continues after plucking, and there may be several pluckings between November and March–April each year.

More recently, production has been greatly increased by using horizontal sheets of plastic netting for the settlement and growth of the alga. These sheets are long and narrow, usually about 50 yards long and 1.6 yards broad. From each of such nets

1,500 to 3,000 fronds can be gathered. The *nori* farmers may drape the plastic nets on which the fronds are to grow over and around heaps of oyster shells heavily infested with the sporangia of *Porphyra,* and so secure a good and uniform settlement of spores.

Losses can be heavy, and arise from two main causes: firstly, when the plants produce spores that are carried away; and secondly, when wave action in bad weather tears off many fronds. Nevertheless, the rate of production of cultivated *Porphyra* beds is comparable with that of the world's most highly productive farmland.

Many other kinds of marine algae are also cultivated. The Japanese, in addition to growing *Porphyra,* make the seafood *wakame* from the alga *Underia.* Simple measures of cultivation include the scattering of rocks over the sea bottom to give a good substrate for the growth of the alga. The red alga *Gracilaria* is cultured to produce agar-agar, which is used in confectionery and other foodstuffs.

There seems no reason to doubt that the cultivation of *nori* and other algae could be done in other countries with similar success, provided a promotional campaign were mounted to popularize their consumption.

Shellfish Cultivation

A bed of shellfish such as oysters, cockles, and mussels, which live by filtering minute algae out of the seawater that bathes them, may produce as great a weight of edible meat per square yard as any other kind of meat-producing unit. In the Dutch Wadden Zee, the figures for animal protein per square yard rank among the highest to be found in the world, and in certain places the sea bottom may be paved with cockles at a density of about 1,000 per square yard. Comparisons are always difficult and depend on technicalities such as "dressed weight," but it has been estimated that cultivated mussel beds can produce more than 1,000 pounds per acre of dry organic substance, or 7,000 pounds of wet mussel flesh, as compared with about 300 pounds per acre of beef on good pasture land.

Oysters and mussels are still raised largely from spat bred naturally. But oysters can also be bred from the egg, so that oyster farming now covers the whole range of true farming. The parent oysters are induced to shed their eggs by the hundred million. When these hatch out as tiny larvae quite unlike the adults, they drift about, aided by minute swimming hairs; but after a short period of this free life, they must find a suitable object on which to settle and metamorphose and start to feed, for they are too

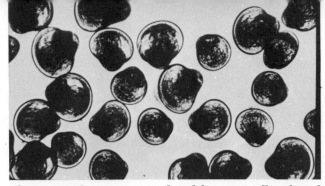

These oyster larvae were produced from specially selected captive parents in a Long Island, New York, hatchery. The water temperature was controlled so that the parents were induced to spawn when the farmer wanted a new generation. The larvae shown here, about ten days old, are approaching the stage when they will settle on some hard surface for the remainder of their lives. (Harold Webber)

Juvenile hatchery oysters from Long Island are suspended in plastic net bags on oyster-shell "cultch." They will be placed on the bottom for their final phase of growth. (Harold Webber)

small to have more than a few days' reserves derived from the egg. The preferred food is minute plant plankton—the μ-flagellates, so called because they are mostly less than one thousandth of a millimeter in size. In the artificial raising of oyster spat, the abundance of these μ-flagellates can be increased by the use of organic fertilizer, such as ground crabmeat, or by inorganic fertilizer, which is used in Norwegian oyster pools and, experimentally, in some bays in Yugoslavia.

Cleaned, empty oyster shells may be distributed over the sea bottom to provide the oyster spat a suitable attachment. But nowadays special artificial spat collectors may be used, such as curved tiles coated with sand and lime. When these devices have collected a good load of oyster spat, they are relaid in neat stacks in creeks and estuaries where the minute foodplants are naturally abundant and enemies few. They soon grow big enough to be scraped off the tiles and relaid for growth to marketable size.

Oyster beds have to be weeded of pests and

competitors: Their worst enemies—starfish—may be swept up by dragging bundles of rope yarns over the shellfish beds.

An improvement in technique is to hang the spat collectors in long bunches strung on wires suspended from rafts or scaffolding. Empty oyster shells used as spat collectors are spaced along the wires by short bamboo rings. When they are well covered with settled spat, the collectors are re-hung on the wires, but now separated by long bamboo distance-pieces. This gives the necessary space for the fast growth of the oysters. This hanging method involves a minimum of handling and also allows the growing oysters to tap the whole water column for food.

Many potentially valuable kinds of clams may occur as deep as 120 feet on suitable areas of sea bottom, and now that scuba divers can stay and work at such depths without very great expense, a profitable extension of the farming of marine mollusks is possible. Recent developments in the northwest of Spain show great possibilities in the raising of mussels suspended from large rafts.

Farming Stock That Can Move

This account has so far dealt with seaweeds and shellfish that cannot move about. To fence in the stock it is only necessary to mark off each farmer's concession; more elaborate barriers are not needed. But for shellfish that can crawl and swim, such as abalones and scallops, some kind of fencing is needed. In the particular case of scallops, Japanese farmers may, by drilling a hole in the shell, grow the animal on suspended wires like the oyster. But prawns (the "shrimp" of commerce), which are valuable enough to be very profitably farmed, are highly mobile, as also are fish. Confining them where they will grow well is a large part of the art of rearing them.

The most valuable commercial prawns belong to the family Penaeidae; most Penaeid prawns are tolerant of brackish water, or at least of rapid changes in salinity. They usually migrate into oceanic water to breed, and the developing young are carried back by wind and current into shallow coastal and estuarine waters. Some are carnivorous; some are herbivorous and feed on the very abundant deposits of more or less disintegrated organic matter (detritus). By far the greater part of the world's supplies of prawns come from the commercial fisheries, chiefly by trawling; but the prawn farmer has the notable advantage of being able to market his stock so as to make the most of good prices.

For example, the *kuruma* prawn in Japan is caught in summer and autumn by the fishing fleets

Shrimp farmers in Japan get high prices for their product, especially during the winter months when the supply is scarce. This large shrimp-rearing pond, with laboratory and hatchery buildings, is located at Aio in southern Japan. (C. P. Idyll)

A Japanese biologist examines net-cages of farm-raised shrimp at Aio. (C. P. Idyll)

155

in such quantities that the prices realized may fall low. Yet in the winter months the price becomes high, because few prawns are then caught. Therefore fish farmers buy live prawns from the fishermen in September and October and keep them alive in tanks for sale in winter. There is always a big cash premium on live prawns.

Ponds for the raising of these prawns are usually made partly of netting, so that the flow of the tide can renew the water. But at low spring tides the prawns may be crowded dangerously in a much reduced volume of water. One of the problems yet to be solved is some economical way of keeping a circulation of oxygen going at such times. Until it is solved, the rate at which prawns may be stocked is the rate applicable to this reduced volume of water at low tides, and the much greater productive volume of water at high tides cannot be fully used.

Thus there is a tendency now for prawns to be raised intensively in cement tanks, independently of the tides. These tanks may be small, with an area of about eight square feet, with water some eighteen inches deep. A layer of coarse sand covers the bottom, and the prawns are able to bury themselves during the daytime. There is a circulation of filtered seawater, and in the winter this may be slightly warmed by passing it through pipes submerged in well water. The prawns are fed trash fish or small shrimps at a daily rate of about 5 to 10 percent of the estimated weight of prawns in the tank. The best time of feeding is at dusk, for then the prawns emerge a few at a time to feed. If feeding were done later in the night, there would be fighting for the food among the many active prawns, with the risk of injury and loss.

From December through February, when prices are at their best, the prawns are marketed alive, packed in refrigerated sawdust in cardboard cartons. So packed, they should stay alive for two or even three days.

Prawns raised less intensively in enclosures in shallow water are likewise fed chopped trash fish to make them grow fast; they can be made to increase their weight fivefold, and the survival rate may be as high as 80 percent.

One of the most valuable of the Japanese prawns has recently been raised artificially from the egg in captivity. The process is being worked out on the full commercial scale, which, if successful, will be a great improvement on the catching of half-grown prawns to raise in ponds, since the supply will be more reliable and the prawns in better condition.

The experimental prawn's natural spawning season in the sheltered Seto Inland Sea of Japan lasts from early May to October, with a maximum between June and August. The mature females, already fertilized, are placed in wooden tanks each about one square yard in area, and supplied with seawater at 82.4° F. Egg laying takes place; the eggs hatch out in about thirteen hours; and after twelve days the young prawns have completed a complicated series of juvenile stages and moults, so that they resemble the adults, though they are still very small. Then they are moved to cement tanks about five square yards in area, with a twelve-inch depth of water; after twenty days they are transferred to the raising ponds, where they grow to full commercial size in about one year.

The first foods of the newly hatched prawns are cultures of suitable diatoms, the eggs and early larval stages of mollusks, or the early nauplius larvae of the brine shrimp, *Artemia*. Once the prawns' larval stages are past, they are fed small shrimps and chopped trash fish.

This artificial raising of prawns is an important discovery, and its use can be expected to spread; but in the meantime most prawn ponds will continue to be stocked from the wild, by capturing young prawns or inducing them to enter the ponds on their own account. An important instance of this occurs in Kerala State in southwest India, where the paddy fields that border an extensive system of brackish-water lagoons may be used to raise a crop of prawns in the off-season for rice. The quantities of prawns produced in this way are large and provide a prosperous export trade in dried prawn meat.

Shrimp-farming research is conducted by the University of Miami's Institute of Marine Sciences in these enclosed ponds at Turkey Point alongside Biscayne Bay. During the daytime the shrimp bury themselves in the coarse sand that covers the bottom. (C. P. Idyll)

Cultures of diatoms (microscopic algae) are collected at the Turkey Point laboratory for feeding to the early larval stages of the shrimp. (C. P. Idyll)

Two pond-raised pink shrimp are attracted by food at Turkey Point. Their daily diet, consisting mainly of trash fish, is usually brought to them at dusk. (C. P. Idyll)

Prawns are raised in special tidal ponds in south China, and Hokkien Chinese migrating to Singapore fifty or sixty years ago brought this technique with them. There are now more than 700 acres of prawn ponds in Singapore.

These are constructed on low-lying tidal mangrove swamps in muddy river basins, which are thus put to good commercial use. The mangroves are cleared, except for the stumps, which are often left in place; and the cleared area is subdivided into ponds by embankments of mud about four feet high, three feet wide at the top, and eight feet wide at the base. At the deepest point, where the original drainage-channel ran, the pond might be seven feet deep,

but it might be only a few inches deep at the landward margin. The area of the ponds varies from one to fifty acres, with an average of about twenty acres. The construction and maintenance of these ponds is a considerable feat of hand engineering. Each pond has several sluices in its embankments, furnished with slots to take screens, sluice-boards, and, when needed, a fishing net.

The cycle of operations is adjusted to tidal conditions and to the habits of the prawns. At rising tides in the mornings the sluice gates are opened and water flows freely into the ponds. This water replaces water drained away during the previous night's fishing; it keeps the temperature of the ponds steady during the heat of the day; and, by equalizing the water levels inside and outside the ponds, it relieves pressure on the mud embankments. At some time before high water the sluices are closed, and they remain closed during the falling tide.

A similar free entry of seawater is allowed at the rising tide in the evening. During the few moments between the rising and falling tides in the evening, however, a long fine-meshed net is placed in slots in the sluice and protected by a screen from floating rubbish such as dead leaves. As the tide falls, water rushes out like a millrace from the pond through the catching net, where the prawns are strained off. After about two hours—depending on water levels and market demand—the sluices are closed, the net emptied, and the prawns sorted and iced for the morning's market.

Tidal conditions allow about twenty fishings a month. Fishing is done at night because the adult prawns burrow into the pond bottom by day, but roam about in the ponds, near the sluices, at night. The juvenile prawns inhabit chiefly the landward, shallower parts of the pond, which are remote from

A fishing net harvests prawns at the sluice gates of a pond in Singapore. (C. F. Hickling)

157

the sluices, in the embankments, and so they are not liable to be caught until they are well grown.

Larval and juvenile prawns are swept into the ponds with the inflowing water of the day tides, since, unlike the adults, they are as active by day as by night. A good flow of water into the ponds should be especially sought in March, June, and November, for then the young stages are at their most abundant following the spawning of the adult prawns in the deep water offshore. To maximize the intake of young prawns, water should be admitted through all the sluices, and any excess of water can be discharged through screens on the next falling tide. This discharge of water should always be over the top of the sluice and not along the bottom, because the better-grown postlarval prawns are likely to settle on the bottom of the pond.

These ponds produce fish as well as prawns, and the fish also enter the pond as young stages with the inflowing water. At suitable intervals—for example, at very low spring tides—the sluice gates are closed and the small amount of water left is treated with an infusion of the seeds of *Camellia sinensis,* or teaseed cake. This infusion contains saponin, which does not affect the prawns buried in the mud, but is toxic to the fish, so that they are easily caught. They are an important addition to the earnings of a pond.

To get reliable data on the economics of this useful industry, the Singapore Government operated some fourteen acres of prawn ponds on the full commercial scale. In four years these ponds produced 43,000 pounds of prawns, fish, and crabs, at a rate of 700 pounds per acre annually. The capital construction costs were recouped in two years, and by a rationalization of labor it was possible also to run a piggery on the ponds, and so add to the income they produced.

To maintain or increase fertility, only simple measures are taken. The ponds may be dried out, and leaves and other vegetable waste may be spread. Drying and refilling introduces a fresh supply of nutrients, which causes the growth of the algae that are the primary food for the prawns, and the vegetable trash has much manurial value.

In addition to being an industry in themselves, prawns are a very valuable byproduct of some fish farms, as will be described later. There they benefit from the cultural methods used to increase the fish crop.

Successful experimental work has been done on the cultivation of "shrimp" (Penaeid prawns) in South Carolina; and indeed there must be great areas in the world where this type of seafood could be cultivated, given the know-how, the aptitude, and the capital. Prawns are hardy and adaptable, fast-growing, and not too fussy about their food. They are highly valued everywhere.

Artificial Breeding of Sea Fish

Toward the end of the last century, the numbers and fishing power of the fishing fleets had grown to the point where they were beginning to catch a significant part of the wild fish stocks. One of the first and obvious remedial measures tried was the artificial breeding of sea fish. This was begun in America as long ago as 1868, and by 1917 as many as three billion young fish were annually bred and released in the sea. European hatcheries added their quota in the eastern Atlantic. But by the 1920's doubts arose as to the real value of adding a few billion fish fry to the astronomical numbers already produced naturally, and many of the breeding programs were scaled down or abandoned.

During both World Wars in Europe, the fishing fleets of the belligerents were requisitioned for minesweeping and as armed escort vessels, while the fishing fleets of neutrals were restricted by minefields and other hazards. The reduced intensity of fishing in turn reduced the mortality rate of fish and thus allowed the survival, to a large size and to breeding age, of large numbers of fish that under peacetime conditions would have been caught and killed while small and immature. Yet this great increase in spawning capacity did not result in larger broods of young fish. It seems to be a fact that the sea can only support so many young fish per square mile, and when this number is exceeded any surplus must perish. Prolific and successful broods of young fish, giving rise to one or several seasons of prosperous fishing, can result from small numbers of spawners as frequently as from large numbers; the survival of the broods depends more on the physical and biological conditions of the environment than on the absolute numbers of eggs spawned.

It may be that the pendulum swung too far, for recently there has been renewed interest in marine hatcheries. There may yet be some scope for breeding valuable species of sea fish or prawns with which to stock areas naturally poor in such species.

A very important and successful case of the artificial propagation of a fish is that of the Baltic salmon. These fish breed in many of the rivers draining to the Baltic, including especially several of the Swedish rivers. But these rivers are becoming less favorable as salmon breeding grounds because of

Early apparatus for hatching marine fishes was photographed aboard the *Fish Hawk,* pioneer research vessel of the U.S. Bureau of Fisheries. This floating hatchery produced millions of fry of dozens of species. They were then released into the sea, but it is doubtful that they had any effect on the marine harvest. (U.S. Bureau of Commercial Fisheries)

hydroelectric dams and, in the higher reaches, an important logging industry. Fish passes or ladders are not considered the answer to the problem; thus the undertakers of hydroelectric schemes make restitution by artificially breeding and releasing to the Baltic a number of young salmon equivalent to what would have been produced naturally. This has been successfully done; indeed, in one river the annual commercial catch of salmon has reached 10,000 tons, whereas the best catch from the previous natural spawning was 8,000 tons.

There are now many salmon hatcheries on the Swedish rivers. Ripe salmon ascending from the sea are trapped and held in ponds until they are ready to spawn. Then they are stripped of their eggs and spermatozoa and the fertilized eggs are incubated under carefully controlled conditions. The hatched young are raised in shallow ponds with a suitable flow of water and fed food pellets prepared from fish solubles, vitamins, and trace elements. The food conversion rate may be as high as 1.5:1, and this remarkable rate compensates for the high cost of the pelleted food.

After two or three years the young salmon reach the silvery smolt stage, and then they are released to descend the rivers to the sea. In the Baltic they grow quickly, reaching commercial size in two years.

But the Baltic is international water, and the fishing fleets of several countries fish for salmon there. Tagging experiments have shown that no less than 10 percent of the artificially reared smolts survive to grow to commercial size, and that half of these are caught by the nationals of countries other than Sweden. Though this half-share suffices to cover the cost of the breeding program, the Baltic is being exploited in this situation like a vast fishpond, stocked by one country, but sharing its harvest with others who have not contributed to the cost.

In a more perfect world, all the Baltic states would contribute an appropriate quota of salmon smolts, thus making the scheme a fully shared one—and indeed Denmark has made a start. The case of the Baltic salmon does point up the handicap that the political factor of the territorial waters presents to any extension of fish farming outside them. An advanced state such as Sweden may enjoin its nationals to safeguard a great natural resource; but what private enterpriser would spend capital on a marine fish farm whose produce would go largely to others?

Enclosed Farms for Marine Fish

In the Seto Inland Sea of Japan, which is wholly in Japanese territorial waters, marine farming, though mostly for the production of marine algae, has also developed for the production of edible prawns and fish. The most important fish cultivated is the yellowtail, *Seriola quinqueradiata.* This fish is grown in enclosures bounded by wire fencing or netting; for

159

example, in one farm the area enclosed is 180 acres at high water and fifteen acres at low water. The outer fence, which is made of netting supported on strong concrete posts held in place by massive anchored wire stays, is 360 yards long. All this represents a big outlay of capital, and also large maintenance costs. The deeper boundary of the farm runs along the twelve- to fifteen-foot depth-contour, so that the net and its supporting posts and stays stand in twelve to fifteen feet of water. As the mesh of the net must be fine enough to hold the newly stocked small fish, the initial investment must be high, as well as the cost of keeping the net clear of debris and of repairing storm damage. In tidal waters, and especially at spring tides, masses of floating seaweed, logs, leaves, and branches will be swept against the netting and will soon, by their damming effect, tear the fence down, unless there is frequent clearing, which is very costly in labor.

As was mentioned earlier in this chapter in the case of prawn farming, the large water volume of the yellowtail farm cannot be fully exploited unless some economical means is found of aerating the lesser volume of water available at low tide. As it stands, the population of the farm is limited to the 65,000 yellowtail—about seven per cubic yard—that can survive when the tide is low. At high tide, when the volume of water is more than double that at low tide, the population of the farm is only about three per cubic yard.

Stocked at such densities the yellowtail could never find enough natural food for growth, or probably even for sustenance. These are predatory fish, and they must be artificially fed on cheap chopped fish such as sardines, herring, and *akta* mackerel. The cost of these trash fish, and the cost of transporting them and of preparing and dispensing the food to the yellowtail, must be high. This kind of fish farming, with its high costs of production, can be profitable only if the fish can be raised at a high density and sold at a high market price.

The yellowtail has not yet been bred in captivity, so the young fish that are to be stocked in the ponds must be caught wild. They are found around floating masses of seaweed. When smaller than about five inches, they feed on zooplankton and small fish, but at a larger size they feed exclusively on fish. The young fish are caught with small purse seines or scoop nets and are then sold to the fish farmers.

A valuable extension of this technique is the floating net-cage mentioned earlier in the chapter. This method makes the depth of the surrounding water less important and provides more constant oxygen conditions. The fish, again chiefly yellowtail, are fed chopped trash fish.

In Mauritius, enclosures called *barrachois* were put up after the island was settled by the French two centuries ago. Barriers made of rockwork cut off creeks and bays, creating impoundments in which small fish, caught in the lagoons outside, were placed to grow to an acceptable size. The tidal rise and fall in this oceanic island is small; there seem to have been no management measures in the *barrachois,* and the fish that flourished best was the predatory barracouta, which feasted on the smaller fish. The enclosure technique never seems to have been taken seriously, and the *barrachois* have long been out of use. Yet this kind of enclosure could make a useful contribution to food supplies, since not only fish, but oysters and clams, could be raised in them, given some interest and management. Where coral rubble is cheap and ready to hand, such enclosures would not be costly. Similar enclosures have been used for centuries for the capture of fish and prawns on some Pacific islands.

Farming Herbivorous Fish

By far the greatest area of marine fish farms in the world today is devoted to the culture of fish that are herbivores, either wholly or to a large degree. These fish are the milkfish, *Chanos chanos*—a tropical fish resembling a large herring—and the grey mullets, species of *Liza* and *Mugil*, of circumtropical and temperate distribution. It would be wrong to believe that the farming of these fish is on a small scale. There are hundreds of thousands of acres of marine ponds, chiefly in Indonesia, the Philippines, and Taiwan, but also extensively in the Mediterranean and Black Seas.

Unlike the farming of carnivorous fish, which use protein food to obtain protein, the farming of these herbivores creates new animal protein; and whereas the farming of predators must consist of selling at a high price fish raised at high cost, the farming of nonpredators gives good to moderate yields at a low labor cost. The food of nonpredators, in fact, is grown in the pond itself. This food consists of the algal felt that grows on a firm substrate on the bottom or on other suitable surfaces in the water. It is a mixture of filamentous green and blue-green algae, with a covering of diatoms and smaller algae, and containing a world of small animals of surprising abundance. For example, an algal felt growing in fresh water was found to contain—in an area of only about eighty square inches—3,550 midge larvae,

11,000 small crustacea, and large numbers of insect larvae and mites. All of these organisms, plant and animal, are good food for fish.

Therefore success in the profitable raising of herbivorous fish depends on the production of a good algal felt, and one that will renew itself as it is browsed throughout the fishes' growing season.

The growth of the algal felt is antagonistic to that of plant plankton. If there is a dense growth of plant plankton, the algae are shaded out. Conversely, a vigorous algal growth uses up the available plant nutrients, and thus suppresses the plankton. But as plankton is of no value for the feeding of growing milkfish and mullet, cultural measures aim at favoring the algal felt at the expense of the plankton. This can be done by having shallow ponds, less than half a yard deep, so that even if plant plankton begins to grow, it cannot grow thickly enough to stop the growth of the algae, which are thus given the chance to gain the upper hand. Secondly, slow-dissolving fertilizers are used, which also favor the bottom-living plants.

The nature of the pond soil plays an important part in the abundance of the algal felt. For example, in Taiwan it is found that sandy soils are poor producers of algal pasture, while clay soils are good producers. In Indonesia and the Philippines the experience is the same: Silty loam is the most productive.

Organic matter present in the pond soil chiefly as detritus is to some extent directly used as fish food. It is also a slow-acting fertilizer and conditioner of the soil; and it may have particular importance to the function of the filamentous blue-green algae, many of which have the valuable faculty of converting atmospheric nitrogen into protein in the presence of organic matter, phosphate, and trace elements. In fact, a surplus of such converted nitrogen becomes available for the nutrition of other plants, such as the green algae and the diatoms that grow on the algal felt. The newly fixed nitrogen also becomes available as the algae decay or are eaten. This must be an important factor in the rate of production of this algal pasture for fish—a rate of production that in Taiwan may be as high as 25,000 pounds per acre in ponds where the organic content of the soil is 4.17 percent, but about 1,330 pounds per acre on soils with an organic content of 1.23 percent.

It will be clear that measures taken to increase the rate of production of algal felt will increase the production of fish, and there is thus a very close analogy with land farming. So far, only very simple measures are used. These include the drying out of the ponds at intervals so as to induce some mineralization of the organic matter in the soil; repeated filling of the ponds with a shallow layer of seawater, which is allowed to evaporate and deposit its nutrient salts; the spreading of slow-acting organic fertilizer such as leaf compost and rice bran. The latter may also act directly as food for the fish, especially late in the season, when the algal pasture may have been consumed or destroyed by pests or bad weather. Research on cultural practices is continuing, and it may be expected to make the same contribution to improving algal pasture as agricultural research has done for grassland.

In the *valli*, or brackish-water lagoons, of northeast Italy, the fish cultivated include the carnivorous eel as well as the grey mullet. The eels find good feeding in the insect larvae that flourish in the very organic soil, but they also take toll of the grey mullet while these are small.

The earlier *valli* were enclosed partly or wholly by open fences. But since success in managing this type of fish farm depends on the correct mixing of fresh water and seawater, the more modern *valli* are wholly enclosed by embankments of earth or masonry, with sluices through which both fresh water from a canalized river and seawater from the lagoon can be admitted as required.

By allowing a slow current of water to flow out of the sluices in the spring months, when the young eels and mullet are migrating in from the sea, great numbers of these can be induced to move into the *valli*, and they remain there when the sluices are closed. But nowadays the natural migration does not supply enough fry, and valliculturalists have to buy additional young fish.

The soil of the *valli* is a very soft and very organic mud, rich in plant and animal debris, and carrying a dense population of living algae, bacteria, and soil animals, including insect larvae and small mollusks. As in the cases of Taiwan, Indonesia, and the Philippines, the higher the organic content of the soil, the better the crops of fish raised. The highest organic content is found in the more saline conditions, in the Italian *valli*.

As yet, no cultivation is done in the *valli*, and fertility remains high because entering seawater is richer in organisms than the water that flows out. From individual *valli*, each of 750 to 1,250 acres, fish crops of the order of 190 or more pounds per acre are produced, so total production is substantial.

The organic matter in the soil can give rise to an oxygen deficiency in the water, which can cause a fish-kill. But there is, in the northern Adriatic, a

Traps for mangrove crabs, which damage pond levees but are valuable seafood, are carried on a raft in the seawater supply canal of a milkfish pond in Taiwan.

Bamboo windbreaks protect a marine milkfish-raising pond in Taiwan. (C. F. Hickling)

modest rise and fall of tide, and by manipulation of the sluices to admit fresh or salt water, conditions of stagnation can be avoided.

The marine fishponds of Indonesia are shallower than the Italian *valli*, so deoxygenation is less of a problem. In Indonesia also, poor sandy soils give crops of only 125 pounds per acre of fish; good ponds on clay soils produce over 540 pounds per acre.

In Taiwan, where this kind of farming is best developed, the ponds are only twelve to twenty inches deep. Here there is a distinct fall in temperature during the winter, and at this time the fish are kept in deep channels protected by thatched fencing from the cold northerly winds, or warmed by well water. The drained and dry production ponds are then subjected to cultural operations. In Indonesia and in the warmer southern parts of Taiwan the milkfish is cultivated; in the colder parts of Taiwan the grey mullet is cultivated. Both fishes have the same feeding habits, and the same technique is used for both. The milkfish has not yet been bred in captivity, but the gray mullet has been hormone-induced to breed and a successful commercial technique is being developed in several countries. At present, in the case of both fishes, the young are caught on the beaches, and in creeks and lagoons, by specialist fishermen who sell to the fish farmers.

162

In Taiwan it has been found that the rate of food conversion of algal felt to fish is about 12:1; thus an annual crop of 25,000 pounds per acre of algae gives a fish crop of about 2,320 pounds per acre—an impressive result. But the quality, as well as the quantity, of the algal felt is important. The filamentous blue-green algae and the diatom flora that grows on them are the best and most nutritious food for milkfish of all ages; but the wiry, filamentous green alga *Chaetomorpha* is poorly digested, and phytoflagellates cannot be used directly as food.

Slow-acting fertilizers such as compost or rice bran can be profitably used, but not the familiar artificial fertilizers such as superphosphate. As said earlier, these would encourage the growth of phytoplankton, which would shade out the algal pasture. In the Philippines, rice bran is used in the nursery ponds to produce the most favorable conditions for the growth and survival of young milkfish. In Taiwan, the teaseed cake used as a fish toxin also acts as a pond-soil fertilizer.

Prawns are a very important cash-crop in the Indonesian marine fishponds; indeed, they may earn nearly as much money as the milkfish. Because they are being cropped all the time, they give the pond owner a steady income. The prawns, which are mostly herbivorous or feeders on detritus, benefit from a good algal pasture. The larvae of the midges, or Chironomids, find in the algal pasture of the shallow marine fishponds a very favorable habitat. In Taiwan it has been proved that by devouring a great quantity of the pasture, they compete with, and so reduce, the crop of fish; and their burrowing activities loosen the topsoil and may destroy the algal pasture. Research into suitable insecticides has shown that they can be controlled and that a notable increase in the fish crops can result. The economics of this treatment are being worked out.

The Future of Sea Farming

The future of farming the sea can be looked at in two ways: either on the assumption that marine fish farming will continue to be developed by private enterprise, or on the assumption that it will be developed internationally by the replacement of private competition by a planned social effort. In the latter case, the imagination is free to range wide and deep, for a civilization prepared to spend many billions on putting a man for a few precarious hours on the moon could raise the money to alter the ecology of the sea to its advantage.

But judging by present trends, the day of planned and truly international collaboration—able to raise the vast resources of money and skills needed to farm the sea for the benefit of mankind—is not close at hand; and this chapter will therefore consider the future of marine fish farming as a projection of the present. The first point to make is that the greatest future developments are unlikely to lie along the lines of the intensive culture of carnivorous fish and prawns, so effectively done in cages in Japan, for example, and proposed for carnivorous flatfish in Britain. A hungry world will find it cheaper and less wasteful to convert trashfish into fishmeal pellets for poultry and other livestock, or to render the trashfish into concentrated protein form suitable for direct human consumption. However, more feed should become available for use in fish farming with the successful mass production of protein from alkanes, waste carbohydrate materials, etc.

Otherwise emphasis will bear more on algal farming, by which the fish food is synthesized in the ponds from sunlight, carbon dioxide, and nutrient salts. Much of this algal production will be for direct human consumption, suitably prepared. Very high rates of production, equivalent to eight to seventeen tons of dry matter per acre per year, have been obtained in pilot-scale trials. There will also be the intensive cultivation of filamentous algae for the on-site feeding of herbivorous fish and prawns. Algae with special efficiencies in growth and digestibility will be produced by selection, and new and more efficient culture techniques will be worked out; for instance, the area of substrate on which the algal pasture can grow could be increased by lattices of flat surfaces.

Brown algae are frequently extremely abundant on coasts and may have a high harvest rate. These are already raw material for the industry that produces alginate, which plays a part in human nutrition as a food additive. Seaweed meal is also fed livestock as a constituent of balanced rations. In the future these prolific seaweeds could be more widely cultivated for direct human consumption, as some of them are now in Japan.

Then there is the other class of marine plant-production, namely, that of the phytoplankton. The fertilizer trials made in Scotland and Yugoslavia have illustrated the great increase in production that can be induced. However, the greatest future lies with the pumping of nutrient-rich waters from the depths of the oceans into confined areas such as atolls. It has been estimated that the lagoon of a

A Petersen disk tag is attached through the back muscles of a flounder. (Robert K. Brigham, U.S. Bureau of Commercial Fisheries)

single large atoll such as Kwajalein could then supply, through the products of aquiculture, enough marine animal protein to sustain 10 million people. Great expertise would be required to obtain such production, but the potential definitely is there.

The culture of phytoplankton-feeding fish also should be studied. Fish that feed on phytoplankton, including such species as pilchard and shad, though valuable as food, are cheap in the marketplace. They would hardly pay to farm unless, like the Japanese *kuruma* prawn, they could be grown for marketing in the off-season at enhanced prices.

At the present, only a very small part of the estuaries and shallow seas are under cultivation, and the possibilities of expansion are so great that there is no need as yet to invoke the thrills of science fiction. But what are the handicaps to this expansion? Firstly, there are the human handicaps: lack of know-how, lack of aptitude and tradition. Cases are known to this writer in which the know-how was provided, but the aptitude was absent and no progress followed. Another problem is the lack of capital. Marine fish farms, to be established in shallow waters whose ownership may be in doubt, might not impress bank directors as good loan risks. In the Philippines, where marine fish farming is well known and firmly established, the Agricultural Bank makes advances for fish farms, and a consequence has been the great expansion in the acreage of these farms since World War II.

Lack of local demand and of marketing facilities are further disincentives to fish farming. The answer would be promotional campaigns for the novel seafoods that marine fish farms can produce. The point to take is that these handicaps are socioeconomic rather than biological, though perhaps the more difficult for that.

163

8 MINERAL RESOUR

by Robert S. Dietz

The roles of the sea in relation to human life are many. Life originated in the sea, and without it all life would cease. It acts as a great thermostat and heat reservoir, leveling out the extremes of temperature that would prevail without its moderating influences. The sea's surface provides an avenue for the least expensive mode of transport known to us. The shore provides a playground, and the open sea is the eventual repository of all human waste. Ocean fisheries are a major source of food. Less appreciated is that the sea is a major storehouse of minerals, both today and especially for the future. It serves as a potential foundation for our increasingly industrial society. But while the undersea world is our nearest frontier, it also is the most distant. It lies at our feet, but it remains unknown—virtually unexplored and unused despite its vast resources.

Russia's first rocket into space, the memorable *Sputnik* of 1957, triggered a great space effort. In the United States a $5 billion-a-year program for space has evolved, with almost all of the funds being governmental, except those relating to communications. The oceans have been eclipsed for the present by the excitement of the race to the moon and planets, but in the long run they will have to provide man with the substances he needs to maintain his culture. The cost of a rocket to the moon exceeds the price of its own weight in gold, so we cannot expect such rockets to pay for themselves with the material they bring back. Their payloads will consist of lunar samples and scientific data, and while the value of such things is not to be deprecated, man cannot live on scientific findings alone.

There is already a substantial industrial involvement in oceanography, and it will become increasingly larger. Eight billion dollars a year are now spent on ocean resources, excluding surface shipping and naval military expenditures, and the prediction is that this amount will triple over the next decade. The recovery of offshore petroleum dominates this ef-fort, but other important facets are offshore mining, fishing, governmental research and development, desalination, and salvage.

Optimistic predictions are frequently made of man's exploiting the sea for its "unlimited riches." Artists in their drawings for brochures show undersea cities frequented by fish-men and midget submarines. Such drawings have a certain dreamlike quality, uninhibited by the practical considerations of human frailty (we are physiologically unsuited for living in the sea) and of the great engineering difficulties involved in making constructions that withstand high pressures and the corrosive saline environment. A credibility gap exists between such "brochuremanship" and practical realization, but we cannot write off the oceans, for the history of technology reveals that bizarre contemplations are oft times achieved.

Prior to the past few decades, mankind was able to wring but little of value from our vast oceans. Except for a rather small harvest of fish, a few minerals, and some pearls for his adornment, the salt sea has supplied only a few of his manifest needs. We cannot accept the idea that this immense ground will never offer more. In fact we are now witnessing a great leap forward in ocean technology, which appears to be only a beginning.

The ocean environment is divided into two major realms, which may be treated separately when considering its natural resources. First is the sea's water —the liquid volume, or "inner space," of the ocean, which scientists term the earth's hydrosphere. The water itself, as well as its salts, is a mineral resource of great value—although of a special sort—and this value is greatly enhanced when the salt water is converted to fresh water. Even the very motion of the restless sea may potentially be harnessed and converted into power. The second major realm of the ocean environment is the seabed, or the marine lithosphere, which in turn is subdivided into: (1) the

CES AND POWER

shelf domain, which is simply an extension of the geological features of the continents; (2) the deep-sea sediment domain, where detritus and chemical precipitates have been laid down through interaction with the salt water; and (3) the underlying oceanic crust, which is an exposure, so to speak, of the earth's upper mantle, and which makes accessible a geologic realm quite unlike anything known on the continents.

In any meaningful discussion of marine resources these diverse realms must be recognized, for each represents a different aspect of the nonbiological natural resources of the sea.

Minerals from Seawater

The ocean, including its boundaries, has many parts. But when we think of the ocean, the realm that first comes to mind is the sea water itself—a vast body of salt water covering seven-tenths of the earth's surface. The oceans average two and a half miles in depth, with the deepest point being seven miles down, and they contain 330 million cubic miles of water.

Seawater is a solution of great complexity. Just as the earth is a microcosm of all the elements in the cosmos of matter—whether planet, star, nebula, or galaxy—so the ocean, in turn, is a remarkably complete sample of the earth: About seventy-seven of the ninety-eight naturally occurring elements on the earth have been detected in seawater itself, or their presence there inferred by being contained in the creatures that inhabit it. Chlorine is the commonest (nearly 2 percent) while radon is the rarest at a mere 6×10^{-16} parts per million. The oceans probably contain all of the naturally occurring elements, but their detection awaits more precise methods of analysis.

The oceans are 3.5 percent dissolved salts, equivalent to 165 million tons of salts per cubic mile.

Thus the oceans of the world represent a storehouse of about 50 million billion tons of dissolved substances. The chloride represents 54.8 percent of the total salts, sodium 30.4 percent, sulfate 7.5 percent, magnesium 3.7 percent, calcium 1.2 percent, potassium 1.1 percent, carbonate 0.3 percent, and bromide 0.2 percent. Although the sea's salt solution is believed to contain at least minute traces of every element, these eight substances in solution account for more than 99 percent of the salts; all other elements total less than 1 percent. Many billions of tons of boron, copper, manganese, uranium, silver, and gold are also present. These seem like enormous quantities, but the oceans are so vast that seawater is a lean broth of such metals. Direct recovery of most of the substances present in trace amounts may never be feasible. It is the major elements that count.

Magnesium, iodine, bromine, common table salt, and fresh water are the five important mineral resources now taken from seawater. Other minerals as yet remain beyond man's ability to extract commercially.

The idea of utilizing the salts of the sea is not new; the first extraction of common table salt (sodium chloride) is lost in history. But perhaps the first complex substance to be taken from the sea was the dye known as Tyrian purple. This dye was a bromine derivative that the ancient Phoenicians found in concentrated form in the body of the purple snail, *Murex*. They obtained the dye by crushing the snail shells and drying the mantles by exposing them to the sun in vats. Varying in hue from scarlet to deep purple, depending upon the period of exposure, the dye was used for the attire of kings, eventually becoming a symbol of wealth and power. This resource of the sea was one of the foundations of Phoenicia's commercial empire and even gave the country its name, the Greek word for crimson or purple being *phoinix*.

Since sodium and chlorine account for 85 percent

Private industry spends several billion dollars a year on the ocean's resources, with the largest share going to off-shore petroleum. These oil rigs, with helicopter landing pads, are located in the Gulf of Mexico. (Mick Church)

of the sea's dissolved salts, it is not surprising that sodium chloride, or common salt, was the first compound to be removed from the sea by man. Its use doubtlessly goes back to the cavemen of prehistoric time, but the first mention of its extraction in the written record goes back to 2200 B.C. in Chinese writings. In·his *Meteorologica*, Aristotle spoke of the origin and usefulness of the salts of seawater and of a method for their extraction. Simple solar evaporation was the ancient method practiced by the Egyptians, Romans, and Greeks—and in the Orient as well. The solar extraction of salt commenced on the Atlantic coast of North America in 1680 and on the Pacific coast in 1852. Lacking a hot, dry climate, the East Coast industry was short-lived, but that along the West Coast thrives today.

The extraction of crude soda and potash from seaweed ash was accomplished in Scotland in 1720, and iodine was separated from seaweed a century later. Magnesia was first prepared from seawater in the Mediterranean at the end of the nineteenth century. In 1923 magnesium chloride and gypsum were first produced by means of solar evaporation of concentrated brine from San Francisco Bay. In 1926 the first seawater bromine was recovered on a small commercial scale by chlorinating the San Francisco Bay brine, then steam stripping, condensing, and purifying the product. In 1931 the production of potassium chloride by evaporation of brine from the Dead Sea was begun. In 1932 bromine was produced on a commercial scale from the residual liquors of the potassium plant. In the thirties, a sharp increase in the demand for bromine as a constituent of the gasoline antiknock compound ethylene dibromide occurred; new bromine plants were built first at Kure Beach, North Carolina, and then at Freeport, Texas. Since the production of bromine from subterranean brine wells has not been able to keep pace with growing demand, 80 percent of the United States' bromine is now derived directly from seawater.

Magnesium metal and magnesium hydroxide are two other important products mined from seawater by a process quite different from that used in the extraction of bromine. Magnesium is taken out of the seawater in an alkaline, instead of acid, condition and is removed by precipitation, rather than by blowing out. From an oceanographic or climatic standpoint, the location of a magnesium plant is not critical. The seawater that must be processed per pound of magnesium is only 5 percent of that required per pound of bromine, and water temperature has little effect on magnesium recovery. More important is a location favorable to the supply of raw

materials and cheap power. The convenient availability of lime, which is used in the process, and abundant and inexpensive fuel are essential for competitive operation. Therefore, the Dow Chemical Company has located its magnesium-extraction plant at Freeport, Texas. About 95 percent of the United States' supply of magnesium metal is extracted there.

The large-scale production of bromine and magnesium from seawater is not an isolated incident in scientific progress over the past few decades. It is indicative of the growing use of the sea for fulfilling more of our industrial needs. As the sophistication of chemical technology grows, more and more substances will be usefully derived from the oceans. The handling of seawater by pumping is easier and cheaper than the majority of mining methods, and the substances to be extracted are already in solution, where they can be chemically manipulated. All these things suggest that in the future there will be a more complete utilization of the resources of the seas.

The most obvious (and the traditional) way of extracting salts from the sea has been by evaporation,

The solar extraction of salt, begun on the West Coast in 1852, is thriving on the shores of San Francisco Bay, where the world's largest solar evaporation plant is located. By a system of levees the seawater is circulated through shallow ponds averaging five inches deep. (Leslie Salt Company)

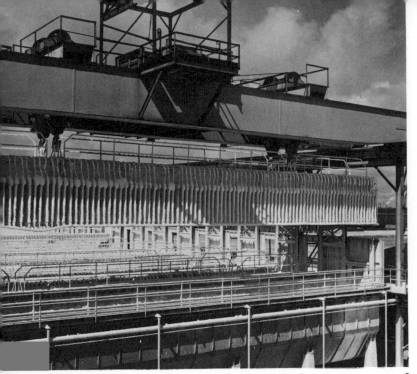

Nearly all the magnesium used by the United States is extracted from seawater at the Dow Chemical plant in Freeport, Texas. (Dow Chemical Company)

Magnesium is collected in shallow settling ponds at the Dow Chemical plant. Using this process, one pound of magnesium can be extracted from 142 gallons of water. (Dow Chemical Company)

either by fires or, in desert regions, by using the sun's warming rays. Both methods are laborious and involve handling large volumes of seawater. Some method of direct extraction is sorely needed. Ion-exchange resins offer some promise, as these have already emerged from the experimental stage and are used commercially—for example, to soften water.

These resins are substances with the ability to capture certain ions (elements or compounds in solution that are electrically charged) while releasing others. Another possible approach is the use of a semipermeable membrane—a "sieve" of such small mesh-size that certain ions can move through it while others cannot. These are but two possible methods that might be used to bypass the customary evaporation process, which is too costly in its use of power.

If man could attain the chemical sophistication of marine animals, new vistas would open up. A promising approach for ocean mining would be to have the creatures of the sea extract minerals, or to look to the bottom oozes, where slow chemical enrichment has proceeded naturally for countless millions of years. More adept at practical chemistry than man, plants and animals exhibit a fantastic ability to concentrate certain elements, many of which are known to exist in the sea only because its creatures have successfully extracted them. The existence in the sea of vanadium, for example, was only realized after chemists discovered it in certain sluggish and soft-bodied creatures like sea squirts and tunicates, which effect a 50,000-fold concentration of this element. It is unlikely that these animals will ever of themselves contribute to the supply of vanadium, but if chemists could learn their technique they might be able to use it to extract these metals from seawater. A single oyster filters several tens of gallons of water each day, and in the process selectively extracts many substances along the way. Lobsters and crabs effect a great concentration of copper, for example, as this element is needed for their respiratory pigments, playing the role in oxygen exchange that iron does in the hemoglobin of human blood.

Iodine can be easily extracted from kelp, but not from seawater directly, and the microscopic radiolaria of the genus *Acantharia* build their intricate and lacelike skeletons neither of lime nor of silica, but of strontium sulfate. Unfortunately skeletons of these free-floating, one-celled animals appear to dissolve rather readily, for their remains, which might conceivably create a useful mineral deposit, are rarely found in the radiolarian oozes laid down on the deep-sea floor.

Hot Brines of the Red Sea

Unlike the rock in the earth's lithosphere, which varies widely in the minerals it contains, the salt water of the open sea is generally homogeneous. There is usually little to be gained by mining one parcel of seawater rather than another, or even by utilizing one ocean rather than another. Being fluid, the oceans are well mixed. Fossil seawater is another matter, for when seawater is trapped in some ancient, porous rock stratum, it may undergo transformations that result in an enriched and economically exploitable brine. Such ancient brines are now pumped from deep wells in Michigan for the recovery of bromine and iodine.

In 1963 oceanographers on the British ship *Discovery II* discovered an exception to the uniformity of the untrapped sea. In the Red Sea they found localized pockets of hot, heavy brine. The largest of four known pools is 8 miles long, 3 miles wide, and 6,800 feet deep. The temperature of the brine is 133° F. and the salinity is 317 parts per thousand, or nearly ten times that of normal seawater. The water is completely without any dissolved oxygen, and the underlying sediments are black, amorphous iron oxides rich in other minerals such as zinc, lead, and silver. Heat-probe measurement indicates that the underlying earth's crust is ten to twenty times hotter than normal. Today this discovery of hot brine is unique, but it seems likely that similar pools will be found around the world in other parts of the world rift system, of which the Red Sea cleft is but a small offshoot.

In 1967, on her global expedition, the U.S. Coast and Geodetic Survey ship *Oceanographer* found that the brine pools could be located by the telltale scattering of sound on the echo-sounder graph, or echogram. Apparently organic detritus drifting down from the sunlit surface settles along the midwater contact between normal ocean water and the brine pool. This detritus layer acts as a soft reflector for sound, so that an echo bounces back to the surface. On the echogram a faint, thin, dark line appears, which is a tip-off that there is some unusual water boundary below.

Hot water normally rises and mixes, but the brine in the deep depressions of the Red Sea is so dense that it remains isolated like pools of mercury. The origin of these intriguing brine pools is not entirely known, but the following explanation seems likely. About ten million years ago, Africa and Arabia were joined together, but then they began to be split apart by rifting, sea-floor spreading, and continental drift—a process that is still going on. This great geologic turmoil is taking place above a hot spot in the earth's mantle beneath the earth's granite crust. Accompanied by earthquakes and the spewing out of hot lava, hot springs have gushed up from the fissures in the ocean floor. The hot water, in turn, has leached buried salt beds and extracted metallic ions from the crust, creating a broth of "liquid ore" that one day will be exploited.

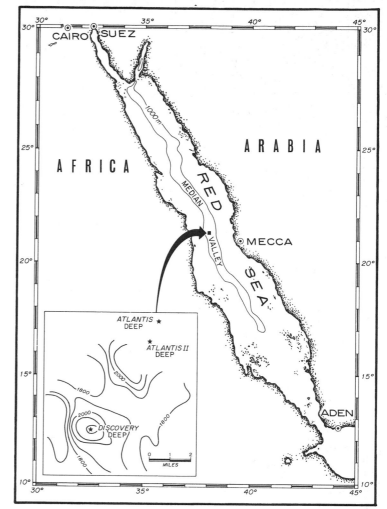

Although the ocean's waters are remarkably homogeneous, occasional brine pools are found where water of very high salinity has been trapped in ancient rock strata. In the brine pools found deep beneath the Red Sea the salinity is nearly ten times that of normal seawater. Heat probes indicate that the earth's underlying crust is extremely hot in these places. (Woods Hole Oceanographic Institution)

On her global expedition in 1967, the U.S. Government's research ship *Oceanographer* showed that brine pools could be located on echograms by the appearance of scattering layers that indicated organic matter separating the hot brine from the surrounding ocean water. In the echogram shown here the dark area at the top indicates an outgoing sound pulse; the thin, unbroken line, a scattering layer; the light areas below it, a brine pool; and the lower dark area, the ocean bottom.

Gold from the Sea

Although the sea is bountiful and its mineral resources abundant, the chemical mixture of its salts is not entirely a happy one. To compute the amount of sodium chloride (common table salt) in the oceans is an exhilarating exercise but not economically meaningful: There is enough to cover all the land on earth to a depth of 150 feet. But while common salt is, indeed, plentiful, the estimated quantities of many metals that we would like to recover from the sea are becoming vanishingly small. Gold is a case in point.

The "oceans of gold" have long been a dream of the alchemists. In 1886 the French Academy of Sciences announced that gold was present in ocean water. The waters of the English Channel supposedly contained about 0.002 ounce per ton. At the turn of the century, the brilliant Swedish chemist Svante Arrhenius declared this estimate to be ten times too high, but his analyses still showed appreciable quantities of this precious metal. Even by Arrhenius' figures the oceans of the world held some eight trillion tons of gold. This would have been enough, if it could have been extracted, to make everyone on the face of the earth today a millionaire. Perhaps, some people thought at the time, the gold could be recovered by running seawater over a pool of shimmering mercury and letting it amalgamate with this quicksilver. This was a fascinating prospect, and several patents on methods of gold recovery from the sea were issued before 1920, but not even one pinch of gold was extracted.

In 1921 the Allied Reparations Commission made demands against Germany amounting to the staggering sum of 132 billion marks, payable only in gold—50,000 tons of it! Germany turned her most brilliant chemist, Fritz Haber, to the task of finding a new source of cheap gold. Haber alone had almost won the war for Germany by his synthesis of ammonia from the nitrogen in the atmosphere. This made it possible for Germany to continue manufacturing explosives during the Allied blockade without relying upon the import of Chilean nitrates. Hoping that Haber could perform yet another miracle, Germany turned to the sea and mounted the famous *Meteor* deep-sea expedition, ostensibly for pure oceanographic research, but with the compelling motive of trying to learn how the sea might be parted from its gold. The expedition succeeded in discovering the anatomy of the South Atlantic Ocean—its general bathymetry, current systems, and water chemistry—but not in finding a cheap source of gold. For ten years Haber meticulously assayed water samples from all parts of the seven seas to the limit of human accuracy, but he discovered only bitter disappointment. His richest samples came from the central South Atlantic, but even there he could find only one percent of the amount of gold claimed by Arrhenius. There was not even a penny's worth in ten tons of ocean water.

Although Haber failed in extracting any gold, chemists of the Dow Chemical Company, in an exercise in pure research, eventually succeeded, but hardly on an economic scale. After laboriously processing about fifteen tons of seawater they finally isolated about 0.0003 ounce of gold worth about one-hundredth of one cent. This constitutes the bulk of the gold that has been unlocked from the oceans to date. The miracles of science seem to affirm that although we cannot dismiss *anything* as being impossible, this does not mean that *everything* is possible. Parting the sea from its gold economically seems forever beyond the grasp of man.

The value of the gold in a cubic mile of seawater is impressive: $100 million worth!—a figure often cited by promotional zealots. But a cubic mile of water is also an impressive volume of water, which could never be processed by usual chemical methods. More telling is that a cubic mile of ordinary granite, the most common rock on earth, contains 500 times as much gold. Admittedly, water, with its ions in aqueous solution, is easier to handle and process chemically than solid rock. On the other hand, one can "high-grade" a granite batholith, extracting gold only from its quartz veins and leaving the rest as waste. Man will never attempt to mine average granite for gold, but he will search out the enriched zones and lodes.

Mining the Ocean Floor

The oceanic realm offers far more in the way of natural resources than just the salts in solution, since it overlies more than two-thirds of the earth and covers crustal rocks often quite unlike those that comprise the continents. In some respects we may compare the hydrosphere to the atmosphere, from which only a few resources, such as liquid oxygen and nitrates, are obtained. The mining man regards the water, like the air, as simply a cover over the riches below. The ocean floor below, and not the salt water *per se*, offers the brightest prospects for future mineral exploitation.

The early oceanographers of H.M.S. *Challenger*, returning in 1876 after her global cruise, related that the ocean floor was a vast volcanic province, a conclusion that has stood the test of time. Today geologists regard the ocean floors as windows that pierce the granite rocks of the earth's sialic crust, exposing the upper mantle of the planet. The ocean floor is a realm of black, iron-rich and magnesium-rich basic and ultrabasic rocks—relatively low in silica (quartz), the common rock-forming mineral found in granite—that have been spewed from volcanic fissures to form basalt lava flows, plus injected rocks termed serpentines and peridotites. With this knowledge geologists can predict that the ocean floor will be rich sources of certain metals, such as nickel, copper, chromium, and the platinoid metals, which have affinities with ultrabasic rocks, but impoverished in many other metals, such as gold, silver, and uranium, which are associated with acidic granitic continental rocks. Also, the rocks of the sea floor are young, and presumably they totally lack the ancient Precambrian shields, which, like the highly mineralized Laurentian Shield of Canada, are the continental storehouses of minerals. Sea-floor rocks also seem to lack variety and the heterogeneity associated with a complex history of differentiation, so that mineralized zones may be rare. Volcanic islands like Hawaii are poor in mineral resources, but similar basalts are the commonest rocks of the abyss. One can predict a dearth of many other substances besides minerals; for example, coal seams will never be found beneath the deep-ocean floor, although they do commonly extend beneath the continental shelves.

Today minerals are already being recovered from the surface of the ocean floor: diamonds from off South Africa, tin from the Indonesian shelf, and sands of magnetic iron ore from the Japanese shelf. But the shelf waters in which such operations are carried out are rarely more than 100 feet deep. None of the substances being recovered are truly ocean minerals, but simply drowned extensions of the coastal plain or shoreline deposits, submerged by the rising seas since the Ice Age. Deep-sea mining remains a challenge of the future, as does mining those substances that are truly of the sea—such as the nodules of phosphorite and manganese oxide born by deposition from the broth of the sea. Doubtless the breakthrough will come soon and the immediate goal will be the recovery of the latter two mineral deposits.

The phosphorite nodules were discovered by the *Challenger* scientists, who did not miss much on their historic reconnaissance of the world ocean. They found these nodules at several sites on the sea floor but most notably on the Agulhas Bank, which juts out from the southern tip of Africa. Phosphorite nodules are objects of intrinsic beauty such as would enhance any mineral collector's cabinet. They are light brown and smooth and have a glazed appearance as though varnished. They come in a variety of forms but are usually ovoid, with bulbous protrusions and deep cavities that make them favored habitats for underwater creatures. These concretions are hard, tough, and internally of earthy texture and color. Commonly they are loaded with fossils, ranging from the microscopic foraminifera to sharks' teeth and the phosphatized bones of sea lions and whales, mostly of extinct species. Green pellets of the mineral glauconite are scattered through the mass, and phosphatic oolitic pellets as well, for which the curious name "sporbo" has been coined—an acronym meaning "small, polished, ovoid, round, black objects." When phosphorite nodules are poured out on a ship's deck from the dredge, it is always possible to reconstruct their original orientation on the sea floor, because their tops are discolored with manganese oxide. The nodules from any particular locality have a group or family resemblance that makes them distinctive to the specialist. They have been found only on offshore banks that rise from the continental shelves. Thus, they are a resource of the nearshore shelf environment and not of the deep sea.

The most extensive known phosphorite deposits lie off the western coast of the United States. These nodules cover numerous offshore banks off southern California. One estimate places the quantity of rock phosphate there at one billion tons, blanketing an area of 6,000 square miles. The phosphorite deposits off California began forming during the Miocene epoch about fifteen million years ago, but the precipitation has been intermittent and the deposition is continuing today. Being far offshore and on the tops of banks to which no sediment from shore can be carried, the most recent layers of these phosphorite deposits remain exposed to chemical interaction with the seawater.

Phosphorite nodules are "things of the sea," for apparently they slowly precipitate directly from seawater in a somewhat special milieu. Regions of strong upwelling waters are the key, for there planktonic life is profuse, as the upwelling constantly brings up new mineral fertilizers into the depleted surface waters. Dying organisms settle to the bottom and upon decay release into the sea microscopic or colloidal particles of phosphate that later precipitate to form the nodules. The overall process is much like the growth of crystal from a saturated solution or the precipitation of sugar candy crystal, except that in the case of the phosphorite nodules, colloidal particles drifting in the seawater are precipitated, rather than ions in solution.

Phosphorite is an important source of phosphorus, which is a vital element in the makeup of all living things. Soils soon become barren if their phosphorus is not replenished either naturally or by the addition of fertilizer. With the world's exploding population, the need to increase farm yield by intensive farming—that is, mainly by adding fertilizer —is becoming ever more vital. The market for phosphate rock is flourishing, and one day it may become our most important natural resource, except for water. Animal bones, fish, and guano were once the main source of phosphate, but that day has long since passed. In the United States phosphate now comes from ancient marine deposits in Florida and Idaho, but soon the sea floor will provide a new source.

The phosphorite deposits on the sea floor off southern California are on the verge of being economically exploited. In 1961 an area richly covered by these nodules named Forty-Mile Bank (it lies forty miles off San Diego) was leased from the Federal Government by Collier Chemical and Carbon Company of Los Angeles. A detailed survey was made, using underwater television and close-spaced sampling. But a full-fledged mining operation never materialized, one reason being the hazard imposed by many unexploded naval shells, because this region had been used as a firing range.

As well as discovering phosphorite nodules, the *Challenger* scientists were the first to dredge up nodules of manganese oxide from the deep-ocean floor. These black, hydrous, iron-manganese concretions fascinated the *Challenger* oceanographers, for they found that the manganese had sometimes accreted around a nucleus of a shark's tooth or a whale's earbone. It remained for Alexander Agassiz aboard the *Albatross* at the turn of the century to show the abundance of the manganese nodules, for he picked them up at every site he dredged off western South America, over a region as large as the entire United States.

Even so, scientists had no concept of the concentration of manganese nodules until the advent of the sea-floor camera after World War II. It could then be seen that these bulbous nodules were strewn across the red clay plains in an almost continuous layer of what looked like dirty potatoes. Potentially, this concentration is a tremendous resource; the manganese deposits from a single seamount would satisfy the world's requirements for a hundred years. But for the marine biologist and the geologist the profusion of manganese nodules is a curse, for they are too easily picked up. They clog the geologist's dredge so that the more precious rocks are missed, and they macerate the delicate specimens of sea-floor fauna that the biologist wants to collect. Manganese incrus-

Manganese nodules, which abound on the ocean floor, were first dredged up by scientists on board the *Challenger*, shown here sifting bottom deposits. The deposits from a single seamount would satisfy the world's requirements for a hundred years.

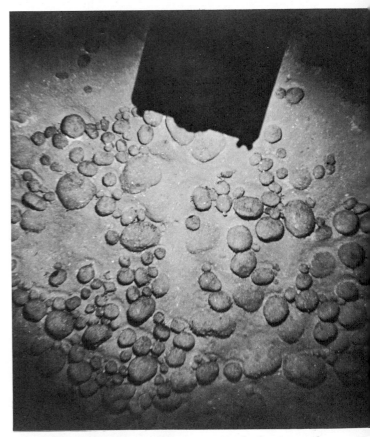

Potato-like manganese nodules, previously known only from dredge hauls, were first photographed in 1948 at a depth of 3,026 fathoms near Bermuda. (David Owen, Woods Hole Oceanographic Institution)

A large manganese nodule from the northern Pacific is shown here; the ore encrustation surrounds a center of volcanic ash. The manganese layer builds up slowly, averaging 0.01 millimeters per thousand years. (Official photo, U.S. Navy, Robert S. Dietz)

tations also promise to annoy future aquanauts by making it impossible to discern the true nature of rock outcroppings on the ocean floor.

Like phosphorite nodules, manganese nodules are products of the sea, for they are precipitated directly from seawater, building up accretionary layers like an onion. Fresh volcanic flows, after they pour out over the ocean floor, slowly weather, and in so doing they release colloidal manganese dioxide, which drifts along in the sea but is eventually re-precipitated to form the bulbous nodules. The rate of growth is extremely slow, about 0.01 millimeter per thousand years. One large nodule ($1 \times 2 \times 3$ feet) recovered in a tangled wire rope from the northwest Pacific was computed, according to radiometric dating, to have taken sixteen million years to grow. But even at this slow rate of accumulation, manganese nodules are a renewable resource, for man could never use up the supply at a rate faster than they are growing. Concentrations of manganese nodules as high as 100,000 tons per square mile are not uncommon.

Manganese metal is an important resource for the manufacture of all steels, but especially for the making of manganese steel, a hard but not brittle variety with 12 to 14 percent manganese. The purest manganese nodules assay as high as 80 percent manganese dioxide, which makes them attractive as potential ores. Probably even more important are the values of other metals associated with these nodules —notably copper, nickel, cobalt, zinc, and molybdenum. Concentrations of as much as 2 percent cobalt, 1.9 percent copper, and 1.6 percent nickel have been assayed. Apparently manganese in the sea has some special electrical attraction whereby it selectively absorbs or "scavenges" these elements present only in trace amounts in seawater.

A prime target area for the future exploitation of manganese nodules is the Blake Plateau, a deep terrace adjacent to the shelf off the southeastern United States. The coal-black nodules in this region cover hundreds of square miles. In any particular area they are evenly sorted by size, some being as small as golf balls and others as big as Idaho potatoes. They are of low density and completely unattached to the bottom, so that they could be readily sucked up by some sort of underwater vacuum cleaner. Most manganese nodules lie at depths of a mile or two, but these are only 2,400 feet down. The seas of the Blake Plateau are quiet, and a market is close at hand, so that conditions are favorable for their early exploitation.

Many of the nodules are too big to fit into the mouth of an oceanographic dredge, and we know of their size only by inadvertently having snagged them with wires and cables. From a depth of 18,000 feet in the Philippine Trench, a British cable ship in 1955 brought up a one-ton mass captured in a snarl of an old telegraph cable. Lacking appreciation for the scientific worth of their esoteric prize, the crew cut it loose, letting it plunge back into the sea. Another large specimen, also inadvertently snagged, now rests in the museum of the Scripps Institution of Oceanography at La Jolla, California. A grab sampler sent down to collect sediment from the bottom hit rock and therefore came up empty, but neatly tied above it in a tangle of oceanographic wire was a 200-pound manganese nodule three feet across.

Ocean mining offers many attractions vis-a-vis traditional land mining. Substances like the manganese and phosphate nodules are available without the need for the removal of overburden or the use of explosives to break up the country rock. Vertical lift is all that is required, without any horizontal transport on conveyor belts, since a barge offers a mobile platform that can be constantly repositioned. There are no shafts to sink, no tunnels to drive, no town sites to construct. The exact quantity of ore can be computed in advance of mining merely by scanning the bottom with an underwater television camera. The ore ends up loaded aboard a ship—the cheapest mode of transport known to man—ready for delivery to any port in the world.

In 1971 a new type of research ship called the *Alcoa Seaprobe* was launched for assaying ocean floor mineral resources and for underwater archaeology. The ship is equipped with sonar, an underwater television camera, spotlights, and 20,000 feet of drill pipe for coring the deep ocean floor, as well as "super

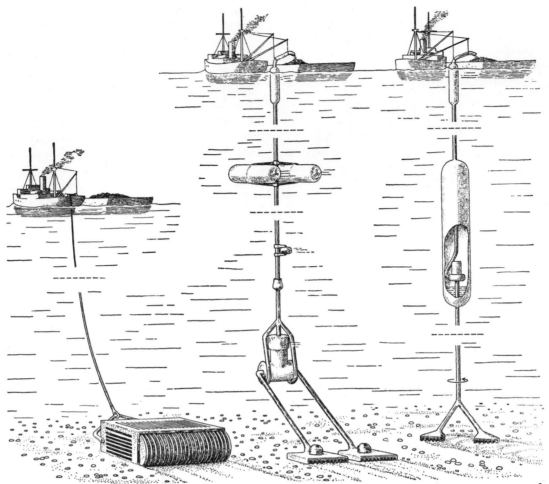

Manganese nodules might be swept or sucked from the ocean floor in several ways, as proposed here. At left is a simple carpetsweeper dredge, which would be dragged along the bottom. The middle dredge would work like a vacuum cleaner, with motor and pump. A similar dredge, the machine at right, would have suction heads that sweep in circles. (From "Minerals on the Ocean Floor," by John L. Mero. © December 1960, *Scientific American*.)

tongs" for retrieving ancient wrecks. The Mediterranean Sea alone may be littered with as many as 15,000 hulks dating from Phoenician times.

One probable objective of such a ship would be the manganese nodules, which are essential in the making of steel. More than $100 million worth of manganese is presently imported into the United States each year, since there are no high-grade manganese ores in this country. The copper, nickel, and cobalt impurities in the nodules can be leached out with acids and offer an additional incentive. Local undersea sources would be valuable in the event of war or restrictive political conditions that cut off the usual foreign sources.

We would not learn much about the mineral resources on land by floating over it in a balloon and occasionally lowering at random a grab or dredge to sample the surface. Similarly we cannot expect the exploitation of the deep sea to be quickened until we can scrutinize the sea floor close at hand and do carefully controlled sampling. In search of mineral resources on the deep-ocean floor, the oceanographer is now going down *into* the sea in ships. It took the loss of the nuclear submarine *Thresher* in 1963 and of the hydrogen bomb off the coast of Spain in 1966 to dramatically reveal our ineptitude and lack of capability for working deep beneath the sea. Continuing development of Deep Submergence Vehicles (DSV's) will not only expand civilian deep-sea ability but will enhance and sharpen the capability of the military submarine, which must also run ever deeper. The capital ship of World War I was the battleship; of World War II, the aircraft carrier; but now it is certainly the submarine.

When the DSV's really come into their own, the exploitation of offshore mineral resources will be accelerated. A decade ago the United States had only one underwater research vessel—the well-known *Trieste*, operated by the U.S. Navy. Now there are about a score in operation or under construction. This country's oceanographic fleet contains ten times as many surface survey and research ships as DSV's, which are more costly to develop and support. Most DSV's so far require the support of a surface mother ship, but the trend is to operate ever more independently, in deeper water, and for longer periods of time. Although the French, under Jacques-Yves Cousteau, pioneered in this field, the United States now has a clear lead, with about a score of DSV's in operation. Even the Russians have been remarkably slow in DSV development. Their first entry in the field, the *Sever II* (the *Sever I* being a World War II submarine converted for fisheries research) was promised for about 1967 but appears to have not yet materialized.

Petroleum Beneath the Shelves

The tapping of the continental shelves for their oil and gas is by far the greatest success story to date in exploiting the ocean's resources. The petroleum industry is rapidly moving offshore, a transition which is now in full swing. In 1968 oil companies bid more than $600 million for leases in a single offshore auction of Federal lands off Texas, while a major oil company paid $240 million for the lease to a portion of Santa Barbara Channel off southern California. Water depths there attain 1,300 feet, which is beyond the reach of today's drilling technology—but presumably not of tomorrow's. In the United States the value of offshore oil production already exceeds that of the fisheries, although offshore drilling is largely a post-World War II technology.

Ten years ago only a few countries had any interest in the offshore, but in 1967 there were seventy-five undertaking offshore exploration for petroleum, and about a score are already producing offshore. World offshore oil reserves on the continental shelves are presently about 80 million barrels (or 20 percent of the world's total of 400 billion barrels). Offshore wells are now pumping about 5 million barrels a day (or 16 percent of the world's output). The petroleum industry has about $10 billion now invested in the offshore and will probably spend another $25 billion over the next decade. In the United States important amounts of oil and gas are now being produced off California, Texas, and Louisiana. In other parts of the world large contributions are being made in the North Sea, in the Persian Gulf, and off southeastern Australia.

The real movement offshore is a development of the past two decades. Earlier drilling was confined to offshore extensions of onshore pools by placing derricks on piers. The Elwood Field near Santa Barbara, California, where drilling commenced in the 1920's, was the first such development. Among the prophets of the offshore industry, consulting geologist Lewis A. Weeks stands the highest, for not only has he long advocated and given geologic fabric to continental shelf exploitation, but he also put together the venture that resulted in the bonanza find in the Bass Straits between the Australian mainland and Tasmania. Another such pioneer is Thomas Baldwin of Los Angeles, who initiated the search for offshore oil by diving off California.

It is becoming increasingly apparent that marine geology and oceanography will form the basis of new-growth industries. Exploration and development methods undreamed of a few years ago are being investigated, and many of these are concerned with a thrust into deeper and deeper water. To date only about a dozen wells have been drilled in waters of 500 feet or more, but within ten years drilling in water 3,000 feet deep or more will be common. The development of completely submerged operations in the near future, including undersea living quarters as well as work chambers, can be confidently predicted.

The reason for the attractiveness of the continental shelf is not hard to find. Young sediments, less than 150 million years old (Tertiary and late Mesozoic), predominate offshore, and 90 percent of all petroleum is found in such rocks, which were laid down in the last quarter of the earth's oil-bearing period. In fact, about 30 percent of the present world oil reserves is in sediments accumulated during the last 6 percent of the earth's oil-bearing period.

This is a natural consequence of the origin of oil, which is derived from the diatoms of the sea, whose dead remains are entrapped in marine sediment. Oil tends to become lost with time, owing to the vicissitudes of geologic history, so that the younger sediments offer the most promising oil reservoirs. The world has undergone only modest changes since mid-Mesozoic time, one of which has been the laying down of a thick cover of sediment on the continental shelf and slopes. Here sand layers saturated with petroleum lie fallow, awaiting the driller's bit. Already 15 percent of America's supplies comes from the shelf regions of the Gulf of Mexico and California. This offshore contribution probably will grow with time, so that by the turn of the century more than half will be produced from beneath the seas.

Offshore oil recovery today is limited to the shallower portions of the continental shelf, where the depths exceed no more than 300 feet. Exploitation of the far offshore areas and the deep sea remains a challenge to future technology; it will be necessary to use other methods than bottom-fixed drilling platforms. Just what petroleum potential the abyssal oceanic regions hold is a moot question. The vast, red, clay-covered expanses of the Pacific Basin presumably have little oil. Red clay is poor in organic matter, for this sediment is produced beneath "blue water," the desert color of the sea where diatoms do not thrive. Also, the oxidation rate in this realm is relatively high, so that any organic matter that does reach the bottom is soon "burned" and returned to the broth of the sea. On the other hand, the diatom oozes, forming a world-girdling belt beneath the Antarctic Ocean, may offer a vast potential for oil. The Atlantic Basin may also contain oil reservoirs. In 1968 the U.S. Navy research ship *Kane* detected domelike structures, apparently identical to those associated with the oil fields in the Gulf of Mexico, on the floor of the Atlantic, about four hundred miles west of Senegal on the bulge of Africa.

But the greatest untapped storehouse of oil doubtlessly must be within the great prisms of sediment that lie at the bases of nearly all continental slopes, with the Pacific margin being the principal exception. All the conditions for oil accumulation seem to be right in these areas. The basic geologic requirements for oil pools are source beds rich in marine organic matter and sedimentary layers that are suitably porous for providing a reservoir and a trap—a cap rock, pinch-out, or some other geologic situation that permits the oil to migrate so far, but no farther, or that puts the cap on the bottle, so to speak. A further condition is that the sediments not be altered or metamorphosed by folding or geothermal heating, as this would destroy the oil or permit its escape through fractures. The sedimentary covers of the continental rise are commonly a few miles thick and constitute the greatest sedimentary bodies on the globe. The sediments are young and quite rich in organic matter. They have been laid down by avalanches of mud and sand shed from the continental block, so they contain numerous lens-like sand bodies that would make fine reservoirs and traps. No oil company has yet attempted to test the continental-rise sedimentary covers, as the technological difficulties in drilling in the deep sea are formidable. We are content today to transport oil 10,000 miles or so from the Persian Gulf rather than to try tapping it two miles down beneath the sea.

The mineral fuels are far more important as natural resources than the minerals and ores one usually associates with mining. South Africa is often thought of as being rich in mineral resources and England as being poor. Actually, the reverse is true, for the value of the coal produced in England far exceeds that of the gold and diamonds mined in South Africa. All the continents of the Southern Hemisphere are poor in the fossil fuels, because the conditions for their accumulation during the Paleozoic era apparently were poor, except along the continental margins. Probably all of these continents were then joined together into one supercontinent, called Gondwana. However, with the breakup and drifting apart of Gondwana beginning about 150 million years ago, the condition for oil accumulation became highly favorable in the new deposits laid down along the continental margins. Thus, the oil potential of the southern continents is largely in the offshore areas. The breakthrough discovery of great oil reserves in the Gippsland Basin off southeast Australia provides a case in point. Several large oil and gas fields have been discovered that promise to transform Australia from a "have-not" nation for oil to one that will export it.

In the Gulf of Mexico off Louisiana and Texas, oil exploration and exploitation is booming. Huge drilling platforms, some as high as the Washington Monument, are assembled on land, towed out to sea, and upended over the drilling site. The oil play there involves salt domes that hang like inverted teardrops in the thick sedimentary prism. The early geologic history of this region is shrouded in mystery, but it seems likely that the Gulf of Mexico did not exist before Mesozoic time. Then either a fully enclosed basin (like the Caspian Sea) or a partially enclosed one (like the Mediterranean Sea) came into being. During the Jurassic period, a geologic heat age, evaporation far exceeded precipitation, so that the Gulf became ever more briny. Eventually the point of supersaturation was reached, so that salt crystallized out and settled to the bottom, as in the Dead Sea of Israel and Jordan today. The searing heat persisted until a deposit a mile thick, the Louann Salt, was laid down over the bottom of the entire basin. Then mud and sands a few miles thick, mainly from the ancestral Mississippi River, were deposited on top of the salt. The salt became unstable under this enormous load and, flowing like ice, was squeezed into thick lenses. These pockets, being lighter than the surrounding sediments, had buoyancy, so that they formed into inverted teardrops, usually less than one mile across, that pushed up like long fingers toward the earth's surface. Geologists have described these salt domes as driven up through 5,000 to 15,000 feet of sediments by earth pressures, like nails through a

An increasing percentage of the United States' petroleum supply is mined from the shelf regions of California and the Gulf of Mexico. This rig, which can drill six wells from one position, is topped by a derrick fifteen stories tall. (Shell Oil Company)

Huge platforms, some as tall as the Washington Monument, are assembled on land, towed out to sea, and up-ended over a drilling site. These tugboats are guiding the British *Sea Quest* to its station in the North Sea. (British Petroleum Company)

A biological bonus of the offshore oil industry, this school of perch is attracted to the encrustations on the guide cups and drill pipe at the base of a drilling platform.

(Charles H. Turner, California Department of Fish and Game)

board. These domes rarely breach the surface, but geophysicists can detect them by anomalies in the earth's gravitational and magnetic fields. The margins of the salt plugs, where the sedimentary strata are upturned by drag and where they pinch out against the salt, offer prime prospects for oil pools.

Drilling at sea involves many technological difficulties that are not encountered on land, as emphasized, for example, by the great oil spill off Santa Barbara, California, early in 1969. Natural gas asso-

ciated with oil is under great pressure and will tend to drive the oil before it, causing a gusher or blowout at the well head. The high pressures are generally contained by dense drilling mud, but from time to time matters get out of hand. Uncontrolled ejections of oil are relatively easy to cap or otherwise control on land, but drilling deep beneath the sea, where the well head is rather inaccessible, poses problems.

Having drilled down 3,500 feet beneath the continental shelf off Santa Barbara, the riggers began

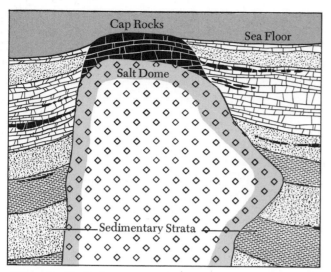

Oil in the Gulf of Mexico is found on the margins of salt domes that rise in the sedimentary strata of the sea floor. The domed upwellings, lighter than the surrounding sediment, are revealed by seismic reflection. In addition to concentrations of oil, the cap rocks often contain commercial quantities of sulfur.

to retrieve the pipe in order to replace the drill bit. During this withdrawal the drilling mud became inadequate to contain the pressure, and the well blew out. An initial attempt to cap the hole was successful, but that led to a great pressure buildup. An expanding mass of oil, rock, and gas opened a fissure, and oil bubbled up 700 feet to the sea surface. Over eleven days some 200,000 gallons of oil flooded to the surface, creating an oil slick covering 800 square miles. Though actual damage to sealife and beaches was smaller than many claimed, offshore drilling received an enormous setback from the adverse publicity.

The highly publicized Mohole Project for drilling a seven-mile hole reaching the earth's mantle beneath the ocean floor is now defunct and has passed into history. This is just as well, for by the time the project was scrapped it had few advocates left among the oil companies, or within the scientific community. This grandiose plan has now been, in a sense, succeeded by a more logical and sophisticated plan to drill many holes into the sedimentary layers beneath the abyssal depths. The project is called DSDP (Deep Sea Drilling Project) and it is cur-

A huge oil slick threatened the beaches and marine animals of southern California when a leak developed at the base of this offshore drilling platform off the coast at Santa Barbara early in 1969. Before the leak was plugged, 21,000 gallons of oil a day spewed uncontrollably to the surface. (Wide World Photos)

This offshore sulfur mine in the Gulf of Mexico has three platforms. The power plant in the foreground supplies hot water, which is pumped down wells to release the sulfur entrapped in the cap rocks of salt domes. (Freeport Sulphur Company)

rently being carried out by the ship *Glomar Challenger* of the Global Marine Corporation. The project is under the overall direction of a consortium of United States oceanographic institutes, with funding from the National Science Foundation. Besides revealing much about the history of the ocean basins, this drilling is expected to provide data on the oil potential of the deep-ocean floor. One such result has already materialized, for in 1968 a drill hole into the Sigsbee Knolls, a belt of submerged hills in the Gulf of Mexico, struck a show of oil.

DSDP is proceeding as Project Mohole should have proceeded half a decade ago, had not Texas politics and administrative ineptness caused the program to become a costly hole-in-the-ground boondoggle—a "rat hole," according to one Congressman. As it is, the ultimate payoff will still be a sample of the earth's mantle, but it will be obtained by drilling in a spot where the mantle outcrops onto the ocean floor rather than by drilling an ultradeep hole. As of 1971, the *Glomar Challenger* had drilled about 200 holes into the abyssal floor in water as much as four miles deep in the Atlantic, Pacific, and Mediterranean. The Antarctic and Indian Oceans will be tackled next. Eventually the DSDP will want to reenter the drill holes and recirculate the drilling mud so that really deep holes can be made in the ocean floor. By any other name, it is still Mohole.

The sea has oil on its surface as well as beneath it. When one gazes across an expanse of sea, it is apparent that the surface is not of uniform texture. Looking toward the sun, one can see that most (but not all) of the water glitters as the dancing capillary waves reflect myriad points of light—actually mirror reflections of the sun. Within the overall glitter the observer will see dark lanes, where the water appears smooth, as though the surface were covered with oil. This is not an illusion; there is oil there, although not petroleum. The oceanographer terms these smooth patches "natural slicks." This oil is derived mostly from the decay of microscopic diatoms—the basic "grass" of the sea, which is so minute that scientists were not even aware of the existence of such organisms until the eighteenth century. Each diatom contains within its cell wall a tiny droplet of natural oil as a reserve food supply and an assistance in flotation. The natural slicks derived from diatoms are new oil, which is eventually lost by oxidation at the ocean's surface; other diatom oil is carried to the bottom, becoming entrapped in the bottom mud as the precursor to fossil fuel.

Of all the energy arriving each year from the sun, only 0.1 percent (one part in 1,000) is captured by photosynthesis, being utilized equally by plants on land and by diatoms in the sea. Of the energy captured by diatoms, only 0.2 percent (one part in 500) becomes trapped in marine sediments as a precursor of petroleum, and the remainder is oxidized by bacteria and returned to the broth of the sea. The entire petroleum supply on earth has been estimated to be about six trillion barrels (or, in terms of energy, the amount that reaches the earth from the sun in about three days); and this has been accumulated over 600 million years since the Precambrian era. We are presently using up this fossil fuel 100,000 times faster than it took to accumulate it over geologic time.

Power from the Tides

Since time began, man has dreamed about harnessing the restless sea. The sea's energy expresses itself forcibly in white-capped waves and swell, but most of all in the surf. Along the shoreline the open sea waves of oscillation are transformed into a moving wall of surging white water—a solitary shock wave. In the turbulence of the surf, mechanical energy is transformed into heat sufficient to raise the water temperature by half a degree. There appears to be no feasible method of taming the surf and harnessing the energy of breaking waves, but the tides

To gather basic scientific data the drill ship *Glomar Challenger* drilled holes in the Atlantic and Pacific ocean floors at depths ranging from 5,000 to 20,000 feet. The derrick towers 194 feet above the waterline. (Global Marine Company)

may prove to be more tractable. Since the tidal pull is only about one ten-millionth that of gravity, and the tides in the open sea rise only one foot, it is evident that the tide wave must undergo natural amplification before its energy can be usefully tapped for power. The combined effect of the continental shelf and a V-shaped estuary often provides the appropriate setting. Tides in excess of thirty feet are found in many places around the world.

Tides are a mixed blessing; they are an engineering nuisance, often requiring the construction of tidal locks, such as at London and Antwerp, to protect ships in port from their influence. At the same time, however, the tides indirectly created the harbors in the first place. The great ports of the world—like London, which is on the Thames—are almost without exception situated on tidal waters. With the advent of steel construction, which removed the limitation of size imposed by wood, seagoing vessels became very large, and consequently the use of the deep tidal harbors became highly important. The river mouths in nontidal regions are clogged with sediment and generally not navigable. Under the scouring action of tidal ebb, however, sediment brought down by rivers is dispersed and carried away. The moving water also flushes and purifies an estuary, preventing the buildup of sewage and poisonous industrial wastes.

Some of the world's highest tides are found in the Bay of Fundy in Nova Scotia, where they attain a range of fifty feet. The high tides of this bay are due primarily to its natural period of oscillation, which approximates the ocean-tide period of about six hours. This brings the period of the stationary wave operating within the bay into resonance with the open ocean tide. Theoretically the rise and fall of water in the Bay of Fundy in one tidal cycle is equal to 200 million horsepower. The amount of

Computerized pulses bouncing off fixed points on the ocean floor keep the *Glomar Challenger* precisely on station while working in water depths up to 20,000 feet. (Global Marine Company)

water entering the bay every six hours is nearly four trillion cubic feet, or as much water as falls as rain over the entire United States in a week.

The tides act as a brake upon the earth's rotation, but our planet's rotational momentum is so great that the length of the day has increased through tidal friction by no more than three seconds in a million years. On the other hand, in the rise and fall of the tide, an enormous quantity of water is placed in motion—and moving water means energy. It is no wonder that engineers with an inventive turn of mind have been fascinated by the prospect of harnessing this seemingly inexhaustible and free supply of energy. To date, however, the utilization of tidal energy has been achieved only on a token scale.

The earliest reference in the English language to a machine exploiting tidal energy appears in the *Domesday Book,* which mentions a tidal mill at Dover, England. And in the Middle Ages, Richard Carew wrote in his *Survey of Cornwall:*

Amongst other commodities afforded by the sea, inhabitants make use of divers creeks, for gristemilles, by thwarting a bancke from side to side, in which a floudgate is placed with two leaves: these the flowing tyde openeth, and after full sea, the waight of the ebbe closeth fast, which no other force can doe: and so the imprisoned water payeth the ransome of dryving an undershoote wheele for his enlargement.

Tide mills have existed for centuries in England and Wales. Among the oldest were those at Bromley-by-Bow, built in 1100, and at Woodbridge, dating from 1170. Both were in use until very recent times. Tide mills of ancient vintage are also found in Holland, and early Dutch colonists built such mills in the early seventeenth century near New York. One of these, near Spring Creek, survived until the end of the last century. The first tidal mill reportedly placed in operation in the United States was built at Salem, Massachusetts, in 1635. In 1790 a small tidal mill was installed on the Tamar River in Devonshire, England; it is still operating, although the original waterwheel has now been replaced by a more efficient turbine. London's water supply was successfully pumped by a waterwheel placed in the tidal stream of the Thames estuary and mounted between the piers of the old London Bridge until its demolition in 1824.

Such tide mills, of course, provide only a token amount of power, but they do utilize the same principles as would be involved in any gigantic engineering scheme. These are a retaining basin, a dam, and sluice gates, with the basin filling as the tide flows in. Among the larger plans that have come under recur-

rent study has been one to harness the tidal ebb and flow of the Severn estuary in England, where 45-foot tides are available, and where 800,000 kilowatts could be generated.

In the United States there has been great interest for nearly fifty years in placing a dam across the Passamaquoddy estuary between Maine and Canada. This bay has a tidal range of fifty feet, one of the highest in the world. A multiple-basin plan at Passamaquoddy actually went into incipient construction in 1935, but work quickly came to a halt when the United States Congress failed to appropriate the funds to implement this public works project of Franklin D. Roosevelt during the Great Depression. Somewhat similarly in 1963, Congress failed to support a 1961 recommendation of a commission appointed by President Kennedy that work commence anew on this tidal power project. Even today the U.S. Army Corps of Engineers continues to update this scheme in the event that a rise in power rates makes it feasible. However, the recent surge in the use of nuclear energy, which is gradually becoming competitive with the use of fossil fuels (coal and oil), probably relegates the construction of such a tidal dam at Passamaquoddy to the distant future.

A major project for utilizing tidal power did reach fruition in 1967 with the throwing up of a dam across the Rance River in Brittany, France, by Electricity of France, a government-owned company. The tides there attain a height of more than forty feet, so that an annual power output of about 500 million kilowatt-hours is obtained—or about half the output of one of the French Rhône River power stations. The twenty-four turbine units placed across the tidal flow have the capability of varying the pitch of their blades, which increases their efficiency. They also are reversible, so that power can be drawn on both the ebb and flood of the tidal cycle. The French now envision damming off and capturing the tidal energy over a large sector of the Brittany coast, an accomplishment that would require a heroic engineering effort.

The Russians, too, have now entered the tidal-power field. In 1968 the Soviet Union announced the opening of an experimental tidal-power station in the Arctic Ocean at Kislaya Bay, a fjord-like inlet just west of Murmansk and near the Norwegian border. This small station is designed as a pilot plant for testing the feasibility of building far more ambitious power plants to supply the needs of the Soviet northern regions. For such future plants, two sites have already been chosen, both of which are in the arctic region of European Russia—on the Kola Peninsula and in the White Sea.

The Oceans' Resources

The sea has provided food for man ever since he began to live on its coastlines. Bones of marine fishes have been found in caves along the Dordogne River in France, where they were discarded 40,000 years ago by men of the Late Stone Age. Neolithic cave dwellers used harpoons, hooks, and nets.

In medieval times fishing became an important industry in northern Europe, particularly in countries bordering the North Sea. Fishermen ventured farther and farther from shore in search of new and richer fishing grounds. By the mid-1500's the Portuguese were exploiting the teeming cod schools of Newfoundland's Grand Banks.

Spurred by the demands of a rising world population, the fishing industry expanded enormously during the nineteenth century. The building of railroads made it possible to transport fresh fish to new and more distant markets. The size of fishing fleets increased dramatically—from 24 to 600 at one English port during a mere 22 years, for example. At the same time, the range and efficiency of boats increased as trawls replaced hooks and as steam power replaced sails.

By the 1880's it had become apparent to some observers that the fish resources of the sea might not be inexhaustible. One of the first to realize that some popular species were in danger of being fished out of existence was C. G. J. Petersen, a Danish biologist who developed concepts of fishery management that are still sound today. To track fish for his studies he invented the "Petersen tag," consisting of two metal (plastic is often used today) buttons that are attached to the fish with a metal pin.

Unfortunately, no fully satisfactory international agreement has been devised to prevent overexploitation of many species in international waters. As a result, the existence of many marine animals, such as certain species of whales, is seriously threatened. International controls have been established, however, in the Pacific halibut and salmon fisheries, with encouraging increases in their yields.

Fish may be farmed as well as caught, and in many parts of the world farming the sea may prove to be more profitable than fishing, just as farming on land is a more efficient way of obtaining food than hunting. The results already being obtained by fish farming are sometimes striking: In Taiwan, for instance, marine ponds produce more than fifty times the per-acre yield of the heavily fished North Sea.

Fish represent only a portion of the wealth of the ocean. Since the earliest times men have also exploited the seas for shellfish, pearls, sponges, and salt. Ocean water itself is rich in minerals. All the natural elements of the land probably are represented in the sea, although some are present in such small quantities that it is impossible to detect them with the instruments of today.

Even the more common elements are so diluted that it is difficult to extract them profitably. There is, for example, $100 million worth of gold in a cubic mile of seawater, but though men have dreamed of mining the sea for this metal, no one has come close to succeeding economically. Today, five important mineral resources —magnesium, iodine, bromine, common table salt, and fresh water—are taken from seawater.

The minerals on and beneath the ocean floor are more easily obtained than those in the water solution. Great expanses of ocean floor are strewn with phosphorite and manganese nodules, some as large as Idaho potatoes. Today they clog oceanographers' dredges; tomorrow they may be vacuumed up for industrial use—especially for their concentrations of copper, nickel, and cobalt.

Private industry is now tapping the continental shelves for oil, pumping some five million barrels every day. A post–World War II development, offshore oil production now exceeds the value of the fisheries in the United States. Drilling platforms, some as high as the Washington Monument, dot the coastal waters of California, Louisiana, and Texas. Outside North America important offshore fields are found in the North Sea and the Persian Gulf, and off southeastern Australia.

For centuries man has attempted to harness the power of the sea's moving waters. William the Conqueror's census of England, the *Domesday Book*, describes a mill that operated by tidal power at Dover. In 1935 the United States considered the feasibility of a dam across the Bay of Fundy, where the 50-foot tides are the highest in the world. But the project was never started.

Today a tidal power plant operates on the ebb and flow of the Rance River in France. Completed in 1967, it attains an annual power output of about 500 million kilowatt-hours. Another tidal power plant recently began operation in the Soviet Union.

The ocean's resources remain vastly unknown and untapped. But the discoveries of the last few years have revealed tremendous new sources of wealth. In our increasingly industrial world, it is almost certain that governments and businesses will turn more and more to utilizing these resources.

High-powered ships and harpoon guns make modern whaling methods so efficient that some species of whales are in danger of becoming extinct. In the days of sailing ships, when harpoons were thrown from small boats, there was often a violent struggle between victim and pursuer. This engraving, from a nineteenth-century painting by Ambrose Louis Garneray, was described by Herman Melville in *Moby Dick* as being the finest representation of whaling he knew. (Lighthouse Gallery, Shelburne Museum, Vermont)

Traditional fishing methods are still used in many parts
of the world. Fish are laid out in the sun to dry on these
grouper boats in British Honduras on the Caribbean coast
of Central America. (C. P. Idyll)

The trawl net is hauled in by hand aboard *Cap'n Bill IV*. This modern side-trawler drags for offshore lobsters in the winter and haddock, cod, and flounder in the summer. (Robert K. Brigham, U.S. Department of the Interior)

Modern fishing boats line the docks in Taiwan. (Chinese Information Service)

A Petersen disk is attached to a spiny dogfish by a biologist of the Bureau of Commercial Fisheries at Woods Hole, Massachusetts. After the fish is tagged its length is recorded and it is released for migratory studies. (Robert K. Brigham, U.S. Department of the Interior)

White sea perch and pile perch congregate inside the remains of a streetcar—a man-made reef placed in sixty feet of water, half a mile offshore. (Charles H. Turner, California Department of Fish and Game)

The sun sets over pearl-cultivation farms in Ago Bay, Japan. (Japan Tourist Association)

Sulfur from the Grand Isle mine, which stands off the Louisiana coast in the Gulf of Mexico, is among the resources being tapped off the continental shelf. (Freeport Sulphur Company)

California giant kelp, a source of commercial iodine, is photographed close up. (Mick Church)

Eight billion dollars a year are now spent on extracting the ocean's resources, with the major part going to off- shore petroleum. Located in Cook Inlet, Alaska, this monopod oil platform rises 306 feet from its base to the top of its drilling rig. (Marathon Oil Company)

The idea of utilizing the power of the tide came to fruition in 1967, when the French government opened its tidal power plant on the Rance River in Brittany. The tides there reach a height of more than forty feet. (French Embassy Press and Information Division)

Fresh Water from Salt Water

Of all of the sea's resources the most priceless is the water itself. Water is recognized as a natural resource—in fact, the most vital of all mineral resources. Water fully satisfies the definition of a mineral, for it is a naturally occurring substance of fixed chemical composition—namely, H_2O. There is nothing in the definition of the term "mineral" (any naturally occurring substance of fixed chemical composition which is a product of inorganic processes) that restricts it only to solid substances.

Water is central to the history of man. Of the world's supply, 97 percent is in the oceans, another 2 percent is tied up in the frozen icecaps, and the remaining 1 percent is found in lakes, the atmosphere, and in ground water. The fresh water of lakes and rivers, so essential to all life, is water in transition, as the flow is ever toward the sea, where the earth's water is stored as salt water that is unfit for many human uses. About ten million billion gallons of water are evaporated each year from the earth's surface, 88 percent of which comes from the sea—enough to lower the ocean surface about three feet if the oceans were not replenished.

In many parts of the world water is often in short, and even critical, supply. Extensive regions of the world lie as barren deserts where rainfall provides less than five inches of precipitation per annum. Mainly for lack of water, Australia, which is as large as the United States, supports a population of only twelve million. A practical approach to solving the agricultural problem would be to develop salt-tolerant, food-producing plants. Such crops could be irrigated with seawater. But the ultimate answer is, of course, desalination, for the 330 million cubic miles of the hydrosphere would provide an unlimited supply for irrigation and human use once the salt is removed.

More than 100 million gallons flow daily through desalination plants around the world. In 1967 the fresh-water pipeline to Key West from the Florida mainland was cut off, and now the entire population of 34,000 uses fresh water from the sea, the first American city to be so provided. A desalination plant on this coral island turns out more than two million gallons of water daily. Plans for the world's first nuclear-fueled desalination plant, scheduled to begin operation in Los Angeles in the early 1970's, have been called off due to rising costs and to the reallocation of Atomic Energy Commission funds.

The seas already contain localized masses of fresh water in the form of floating icebergs derived from glacial ice and also in the form of pack ice, for when salt water freezes the salt is largely excluded. Greenland and Antarctica are the spawning grounds of huge icebergs, which drift in silent and ghostly majesty through the polar seas. The flat-topped tabular bergs that break away from the ice shelves of the white continent attain truly gigantic proportions. In 1893 one such berg was sighted near the Falkland Islands off Argentina. It was 1,200 feet thick and ninety miles on a side, an overall area equivalent to Corsica. Here was a solid block of fresh water of a volume greater than all of the Swiss lakes combined, and far removed from its birthplace. What a boon it would be if such a wandering berg could be nudged into the Humboldt Current, so that it could water the parched deserts of northern Chile, or if a berg could be discovered that might be towed into Perth to water the arid wastes of Australia! Doubtless icebergs will someday augment man's supply of the most vital of all natural resources, fresh water.

The Lawless Seas

The seas are both a new and a lawless frontier. The rapidly expanding technology that is making the exploitation of the ocean floor increasingly more feasible has not been accompanied by the necessary international legal agreements. Through international treaties the legal aspects of mining on the moon and in the Antarctic have been made clearer than the ground rules for mining the ocean floor. Use of the high seas for navigation is rather well established by centuries of precedent, but we urgently need an international "ocean space" treaty that would cover both the sea floor and the sea itself for mineral exploitation purposes. International control somewhere between complete laissez-faire and placing the resources under the jurisdiction of the United Nations seems needed. Possession is now "nine-tenths of the law," and things of the sea belong to "he who can reduce them to his possession"—a sort of finders-keepers situation. Big nations with navies capable of backing up industry with naval defense may wish to keep it this way, but this attitude will certainly lead to dangerous confrontations. More reasonably, nations must work within a framework of international law in opening the sea to profitable enterprise. Private industry will be hesitant to invest its dollars until international, legal, and economic rights have been clearly defined. A step in the right direction recently has been made by the countries surrounding the North Sea, which have arrived at a workable agreement with respect to exploitation of natural gas and oil from beneath these waters. Hopefully this will establish a pattern for other countries to follow.

At the present time the coastal nations are generally conceded to have control of the seabed out to the 200-meter line (668 feet), or just beyond the usual break at the shelf edge. This applies to mineral rights and benthos (bottom-attached animals). On the other hand, the limit to the control of the overlying water generally goes out only twelve miles. Presumably a crab walking along the bottom on the shelf but beyond the twelve-mile limit belongs to the adjacent country. However, if this crab were to momentarily lift all six feet off the bottom, he would become the property of anyone who could grab him.

Senator Claiborne Pell of Rhode Island has called for an international treaty of jurisdictional rights to the sea. Ocean space is characterized by two components—the seabed-subsoil and the liquid segment above it. According to Senator Pell, the liquid portion beyond a country's territorial seas should remain as high seas open to all nations to engage in fishing, aquaculture, in-solution mining, transportation, and telecommunications—but in accordance with existing conventions on fishing and living resources. He suggests that the United Nations control the leasing of the ocean floor beneath the high seas and that it have its own Sea Guard, patterned after the U.S.

Coast Guard, to police the treaty. He would restrict the seabed to "peaceful uses" but permit its use by military submarines and devices for submarine and weapons detection, identification, and tracking.

The legal aspects of sea-floor mining remain tangled and uncertain, for there are few precedents. A case in point was the recent application of a California firm for a precedent-setting lease from the United Nations to investigate the pools of hot brine in the Red Sea. This firm, Crawford Marine Specialists of San Francisco, asked for a 38.5-square-mile, exclusive mineral lease for three years to sample, map, and determine the economic significance of the mineral deposits associated with these brine pools. These deposits now underlie the high seas (unless the Red Sea is considered to be an inland waterway), being about fifty miles from the nearest country.

Future Promise

Noise is defined as unwanted sound, and dirt has been defined as misplaced mud. Many things are useless if out of place, or until a new use is found. History reveals that nothing is a natural resource until man finds a use for it. Until the Wright brothers made their historic flight sixty years ago, lead and aluminum were unimportant. Today fuel additives and batteries consume about half of all the lead produced. Until the development of the A-bomb in 1945, uranium had little value. Today there is great demand for the so-called space metals, many of which were formerly regarded as useless—tantalum, beryllium, and germanium. These metals are now needed for transistors and other solid-state devices used in electronics. The sea is a vast storehouse of minerals, many of which are unwanted today—but, as surely as morning follows the night, the day will dawn when the insatiable appetite of industry for new raw materials will create a demand for their exploitation.

With today's quickened pace of technology, the general promise of the oceans appears to be enormous, even though it is impossible to fully identify just what resources are to be gleaned. The Wright brothers, imaginative and inventive as they were, did not foresee, when they built the first aircraft in their bicycle shop, that airplanes would be able to fly over any ocean in their own lifetime. And certainly the brilliant rocket pioneer, Dr. Robert Goddard, never foresaw the $25 billion Apollo program for carrying man to the surface of the moon. The concept of a frontier has a special meaning to Americans. Today the frontiers of the West have passed into history, but the oceans around us remain as our deep frontier.

Marine geologists prepare to lower boomerang corer to collect bottom samples off the coast of Virginia. (Mick Church)

194

The land is broad, but the sea is wider. The hidden depths remain a wonder world of which little is known. Until men learn to live there, they shall never fully know their planet. The ocean floor remains to be mapped, even in a reconnaissance manner. Until man explores this challenging landscape, he cannot know the economic rewards or benefits it holds for him. But he can be sure that the bottom of the world, the basin into which the ancient sediments have gathered, has many magnificent secrets. The ocean may hold the key to man's survival. With the current population explosion, man is already straining the array of traditional land resources, including—curiously enough—even water. He must urgently seek help from the sea. He must turn to the sea.

9 UNDERWATER ARC

by Mendel Peterson

Men have left their traces in the waters of the globe from the time the first man floated downstream on a log to the present age of great steel ships. Accidents of storm or human miscalculation have littered the ocean bottoms with the wreckage of vessels of all ages, including their cargoes, and the tools and possessions of those who sailed in them. Even the homes of these people have been inundated and preserved beneath the waters.

The men of the Old Stone Age often chose to live in the caves that abounded in certain seashore areas, especially the Mediterranean, and the records of thousands of years of human history are contained in the layers of plant, animal, and human remains—the refuse from everyday living—that were gradually built up in the floors of their homes. Through succeeding ages, as shorelines settled and as the waters of the oceans rose due to melting of the polar icecaps, these caves were slowly inundated, until today they lie as much as 200 to 300 feet below the surface of the sea.

The slow rise of the sea continued as man began to build the first stone dwellings in villages bordering the Mediterranean, and then to construct new buildings on top of the first ones during centuries of occupation. Many of these sites, too, now are covered by the sea, but since they are only centuries rather than millennia old, the depth of the water is measured in tens rather than hundreds of feet. Today, protected by ten to twenty feet of water, hundreds of ancient architectural remains ring the Mediterranean.

Thus, the history of the human species may be traced through the wrecks, the dwellings, and the precious trash scattered about the bottoms of the planet's oceans, and of the lakes and rivers, too. It is the object of underwater archaeology to systematically discover, excavate, study, and publish treatises on these remains in order that we may better understand the lives of our ancestors, both remote and recent. While this discipline is new, some of the diving concepts and tools the underwater archaeologist uses have a history of several centuries.

The Art of Diving

The lives of such early peoples of the eastern Mediterranean as the Minoans, Phoenicians, and Greeks were centered on the sea, which furnished them a livelihood through fishing, trade, and warfare. The natural resources beneath the surface of the sea —sponges, coral, and seashells—also attracted them and they became expert in diving. The earliest divers undoubtedly relied on sheer lung power; later, they learned to use simple breathing tubes as well as inverted pots in which air was trapped and brought below the surface for breathing. Many ancient writers describe the Greek sponge divers, who were able to reach surprising depths—probably seventy-five to a hundred feet—with no apparatus but their own lungs. But the cost in physical suffering and in life was high. Writing in the second century A.D., Oppian said in his *Halientica* that "No ordeal is more terrible than that of the sponge divers and no labor more onerous for men."

Several incidents involving the use of divers or underwater swimmers in warfare are recorded in history. The use of underwater swimmers to cut the anchor lines of enemy ships, for example, seems to have been a rather common tactic. And Thucydides tells how the Athenians used divers during the siege of Syracuse to saw down the subsurface barriers that had been erected to block their vessels from entering the harbor. Alexander the Great is reported to have used similar underwater demolition teams when he besieged Tyre.

As accidents deposited valuable goods on sea bottoms, it was natural that divers should also turn to salvage operations. In Rhodes, the divers were governed by special laws prescribing the percentages they were to receive of recovered goods. In the case

HAEOLOGY

of goods recovered from a depth of twenty-two and a half feet or more, the diver was allowed to keep one-half; on recoveries from twelve feet to twenty-two feet he received one-third. For recoveries at less than twelve feet the rewards declined to a share of one-tenth.

While legends tell us that Alexander the Great descended beneath the surface of the Mediterranean in a device that resembled a diving bell, not much is known for sure about this excursion and none of the illustrations of Alexander's chamber are contemporary. The earliest known illustration of what today is recognizable as a diving suit appears in a manuscript, now in the Munich Library, that dates from 1415. This was a waterproof suit, presumably leather. It had a hood that fitted over the diver's head and was joined to a leather tube extending to the surface of the water, where it floated on two air bladders. Although such a suit would not have functioned, variations on this theme were repeated in literature for several hundred years. Even the great Leonardo da Vinci, who offered an improvement in the form of a valve at the top to prevent entry of water into the tube, failed to appreciate the essential flaw in this arrangement. It is simply not possible for the human animal to breathe on lung power alone in depths of over a very few feet; air must be fed to the lungs at the same pressure as the water on the chest. This basic fact was not realized until the seventeenth century.

The first happy relief from the monotonous repetition of the Munich manuscript was provided in 1551 by Nicholas Tartaglia, an Italian mathematician. In Tartaglia's scheme, the diver stood on a platform, with his head encased in a large glass globe. By cranking a windlass on the platform, the diver could move himself up and down a static line held at the bottom by a heavy weight. This apparatus was as impractical as the leather suit, since water would compress the air in the globe and rise over the nose

of the diver as he descended into the sea. But it suggested the principle of the diving bell, and Tartaglia a decade later illustrated a large globe in which the diver sat. This principle, foreshadowed by the ancient technique of using pots to trap air for the diver, actually may have been applied in Tartaglia's time.

In 1597 Buonaiuto Lorini suggested two systems for furnishing air to divers. The first was a variation of the old leather suit. Lorini seated his diver on a platform, encased him in a leather suit, and stuck his head in a leather helmet, which continued upward to the surface as a large tube some eight to ten inches in diameter. The idea appears absurd now. If the water pressure did not collapse the tube and smother the diver using such a suit, he would very soon use up the oxygen in the bottom of the tube and die in the carbon dioxide he had exhaled. The second of Lorini's ideas, however, was very practical; in fact it represents an early form of the diving bell, and it would have worked. The apparatus was merely a large box, presumably metal, with viewing ports and a platform on which the diver stood. As it was lowered into the sea, the water in it rose up to the diver's chest, compressing the air in the container to the same pressure as the water surrounding it, thus enabling the diver to breathe.

The exact stages in the evolution of the diving bell during this period are obscure, but its development seems to have been fairly rapid. In 1678 Jean Christophe Sturmius described a true diving bell— the first really successful diving apparatus. According to Sturmius' description and an accompanying woodcut, which appeared in the *Journal des Scavana*, published in Paris, the diver stood on a platform that hung by chains from the bottom of the heavy metal bell. Similar or identical bells had probably been used for some time before this description was published. The basic principle of this bell was the same as that suggested by Lorini in the previous century:

The earliest known illustration of a diving suit appears in a German manuscript of 1415. The suit, with leather hood and tubing, would not have worked. (Smithsonian Institution)

Leonardo da Vinci suggested adding a breathing mask, hoses, and a floating air-valve to his diving suit, but failed to recognize that air pressure must be increased in proportion to the surrounding water pressure. (Smithsonian Institution)

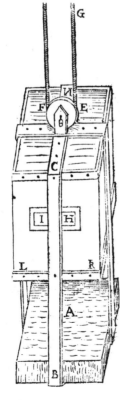

A practical diving bell, invented by Bonaiuto Lorini, is shown in this woodcut of 1597. As the bell was lowered, the water level rose in the box, compressing the air and allowing the diver to breathe. (Smithsonian Institution)

As the bell was lowered, the air that had been trapped in it was automatically compressed by the pressure of water trying to force its way into the bell.

The Sturmius bell had one serious defect; as the oxygen in the bell was consumed, it was necessary to bring the contraption up to the surface to renew the air. The problem of renewing the air in the bell, however, was certainly solved in the next century by Edmund Halley, the English scientist and astronomer, whose treatise on diving bells was published in 1717 in the *Philosophical Transactions of the Royal Society,* and it may well have been solved fifty years earlier by the divers who salvaged the guns of the *Vasa* from the bottom of Stockholm harbor. To extend their working time underwater, the *Vasa* divers probably sank small containers—Halley's version called for the use of small bells—of fresh air below the main diving bell, then bled the new air up into the larger bell through a leather tube. With this development, the diving bell attained what is very nearly its modern form, the most significant later im-

Using diving bells such as this, salvagers raised more than fifty cannons from the gunboat *Vasa* in the mid-seventeenth century. The three-deck ship had rolled on one side and sunk in the Stockholm harbor in 1628. (Smithsonian Institution)

provement being the introduction in the early nineteenth century of pumps to provide a constant flow of new air from the surface.

The raising of the *Vasa's* guns is one of the earliest examples of a major salvage operation successfully completed. This brand-new, 64-gun, three-decker had rolled over and sunk in August 1628, just after leaving the Royal Dock in Stockholm, where she had been constructed. Many of her crew of 133 and the distinguished visitors aboard were drowned in full sight of the thousands who lined the shores of the harbor to see the new ship off. Almost immediately after the *Vasa* sank, efforts to raise it were begun. An Englishman, Ian Bulmer, using two hulks

and numerous large cables that he attached to the *Vasa* with grapnels, merely succeeded in getting it on an even keel. After a period of some confusion, during which several salvagers concerned themselves with the project, a syndicate emerged, headed by a Swede, Colonel Hans Albrecht von Treileben, who received a license to salvage in December 1658. He had to wait four years, until another license expired, before he could begin operations on the *Vasa*, but he used this time to good advantage, engaging in several successful salvage projects involving recovery of guns and even one small ship. In 1663 Treileben and his group, now including two expert divers named Andreas Peckell and Jacob Maule, began work on raising the guns.

Although the *Vasa* by this time had become buried deep in the mud of the harbor bottom, Treileben and his company recovered more than fifty of its valuable bronze cannon by making use of the newly perfected diving bell, along with special long-handled cutting and grasping tools, and by working the divers to the limit of human endurance. When one considers the primitive equipment, the prevailing ignorance of diving physiology, the depth and coldness of the water, and the almost complete lack of visibility, the salvage of the *Vasa's* guns still stands as one of the most remarkable underwater recoveries ever accomplished, a fact especially appreciated by the modern divers who worked for three years to raise the ship itself in 1962.

The development of the diving bell by no means rendered traditional diving techniques obsolete. They continued to be used—and most profitably, too. In 1687, some two decades after the *Vasa* salvage, William Phipps, a farm boy from Maine (then a part of Massachusetts), hauled up a tremendous fortune from a Spanish ship that had sunk on a coral reef north of Santo Domingo. He employed native divers from the West Indies, who used air tubs to perform the salvage—in principle, the same system of diving that had been used by the ancient Greeks.

The ship that Phipps salvaged had sunk sometime in 1643 in forty to fifty feet of water, while homeward bound with a rich cargo of treasure from the New World. A few survivors managed to get away by making rafts of the wreckage, and at least one of them remained in the West Indies. Phipps had already spent several years looking for sunken treasure, and had failed in one well-organized and well-financed expedition, when he met the Spanish survivor and learned from him the location of the wrecked ship. Sailing again with another well-planned expedition, backed by important British

199

peers, Phipps found the site and recovered silver and gold worth 300,000 pounds, along with bronze guns. Phipps became wealthy on his share, was knighted, and later was appointed Governor of Massachusetts.

The Diving Suit

While the development of the diving bell increased the length of time that divers could remain below the surface, underwater salvage operations remained comparatively inefficient. Divers were restricted to the area adjacent to the bell. The farther their work site was from the bell, the shorter was the period of time they could work before having to swim back to the bell for another breath of air. Even when the work site was close at hand, the necessity of frequently halting their activities in order to return to the bell seriously handicapped their operations. What was needed, obviously, was some form of airtight container in which a diver could move around on the bottom and pursue his work without interruption.

Not surprisingly, the first successful apparatus of this sort, invented in 1715 by John Lethbridge, an Englishman, looked more like a mobile diving bell than a suit. In effect, Lethbridge just closed the bottom of a diving bell, suspended it on end, added a glass viewing port, and provided armholes with waterproof leather collars through which the diver could extend his arms to accomplish useful work.

The air in Lethbridge's bell was kept at normal atmospheric pressure, thus eliminating any danger of "the bends"—the formation of minute nitrogen bubbles in the joints, muscles, and nerve-control centers that may cripple or kill a diver if he returns to the surface too quickly after a long time of breathing air under pressure. The diver who used the Lethbridge apparatus, however, still had to contend with water pressure. The pressure of the surrounding water on his extended arms restricted circulation of the blood, thus greatly limiting the duration and depth of the dive. Despite these trying circumstances, Lethbridge recorded dives to a depth of sixty feet for thirty-five minutes.

In attempts to gain more mobility for the diver, Lethbridge's bell-like enclosure gradually began to evolve into a suit toward the end of the eighteenth century. A Parisian named Freminet introduced an apparatus in the 1770's that consisted of a waterproof helmet with a dress strengthened by a metal frame. The diver was supplied with air from a tank, pumped by a bellows through a hose. It is doubtful if this apparatus was really successful, although prints exist

The first successful diving suit looked more like a barrel than a piece of clothing. Invented by John Lethbridge in 1715, it consisted of an airtight diving chamber turned on its side. Holes with leather collars enabled the diver to use his arms outside the enclosure. (Smithsonian Institution)

showing it in use. In the 1790's a German engineer, C. H. Klingert, built, tested, and used an apparatus that combined an armored leather diving dress with a copper helmet and a reservoir of air. The reservoir was really a closed bell with a cylinder and piston extending into the bottom. The diver stood on a platform with a hose connecting the helmet to the bell. As the bell descended, the increasing water pressure forced the piston up into the cylinder, compressing the air in the bell and feeding it to the diver at the same pressure as the water surrounding him. While incorporating most of the elements of later successful diving suits, this apparatus also had several severe defects. The activity of the diver was restricted to the platform; if he stepped off and descended to a point lower than the bell he immediately suffered a squeeze from water pressure. Secondly, as no means was provided to renew the air in the bell, the dives were limited to the time it took the diver to breathe up the original supply of oxygen. If Klingert had provided his apparatus with a good air pump instead of the bell, he would have had a practical diving suit. Two decades later, this was to be invented by Augustus Siebe.

Siebe, a German by birth, was trained as an in-

200

The Siebe closed suit, which covered the diver entirely, was used in the recovery of the guns of the *Royal George*, a British ship sunk in 1782. Divers wore copper helmets and weighted shoes. A pump supplied pressurized air from the surface. (Smithsonian Institution)

Front view of the copper helmet from the closed diving suit perfected by Augustus Siebe in the late 1830's. (The Science Museum, London)

strument maker and gunsmith. He served as an artillery lieutenant in the battles of Leipzig and Waterloo against the forces of Napoleon, and after the war he emigrated to England, where by 1819 he had developed what is known as the "open diving dress." This apparatus included a helmet and a canvas suit extending just to the diver's waist, and a weighted belt and shoes. An efficient pump supplied the suit with air at the proper pressure as the diver descended into the water. In 1837 the "closed dress," which covered the diver entirely, was perfected by Siebe, and modern diving was underway.

The first great salvage job with Siebe's new suits was the recovery of the guns of the *Royal George* and the destruction of her hulk. This 108-gun vessel, flagship of the British navy, had sunk while anchored in Portsmouth harbor in 1782. The ship had been heeled to one side to permit repair of a leak in planks that were normally below her waterline when, through a miscalculation in loading supplies, it suddenly rolled over and sank, carrying down most of its 800-man crew, an admiral, and numerous women and children who were visiting aboard. Early efforts to pass cables beneath the sunken hull and then raise it with hulks failed, as they had in the case of the

Vasa, and the *Royal George* became increasingly dangerous to navigators in the harbor as mud and gravel built up around the wreck.

During the late 1820's and early 1830's, a diver named Anthony Dean recovered some thirty of the *Royal George's* guns, together with many small objects. It is thought that he used the new open dress invented by Siebe, although there is some confusion in the accounts concerning Dean. A contemporary illustration shows him in a closed, not open, suit and dates his work earlier than the written sources describing Siebe's model. If the illustration can be trusted, perhaps Dean, rather than Siebe, should receive credit for inventing the closed suit.

In any event, Dean's salvage operations did not diminish navigational hazards posed by the sunken ship. As a result, in 1839 the Royal Sappers and Miners under Colonel Charles Paisley began work on salvage of the rest of the guns and destruction of the hulk. They used a diving bell but failed initially to achieve either of their objectives. The next year

201

The Denayrouze-Rouquayrol tank, which was strapped to the diver's back, had an automatic valve to control air pressure. The air was replenished through a hose from the surface. (Smithsonian Institution)

Augustus Siebe introduced the colonel to the new closed diving suit and the work began in earnest. By 1844, after a remarkable operation, the guns had been raised and the hulk blown apart with containers of gunpowder fired from the surface. This operation marked the beginning of heavy diving operations as we know them today and a momentous turn in the history of diving.

A few years after the final destruction of the *Royal George,* a new concept of diving made its appearance in France. By 1865 a mining engineer, Benoît Rouquayrol, and a naval officer, Auguste Denayrouze, had invented and placed in use an apparatus utilizing the principle of a back tank and a demand valve automatically supplying the diver air at the pressure of the water surrounding him. The cylinder, which was carried on the back, was resupplied from the surface with a hose that could be detached when the cylinder was full and reattached when the tank required more air. So successful was this apparatus that forms of it were still in use until quite recently, and its principles are incorporated in the aqualung. Except for refinements of gas mixtures, valve arrangements, and minor details of design, the

Siebe closed suit, the Denayrouze-Rouquayrol apparatus, and the diving bell remained the basic deep-diving apparatus until the period of World War II.

Mediterranean Treasures

The invention of the closed diving suit was a boon not only to salvagers, but also to the sponge divers of the Mediterranean. Once they began to adopt the new gear, it probably was inevitable that sooner or later a sponge diver would make an important archaeological find. The floor of the Mediterranean is littered with wrecks, and fishermen have dredged up ancient bronzes and marbles since the beginning of modern times. During the nineteenth century a number of important fragments were found in this manner. In 1900 a discovery surpassing all previous ones was made by a sponge diver using a closed suit. The event marks the beginnings of salvage archaeology.

While returning from Tunisia, where they had harvested sponges, a group of divers under Captain Demetrios Kondos was forced by high winds to anchor under the sharp face of a cliff at the island of Antikythēra. In order that the time should not be entirely wasted, the captain decided to send a diver down to see if there were any sponges on the bottom. Elias Stadiatis suited up and plunged overboard; on the bottom, at 150 feet, he must have thought that he was having visions, for all about him lay sculptures of horses and human figures in marble and bronze. One large, blackened bronze arm protruded from the bottom. The diver pulled it out and signaled his tenders to bring him up.

On hearing what was below, Captain Kondos climbed into his suit and immediately went down with a tape measure. He returned to the surface, recorded some figures on paper, and placed the bronze arm under his bunk. When the wind died down Kondos hoisted sail for home, the island of Symē. There, at a meeting of the elders, it was decided to send Kondos to Athens with news of the find and the bronze arm to prove it.

The Greek government now had laws protecting the archaeological treasures of the nation and was very conscious of new discoveries and the need to protect them. Kondos was warmly and excitedly received in archaeological circles. The government agreed to give the divers a reward for each piece of sculpture recovered from the site and furnished a Greek naval vessel with machinery to lift the heavy statues to the surface. Thus, this became the first Greek archaeological project to be carried out by the Greeks themselves.

During the following year the divers worked with energy and audacity beyond their normal depths and under severe wind and water conditions. One died and two were crippled for life. Today, the prizes they brought up rest in the National Museum at Athens. The bronzes include two statuettes of the Age of Pericles and a life-size youth identified with Perseus, Paris, or Hermes—the experts can't agree which. The divers also salvaged a large collection of marble copies of famous earlier works. The wreck was identified as a ship containing loot bound for Rome in the first century B.C.

The sunken ship also yielded a bronze geared instrument, or the corroded remains of it. It lay unstudied and unrecognized until the 1950's, when Derek Price, a historian of technology, identified it as an astronomical clock, which may also have been used as a computer to predict planetary motions. The only such instrument known from the pre-Christian era, its discovery has radically changed our opinions of ancient technological capabilities.

The next archaeological find in the Mediterranean was made in almost the same circumstances as those at Antikythēra. In 1907 a Greek sponge diver working in 130 feet of water three miles off Mahdia, Tunisia, came upon a row of strange cylindrical objects that he first took to be cannon from a sunken warship. Closer inspection revealed them to be ancient marble columns. They were laid out on the sea bottom in six main groups in parallel rows, forming the outline of a ship. Among the columns were odd shapes that proved to be bronze figures and marble carvings.

When authentic pieces of Greek sculpture suddenly appeared on sale in the Arab market in Mahdia, Alfred Merlin, Director of Antiquities in Tunisia, quickly traced them to the divers and learned of the site. He suggested that an official expedition be organized to examine the wreck. An antiquarian, Solomon Reinach, excited about the possibilities of the wreck, secured financial backing for the project from an American friend in Paris, James Hazen Hyde. With the $25,000 contributed by Hyde, two small ships loaned by the French navy, and support by the Tunisian government and some French ministries, a series of expeditions was mounted between 1908 and 1913 to explore the Mahdia site and recover the sculptures that remained. As in the case of the earlier discovery, the ship was identified as one carrying a cargo of loot to Rome from Greece. It apparently had been blown 500 miles off its course and had foundered in sight of a sandy beach on which it might have been saved.

Through the efforts of a group of Greek divers and one lone Turk, a collection of bronze and marble sculpture was recovered that led Reinach to compare the discovery to that of Pompeii and Herculaneum. Most of the sculptures proved to be of a late decadent style, but a fine winged figure (Eros, or possibly Agon) and a herm of Dionysius signed by Boëthos of Chalcedon elevated the quality of the find. This collection now rests in the Alaoui Museum at Bardo, Tunis.

Freshwater Discoveries

The domain of the underwater archaeologist includes rivers, lakes, and other small bodies of water, as well as the seas of the world. In fact, aside from the finds at Antikythēra and Mahdia, the most significant discoveries of the first decades of the twentieth century were made in fresh, rather than salt, water.

The first, and richest, of these was made by Edward H. Thompson, United States Consular Agent in Mexico, who began exploring the *cenote* at the ancient Mayan city of Chichén Itzá in 1904. Thompson himself learned to use a diving suit in order to explore the *cenote,* a very deep and very narrow sacred well into which the Maya had thrown sacrificial victims and many precious objects of gold and jade. He also obtained the services of two Greek sponge divers and erected a derrick from which a bucket could be lowered into any part of the *cenote*.

Thompson's haul was breathtaking. Bucket after bucket of the black mud from the *cenote* was shot full of ancient artifacts. Gold objects abounded: a jar twelve inches in diameter; more than forty disks, from four to ten inches in diameter; a tiara; many gold rings, and one hundred gold bells of various forms. Jade objects included seven slabs, three to four inches in diameter; fourteen spheres; thousands of points, knives, and beads; and an exquisitely carved figure that has been called the finest small Mayan figure yet found.

This was by no means systematic archaeology, but in the muddy soup at the bottom of the well, it is doubtful that systematic techniques could have been practiced or that they would have been worth the effort. The objects probably settled into the mud in disorder, obviating the value of carefully determining their positions. In any event, the find fully justified Thompson's faith in the sixteenth-century account of the Yucatán by Diego de Landa, bishop of that province, whose writings included a description of the sacrifices. The clincher was Thompson's discovery of human bones in the well. As the treasures came to light, he packed them in diplomatic pouches and sent them to the Peabody Museum at Harvard, where many of them still repose. In 1959 a portion of the collection was returned to Mexico as a gift to the nation.

Less valuable than the Chichén Itzá find monetarily, but perhaps more valuable historically, was the exploration of the Montespan Cave in 1922 and 1923 by Norbert Casteret. On his first entry into this cave, which is on the edge of the French Pyrenees, Casteret took one of the greatest gambles that a diver can take. Coming to a pool where the cave roof slanted down into the water, he stuck his candle on a nearby rock and plunged in, although he had no way of knowing how far the water flowed beneath the rock. He had no equipment and might have exhausted the two minutes' maximum time during which he could have held his breath. Or he might have become entangled in tree branches, or popped up into a pocket of poisonous air. But Casteret's luck was good; the rock barrier was a short one and he surfaced in what he sensed to be a very large cave. The following day, Casteret returned to the cave, this time with a candle and matches wrapped in a bathing cap. Diving beneath the barrier once more, he proceeded to explore a series of corridors and chambers down to the 1,000-foot level, in the process negotiating another barrier that was longer than the first.

The next year Casteret returned with a friend to Montespan. This time, instead of making a right turn after emerging from a large chamber just beyond the first rock barrier, which was now dry, Casteret and his companion turned left, and in so doing stepped into the Stone Age. Following an incline with rock outcroppings that formed a natural staircase, the two reached a large, tunnel-like chamber 600 feet long. Here they found their first evidence of human occupation, a flint tool. Holding their candles high, the two proceeded onward, suddenly coming upon a startling sight: a large bear, formed of clay. Between the front paws lay a bear skull, and the whole thing was covered with a calcareous encrustation deposited by the dripping of water through thousands of years. The explorers surmised that the effigy had been covered with a skin and that the skull had been in place on the neck when the image was ceremoniously attacked with spears. The clay was pierced in numerous places.

As Casteret and his companion moved around the chamber, the light revealed other sculptures, which were in high relief or cut into the walls. These included horses, ibex, asses, bison, stags, and other animals now extinct in this region. Included were lions and tigers, also formed of clay. Only one human

figure was found, a profile incised in the cave wall near the end of the chamber. Casteret's initial venture into the cave still stands as an unequaled example of underwater exploration in the most difficult and dangerous of conditions. The feat was accomplished on lung power and courage alone.

A third important freshwater discovery of this period was that of two Roman galleys in Lake Nemi, seventeen miles south of the Italian capital. The existence of these galleys was known according to a tradition extending back to the early days of the Roman Empire. Several attempts were made prior to the twentieth century to raise them. In 1446, in what apparently was the first such effort, Leon Battista Alberti devised a scheme for hooking one of the ships with grapnels and floating it by means of a raft of barrels. All he succeeded in doing was to bring a large statue to the surface. Then, in 1535, Francesco Demarchi explored one of the ships, finding the grapnels left from the previous century's attempt. Demarchi reportedly used a crude diving helmet. This must have been a small diving tub without an air supply, for the concept of providing diving gear with air under pressure had not yet been developed.

Now in the Smithsonian Institution, the *Philadelphia* is the only intact warship of the eighteenth century in the Western Hemisphere. She was sunk in Lake Champlain during the Battle of Valcour Island. (Smithsonian Institution)

There were no further known attempts until 1827, when Annesio Fusconi, a hydraulic engineer, used a large Halley diving bell to explore the site. While also failing to raise one of the ships, Fusconi did recover a collection of artifacts and fragments, which found their way to the Vatican Museum.

The Italian government finally declared the galleys off-limits, following the incursion in 1895 of a Roman antiquities dealer, who hired a diver to bring up sculpture, tiles, parts of mosaics, and other fragments. Suggestions were made for recovering the galleys by draining the lake, but no action was taken until 1928, when the Mussolini government undertook the project. After many months of pumping, the two galleys finally came into sight, with the result that previous concepts of the construction of ancient ships had to be altered radically. The ships were large even by modern standards—234 by 64 feet, and 230 by 78 feet. They were decorated opulently with marble and bronze columns, mosaics, and marble paving; they probably even had heated baths. The fittings, anchors, and blocks were surprisingly similar to their twentieth-century counterparts in form and function. After such a long history, it is tragic that the Lake Nemi ships were senselessly destroyed by fires set by German troops in 1944.

American Beginnings

The first major underwater archaeological discovery in the United States was the finding and raising in 1935 of the *Philadelphia*, which had been sunk on October 11, 1776, during the battle of Valcour Island on Lake Champlain.

The *Philadelphia*—a gundelow, or bargelike vessel with a single mast—was one of a small squadron of ships hastily constructed by an American army under General Benedict Arnold, and then launched to counter a British force moving down Lake Champlain toward Fort Ticonderoga in an attempt to isolate rebellious New England from the colonies to the south. Although tactically defeated, the Americans gained a strategic victory, since the British force was unable to continue its advance after the battle. Thus the British plan to divide the colonies was delayed for a critical winter season, and the next year, when a British army under General John Burgoyne thrust south with the same intention, it was met and defeated by a stronger American army at Saratoga. Valcour Island was therefore one of the decisive battles of the Revolution.

The *Philadelphia* was found as a result of a thorough search of documentary evidence about the battle by Colonel Lorenzo Hagglund, a veteran salvage engineer, whose recovery of timber remains of the *Royal Savage*, the American flagship at Valcour Island, had led him to believe that the *Philadelphia* might also be discovered. Hagglund's approach was unusual for the field of underwater archaeology. Most discoveries of significant wrecks have been made by accident and then identified by consulting written records. Oddly enough, this is true even today, when the explorer has at his disposal sophisticated electronic equipment for searching cloudy water and muddy bottoms.

After his study of the documentary evidence, Hagglund had only to search the site for a short time before locating the *Philadelphia*. The ship was resting in a depression in the mud bottom, her mainmast erect, in just sixty feet of water between Valcour Island and the New York shore. This is how he described his first view:

Here there is a brown shadow in front of us. We advance toward it, and it takes shape. It is the hull of a vessel. We are approaching the stern. Before us, partly buried in the mud lies the rudder. . . . Now we are abreast of her mast. It is still standing upright. Here are two poles projecting out over the rail. One is twenty-seven feet long and over five inches in diameter at the center. This is her main yard. The other, twenty feet long, is her topsail yard. They had hung on that mast through the century, and finally, when ropes which secured them had rotted away, they came to rest waterlogged, across her gunwales. We are now approaching the blunt bow. Just above the mud line there is a hole in her side through the outer planking, a shattered rib and the inner planking. . . . We have arrived at the bow. In place of a bowsprit, we find a cannon with a peculiar shaped object fixed in the muzzle. This object, now covered thickly with rust, is a bar shot . . . as the Philadelphia went down bow first, this bar shot slid forward and half out of the muzzle . . .

Like the bow gun, the two broadside guns were still standing on their carriages, and they were later found to be loaded. The brick cookstove was found intact with rusting pots and pans on it. Scattered about the deck were remains of muskets, uniform buttons, and other objects of daily use. Recovery operations began immediately. First the guns were raised with floats and a jury-rigged boom. A large floating crane was brought to the site, and on August 9, 1935, after 159 years in the mud, the *Philadelphia* broke the surface of the lake. Subsequently she was cleaned and placed in a building at the edge of the lake, where she was seen by thousands of visitors during the following twenty-five years.

In the summer of 1961 the ship was transported to Washington, where she was placed in the Smithsonian Institution by terms of Colonel Hagglund's will. The *Philadelphia* is the only intact warship of the eighteenth century in the Western Hemisphere, and it is the most significant naval relic of the United States in existence today.

At about the same time that this ship was raised, another important underwater site of Revolutionary times was being explored at Yorktown, Virginia. During the siege of the town in October 1781 (a British army under General Charles Cornwallis had been trapped here by the combined American and French armies of George Washington and the Comte de Rochambeau) French batteries firing red-hot shot sank the frigate *Charon* and three transports moored in the York River. Subsequently, the British scuttled two more frigates to prevent their capture. For the next 150 years the ships lay on the bottom, their memory kept alive only by the fishermen who continually snagged their nets on the sunken hulks.

In the early 1930's Joseph Holtzbach, director of the newly established Mariner's Museum at Warwick, Virginia, became interested in the sites and with Floyd Flickinger, superintendent of the Yorktown reconstruction project, planned a recovery operation on the hulks. With the help of professional divers, a large collection of artifacts was recovered over a three-year period. The murky water and the nearness of the hulks to one another prevented accurate identification of the specimens with any particular ship. Nevertheless, since all the sites date from a single event and are of one origin, the recovered artifacts form a valuable reference collection for dating similar materials from other wrecks. The collection includes bottles; redware pots; barrel staves and heads, some bearing names of contractors who furnished provisions to the Royal Navy; iron guns, including swivels; iron fittings from ships' rigging; a ship's bell; and two iron anchors. The operation utilized a grab operated from a crane—salvage in the crudest sense, but it was a pioneering effort and conducted under extremely poor diving conditions.

Underwater exploration of a much richer area archaeologically, the Florida Straits, was begun in 1938 by Charles M. Brookfield, an ornithologist and student of Florida history. Brookfield and his neighbors were rewarded with some interesting information for having helped two fishermen who had arrived at his dock on Elliott Key in December without food and water. One of the men reported having seen coral- and sand-encrusted cannon resting in the reef and offered to reveal the location for a consideration. Within a day Brookfield, two neighbors, and Walter

Williamson, a diver, were on their way with the two fishermen to the site, some twelve miles south of Elliott Key. Looking into the water through a waterglass—a wooden box with a glass bottom—the party spotted a number of large guns scattered about the coral bottom. Of course everyone concluded that this was the site of a Spanish galleon. Williamson went down in the swirling water and managed to determine that the guns were iron.

In operations during the following year, many guns were recovered. While the coral-sand crust was being removed, a copper coin was found—a halfpenny of England dated 1694. Following this, hand grenades, bar shot, solid iron shot, and a silver porringer were brought up. The guns were found to bear a broad arrow and crowned rose, evidence that led finally to identification of the wreck as that of H.M.S. *Winchester*, a frigate of sixty guns, which sank on the reef on September 24, 1695. The cause of the wreck was scurvy, which so decimated the crew that the ship ran helplessly aground because there were not enough hands to work her properly. Perhaps the most startling discovery on the site was of fragments from a prayer book, still legible after nearly 250 years in salt water.

A New Age in Diving

World War II was the beginning of a new age in diving. When the war started, deep-water divers were still encased in the old canvas dress, heavy copper helmet, weighted shoes, and heavy lead-weight belt that had been standard since the middle of the nineteenth century. Improvements had been made in techniques, and knowledge of diving physiology had expanded remarkably, but the diver still went down to work on a stage and his movements were greatly restricted by the heavy gear. Three principal types of gear were in use: (1) The open helmet, essentially a small diving bell, which sat upon the shoulders of the diver and received a constant supply of air from a surface pump; while useful for work in depths to fifty feet, the diver had to take care to remain upright at all times in order not to spill the air from the helmet. (2) The closed canvas dress, supplied with air under pressure equal to that of the water surrounding the diver; it gave the diver more freedom of movement and was useful in moderate depths. (3) The armored suit, supplied with air under atmospheric pressure, the diver being protected from the water pressure by the strength of the heavy steel dress; this was used for work in depths greater than 150 to 200 feet (a diver in an armored suit salvaged gold from the sunken steamship *Egypt,* which lay in about 400 feet of water).

Two developments during the war changed all this, suddenly freeing divers from such encumbrances and enabling them to swim about with the freedom of a fish. In France, Commandant Jacques-Yves Cousteau and Emile Gagnan developed the aqualung, using the concept of the demand valve—a concept similar to that used in the Denayrouze-Rouquayrol apparatus of the previous century. The result was an apparatus that has a range of depth to 200 feet or more on air and a submersion time of up to two or more hours, depending on depth and on the number and capacity of air bottles. In the United States, the inventor Jack Browne developed a lightweight rubber mask with a manual valve that was supplied by an air hose attached to a surface compressor. With this apparatus the diver has a range of depth to 100 feet. While the length of the hose restricts the diver to a limited area, this area is almost always large enough to permit thorough exploration of an older shipwreck site, and the advantage of a continuous high volume of air increases the time the diver is able to remain underwater and permits him to engage in extended hard physical labor.

The result of these two inventions was nothing less than a complete revolution in diving. Since the end of World War II a new generation of divers has grown up knowing only these improved methods, and scientists, historians, and archaeologists now go regularly beneath the sea to pursue their studies as routinely as they formally followed them in the laboratory, library, and land dig.

Systematic underwater archaeology had its birth in the Florida Straits and the Mediterranean. Exploration of underwater wrecks in the Florida Straits began in the late 1940's with the work of an experienced diver, Arthur McKee, who investigated a number of sites and recovered from one of them three large Spanish-American silver bars dating from the 1640's. This discovery naturally helped spur interest in wreck diving, particularly at the Smithsonian Institution, which acquired one of the bars found by McKee. In 1951, I was invited by Mr. and Mrs. George Crile, Jr., of Cleveland to go on a small expedition to explore wreck sites that they had been probing with Bill Thompson of Marathon, Florida, during the previous two years. Setting out in May, the expedition was joined by James Rand, Jr.; Dan Moore of Cleveland; John Shaheen of New York; and Edwin A. Link, inventor of the Link Trainer for instructing airplane pilots. The expedition proved to be a step forward not only for underwater exploration

in the Western Hemisphere but for diving in general, since it resulted in Link's entry into a field in which he is now making major contributions, especially in the area of deep diving. This expedition also marked the entry of the Smithsonian Institution into a modest program of underwater exploration, which has continued to this day.

Working through the month of June, the expedition devoted major attention to the site of a ship that went down in a channel between two steep coral reefs in the central Keys. With Arthur McKee furnishing the professional diving assistance, as well as many practical ideas that have since been utilized on Smithsonian expeditions, the group probed the rough coral bottom in the channel, recovering masses of encrusted iron and one long iron six-pounder. The symbols on the gun—a crowned rose—and on numerous solid iron shot—a broad arrow—indicated a British vessel. The condition of the gun and the evidence of the crowned rose led us to the conclusion that the ship had gone down before 1750. The name of the reef, "Looe," on British naval records proved to be significant, and the ship was identified as the 44-gun frigate *Looe,* which sank February 5, 1744.

During the operation, I attempted in two short dives to apply one of the standard techniques of land archaeology to an underwater site by laying out a grid pattern to permit systematic excavation of the wreck. This proved to be impossible due to the rough underwater terrain. An important first in underwater archaeology, however, did occur during this operation: An airlift was used to remove the sand that had accumulated on top of the sunken ship. This instrument, which operates like a powerful underwater vacuum cleaner, has since been improved by Link and others, and it is now a standard tool for underwater archaeologists throughout the world.

In 1952, 1953, and 1954, Link supported further underwater explorations in Florida in cooperation with McKee. The base of operations for these investigations was Link's new vessel, the *Sea Diver.* Several sites were explored with sophisticated electronic equipment, including a magnetometer for locating buried iron objects by means of their minute magnetic fields, and a new underwater detector, developed by Link on the principles of the conventional landmine detector. The most notable of the sites investigated was off Plantation Key. This site, originally spotted by McKee, proved to be the wreck of the *Rui,* flagship of the 1733 Plate Fleet. During the summers of 1953 and 1954 this wreck produced significant collections of ship fittings; weapons, including iron guns from the ship's main battery; pewter; tools; silver

Man Beneath the Sea

The undersea world is man's nearest frontier, and also his most mysterious. It lies at his feet, but remains virtually unknown and unexplored. Man, whose remote ancestors lived in the ocean, can return to it only if he brings his own atmosphere with him and if he finds a way of withstanding the pressures that increase with water depth.

For almost the entire span of history man has been restricted to short plunges beneath the surface. By developing their lung power, Greek divers of ancient times were able to gather sponges at depths of seventy-five to a hundred feet; they also salvaged wrecks and served on underwater demolition teams in wartime. In the Pacific, pearl divers have traditionally reached even greater depths, using no equipment but a stone to speed their descent.

The earliest design for a diving suit dates from a medieval manuscript illustration now in the Munich Library. It shows a waterproof suit with a leather tube extending to the surface. During the Renaissance Leonardo da Vinci sketched a diving suit with a control valve to prevent water from entering the tube. Nicholas Tartaglia, a sixteenth-century mathematician, devised a diving platform topped by a large glass globe for the diver's head. None of these designs was practical, however, for even the great Leonardo failed to recognize that the air pressure within a suit or bell must be made to equal the water pressure on the outside.

The first true diving bell was described in a French scientific journal in 1678 by Jean Christophe Sturmius. Two generations later, English astronomer Sir Edmund Halley brought the diving bell to very nearly its modern form. In Halley's device the diver's breathing supply was renewed with air from two auxiliary bells positioned below the main bell.

Although the diving bell extended the time a man could remain below the surface, his mobility was restricted to a platform. This limitation was overcome by Augustus Siebe, a German artillery officer who emigrated to England after the Napoleonic Wars. Siebe built two models, each with metal helmets and canvas bodies: the "open dress," which extended only to the diver's waist, and the "closed dress," which covered his entire body. In each case air was supplied to the diver at the proper pressure from a surface pump. The worth of the closed suit was proved in 1844 in the salvage of the guns of the *Royal George*, flagship of the British navy, which had sunk in Portsmouth harbor more than sixty years before.

World War II brought a revolution in diving when Emile Gagnan and Jacques-Yves Cousteau invented the aqualung. This piggy-back air tank, with a demand valve to regulate pressure, freed the diver to swim about without the restrictions of a surface air supply. Convenient, transportable, and relatively inexpensive, the aqualung not only facilitated underwater exploration, but also helped create a new generation of recreational divers.

Modern divers have at their disposal a wide variety of tools for surveying and excavating the ocean bottom. The airlift, perfected by Edwin A. Link, vacuum-cleans mud or debris that has built up over an archaeological site. Sonar scanners and mud pingers penetrate murky water to reveal materials different from the surrounding sediment. Magnetometers detect the presence of iron, a valuable clue in locating an ancient ship.

Small research submarines, especially designed to explore and salvage the ocean bottom, now take man to the deepest regions. The bathyscaphe *Trieste* set a depth record in 1958 when it descended seven miles down in the Challenger Deep, near the island of Guam in the western Pacific. The little submersibles *Alvin* and *Aluminaut* played important roles in retrieving the lost hydrogen bomb off the coast of Palomares, Spain, in 1966. Deep Sea Rescue Vehicles, now under construction by the U.S. Navy, are designed to match the capabilities of its new submarines.

The final step in man's return to the sea will be accomplished when he is able to live and work in an underwater environment for extended periods of time. Much still has to be learned about the physiology and psychology of undersea living, but knowledge is being accumulated rapidly. A group of six "oceanauts" under Cousteau's direction has lived for three weeks at a depth of 325 feet. The U.S. Navy is submersing teams of "aquanauts" in its Sealab habitat projects. These man-in-the-sea experiments represent man's most ambitious efforts to adapt himself to the last outpost on Planet Earth.

Ancient divers in Ceylon dove for pearls at depths of
more than a hundred feet. The weights sped their descent
to the bottom. (Matthew Kalmenoff)

The effects of water pressure eluded Italian mathematician Nicholas Tartaglia, inventor of this diving platform. His device had one major defect: the water level would eventually rise above the diver's nose as the air was compressed within the glass globe. (Matthew Kalmenoff)

Better known for tracking the orbits of comets, Sir Edmund Halley devised a diving bell with auxiliary air containers connected to the main bell by a leather tube. An improved model of the Halley diving bell, dating from the late seventeenth century, is shown here. (Matthew Kalmenoff)

Divers of the Royal Sappers and Miners, wearing closed diving suits, successfully salvaged the guns of the *Royal George* in 1844—an operation others had failed to accomplish for more than sixty years. The 108-gun flagship had rolled over and sunk while at anchor in Portsmouth Harbor. (Alan Albright, Smithsonian Institution; Lithograph in Mendel Peterson's Collection)

A modern diver performs an undersea welding operation. Hard helmets and surface air are used when the diver needs a protective covering and works within a prescribed radius. (Ocean Systems, Inc.)

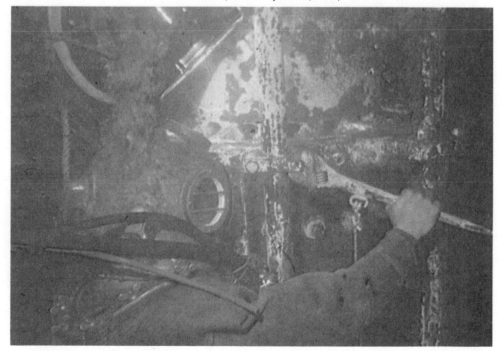

Divers with aqualungs make adjustments on *Deepstar 4000*'s stereo camera. A World War II invention, scuba tanks have a submersion time of up to two hours or more and a depth range of 200 feet. (Ron Church)

A diver excavates a site with an airlift. The outflow is piped to a point clear of the site, then sifted to recover any objects that may have inadvertently been sucked up. (Ocean Systems, Inc.)

A modern descendent of the diving bell, a welding chamber is lowered into Raritan Bay, south of Staten Island, New York. (Ocean Systems, Inc.)

Workboat submarine *Beaver IV* was built to transfer personnel at undersea stations to depths of 2,000 feet. A diver lockout hatch, effective to 1,000 feet, is located on the lower portion of the after section of the hull. (North American Rockwell)

With the second team of aquanauts inside, the Sealab II
personnel transfer capsule is swung onto the deck of the
support vessel. Moments later the aquanauts entered the
decompression chamber. (U.S. Navy)

This cutaway drawing shows a Deep Sea Rescue Vehicle
performing rescue operations atop a disabled submarine.
Designed to answer emergency calls from the deep, the
DSRV's will be transportable by air and capable of attaching to submarine hatches. (U.S. Navy)

A portent of the time when man will live and work for extended periods in the sea, Edwin A. Link's underwater shelter rests on the ocean bottom off the Bahamas. (© National Geographic Society)

coins produced in 1732 and 1733 at the Mexico City mint; and ceramic fragments, including Chinese porcelain brought from the Far East by the Manila Galleon (the term applied to the ship, or sometimes, ships, that sailed between Manila and Acapulco) and then sent by land to Vera Cruz, where they were transferred to the flagship of the convoy bound for Spain. The objects from the *Rui* are now divided between McKee's museum in Florida and the Smithsonian Institution, where they are displayed in a hall of underwater exploration.

In the Mediterranean, the first underwater dig to proceed with some care was that of Grand Congloué—a frowning, cliff-girt island lying off the ancient port of Marseilles. In 1951 a local diver, ignoring the decompression tables, suffered a severe case of the bends. His life was saved by prompt treatment by the Groupe d'Etude et de Recherches Sousmarines, and he expressed his thanks to the group by giving Commandant Cousteau and Frederic Dumas the location of what appeared to be the site of an ancient wreck. The next year an expedition led by Cousteau and Dumas commenced work on the site.

The problems were great, but the solutions were imaginative. The wreck lay at the foot of a precipitous cliff in about 140 feet of water—a great mound of gravel and clay from which the necks of amphorae protruded. On the mound lay boulders that had fallen from the cliff through the centuries. The first day on the site, Cousteau brought up three drinking cups of Campanian ware. This gave Professor Fernand Benoît, Director of Antiquities for Provence, the first evidence of the antiquity of the ship—200 B.C.

At first Cousteau's ship, the *Calypso,* was used as a platform for diving operations. Since it was apparent that the mound covering the wreck was too large to be removed by hand, an airlift was installed on the ship. Working in two-man shifts that averaged fifteen minutes' duration, the divers swept the mound with the airlift's five-inch diameter semiflexible hose, bringing up dozens of intact amphorae each day. The cargo of the ship rested in layers in the order in which it had been loaded, with amphorae from Delos on the bottom and Roman jars and Campanian ware on top, thus establishing the chronology of the fatal voyage.

Since the airlift could be used from the *Calypso* only when the water was calm, the onset of winter weather made it necessary to transfer the base of operations to the island. A working platform with the pumps was set up on the cliff, while shelters on the island provided living quarters. From this base work continued for five years, during which a tre-

Working from his oceanographic boat *Calypso,* Jacques-Yves Cousteau conducted the first archaeological dig on an ancient ship off the French island of Grand Congloué. In the ship he found amphorae and ceramics dating from 200 B.C. (French Embassy Press & Information Division)

mendous collection of amphorae, Campanian ceramics, ship's equipment, and portions of the hull near the keel was recovered. Examination of hull sections revealed that the ancient shipwrights had used many sophisticated construction methods, including lead sheathing, insulated nails to prevent electrolytic damage, and different woods with different characteristics for various parts of the ship. More than 130 different forms of Campanian ware—much of it in almost mint condition—were found, greatly multiplying our knowledge of this ware. In sum total, this dig—the first comprehensive archaeological dig on a large merchant ship—increased tremendously our knowledge of Mediterranean trade of the second century before Christ and provided scholars with important material that will be studied for years to come.

The Bermuda Treasures

The relatively low cost of aqualung equipment helped make diving a popular sport in the postwar years, with the result that amateur divers began making finds at a rate that startled—and sometimes frustrated—professional marine archaeologists, who not unreasonably feared that sites would be plundered instead of systematically excavated. The vigorous community of amateur scuba divers, nevertheless, plays an important role by providing a wide-ranging and efficient detection apparatus for its professional counterpart.

The most important of the many amateur finds of the postwar period was that of a young Bermudian, Edward Bolton (Teddy) Tucker, and his brother-in-law, Robert Canton. While salvaging metals from modern wrecks in 1950, Tucker had found a sand hole that contained five encrusted iron cannon and a huge copper pot full of lead musket balls. These he raised and sold to the museum of the government of Bermuda. Returning to the site late in the summer of 1955, he discovered gold bars, ingots, pearl-studded buttons, and an emerald cross, as well as additional military equipment and other artifacts. I first learned of their discovery in the fall of 1955, when *Life* asked me to inspect and, if possible, authenticate photographs of gold bars and ingots, pearl-studded buttons, and an emerald-studded pectoral cross that the magazine had just received from Bermuda. When the pictures were laid before me

in the New York office, I knew immediately that this must be the most remarkable underwater discovery since William Phipps recovered the cargo of a sunken Spanish galleon in the 1680's. The color transparencies of the gold bullion clearly showed the marks that should have been there; the pearl-studded buttons were of the right design and workmanship; and the emerald cross fairly leaped out from the transparency, the emeralds flashing green fire. I assured the editors that all the evidence indicated the objects to be authentic, and I immediately determined to go to Bermuda to see these wonders.

In November I joined Mr. and Mrs. Link and flew to Bermuda. Here, in a private home in Paget, we saw the great treasure of Spanish gold and jewels: a forty-ounce bar of gold; two round gold ingots, four inches across and half an inch thick; two cube-like sections of gold bar; several single and triple pearl buttons; and, dominating this impressive and glittering array, the emerald cross—breathtaking in the fine color photograph I had seen in New York, unbelievable in actuality. The finely wrought cross,

Edward Tucker operates a small airlift at the site of a sixteenth-century ship he discovered on the northern reefs of Bermuda. The ship yielded a treasure in gold, gems, weapons, and marine artifacts. (Smithsonian Institution)

some four and a half inches high, was studded with seven fiery emeralds—symbolizing the seven last words of Christ—of the finest color and luster. Three small gold nails once hung from the arms and foot of the cross; two were still in place. The back of the cross was riveted on, and when Tucker carefully removed it, a small cavity that once held a holy relic was revealed. The gold inside was stained a reddish tone from some organic matter. At the time I did not know the monetary value of the cross, but I did realize that it was great enough to make it the most valuable single object to have come from the sea in modern times.

After drinking in this overwhelming sight for an hour or so and hearing Tucker's fascinating tale of how the treasure was found, we went across the harbor to a storage building and there saw some of the artifacts from the vessel: pewter, remains of matchlock muskets and arquebusses, a breastplate, black palm-wood bows, and a beautifully carved staff of office for an Indian chief—one of the very few to have survived, a prize of some ethnologically inclined collector who died in the wreck. Navigators' compasses and brass-enclosed time glasses were impressive testimony to the skill of the instrument makers of the time. A heavy brass sheave, or pulley, which once rested on a shaft at the head of the mainmast and received a line used to run up the top mast, had survived unscathed and only lightly corroded. A pewter cylinder with a lock-on lid pierced with a one-inch hole was a mystery. Tucker said that rotted wood was found in the cylinder. I guessed a lighting implement but couldn't explain the wood. Only after years of puzzling over it was the implement identified when I saw a similar one exhibited in a Smithsonian hall. It was an enema pump with the nozzle missing. We also observed the cast-iron guns from the wreck—some of the earliest to come from western waters—and a huge copper pot full of musket and arquebus balls. The collection taken as a whole is one of the most important of the period to have come from any site, wet or dry. The fact that it was from a single ship added tremendously to its value to historians, for here was a neat slice of late sixteenth-century maritime life, history in three dimensions, dropped into the sand and preserved for modern historians to study.

This initial discovery started an interest in underwater exploration that continues in Bermuda to this day; each year new and important finds come to light. In 1957 the author began working with Tucker and Canton on Bermuda sites, and this collaboration continues to the present.

In 1958 Tucker discovered the site of the *San*

This jar was found in the *San Antonio* wreck off Bermuda. Like the ancient Greek and Roman amphorae found in the Mediterranean, it does not have a flat bottom and must be wedged or racked in position to stand up. (Smithsonian Institution)

Antonio, a large Spanish vessel that was wrecked in a storm in 1621. Four years of exploring the site produced a significant collection of ship's equipment, armament, cargo, and personal effects of the passengers and crew. The sinking of this ship is well documented in the archives of Bermuda. Going aground in September during a violent storm, it sank by the bow, but the stern stayed above water. The crew mutinied and escaped in the only boat. The passengers and officers spent the succeeding night in making a large raft from the ship's timbers, on which they floated to safety ashore the next day. The Bermudians immediately invaded the wreck, removing everything in sight, and the Spaniards ashore were fleeced of the valuables they had saved for the expenses of getting them back to Spain.

For two years, I explored the site with Tucker and examined the collection he found. While gold and jewelry—including a fine emerald ring—was discovered, perhaps the most important historical material was furnished by the traces of ordnance stores and the cargo the vessel carried. New types of wired lead balls for the small arms were discovered, as was an unknown type of wooden case shot for the heavy

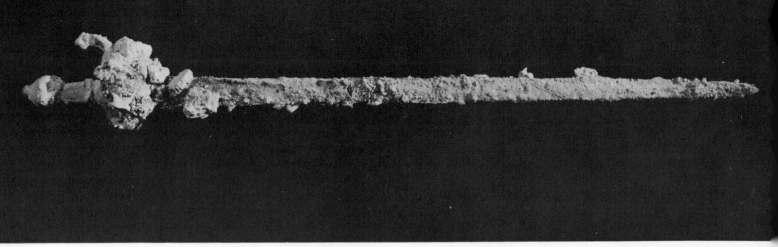

A Spanish battle sword was raised in this beautifully encrusted condition from the Tucker treasure site. To protect it from oxidation, it has been placed in a clear plastic block. (Smithsonian Institution)

guns. Evidence of the armorer's work aboard in the casting of lead balls, knapping of flint, and assembling of the wooden case shot gave us a new insight into an obscure phase of naval warfare. The cargo was represented by finds of indigo and cochineal dyes, tobacco rolled in cones around sticks, lignum vitae, and tortoise shell. The most surprising item of cargo was a shipment of several thousand cowrie shells, which was being sent from the southwest Pacific to Spain by way of Manila, Acapulco, Vera Cruz, and Havana. The ultimate destination of the shells would have been Africa, where they were used as money, and with them the Spanish would have purchased slaves.

In 1962 an exploratory probe was made into a site discovered by Tucker near the *San Antonio*. Ballast stones were found under four or five feet of sand, and there was evidence of a solid ship's bottom below that. One small clump of silver coins, almost completely converted to silver sulfide, was found. Upon examination, the coins proved to date from the middle 1500's. In 1963 a thorough study of the site was

The reef flats off Bermuda are one of the world's great graveyards for ships. A mid-seventeenth century chart of the island is shown here. (Smithsonian Institution)

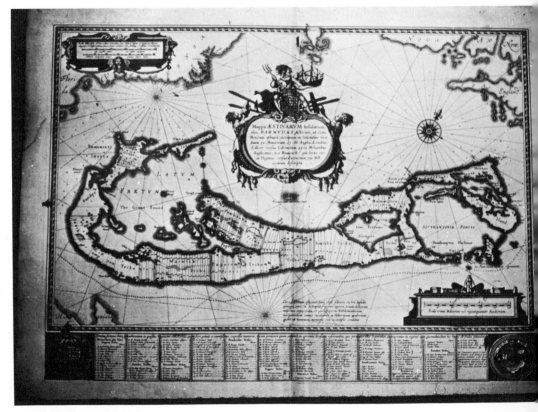

An airlift sucks up covering layers of mud and sand in the coral-reef area of Bermuda. (Alan Albright, Smithsonian Institution)

Divers with aqualungs explore a wreck site in the Bermuda reefs. (Smithsonian Institution)

made by an expedition sponsored jointly by the Smithsonian Institution and the National Geographic Society. The overburden was removed, the timber remains carefully measured, and a plan drawn. New equipment for the precise measuring of curves underwater provided an accurate means of determining the exact elevation of the timber remains at any given point. While the site yielded very few artifacts, the study of the timber remains provided valuable information on ship construction of the period.

Since the excavation of this area, the exploration of wreck sites in Bermuda has continued, aided by the introduction of sensitive electronic devices to survey the entire reef area. In 1967 the Explorer's Research Corporation sponsored a comprehensive electronic survey of the reef flats and discovered and identified the English ship *Warwick*, which sank in Castle Harbor in 1619. The timbers of this ship are preserved from the keel to one gunwale and lie protected by deep silt awaiting further study.

Northern Europe

Underwater archaeology has by no means been limited to the warm waters of the Mediterranean, Florida, and Bermuda. Since the mid-nineteenth century, for example, there have been historic finds around the island of Anholt, which lies in the strait between Jutland and Sweden. In 1847 eight ancient iron guns were recovered, five of which are now in the Tojhusmuseet in Copenhagen. Ninety years later a diver looking for wrecks for salvage found the site of a wreck that was about ninety feet long by twenty-five feet wide. He recovered six breech-loading iron pieces, several cast iron muzzle-loading guns, quantities of cast iron shot, and a large number of wrought iron bars. It was later concluded that the breech-loading pieces were the ship's battery, while the muzzle-loading guns, iron shot, and bars were cargo bound for Denmark from England (a lead seal identified as English had been found on the site). Frederick II of Denmark is known to have bought English cast iron guns from an English merchant named Foxall between 1563 and 1570. At this time the English founders had perfected iron gun casting and had become suppliers of heavy artillery to continental Europe. In 1942 a third find of guns was made in the Anholt area, and these are also in the Tojhusmuseet.

While none of the Anholt sites was systematically excavated and described, these finds served as a prelude to one of the most important and dramatic feats of modern underwater archaeology: the raising of an entire ship, the *Vasa*.

During the centuries following the salvaging of its guns in the 1660's, the location of the *Vasa*, and even its existence, had gradually been forgotten. For almost three hundred years the ship lay unremembered in her grave until a young Swedish government employee, Anders Franzen, began an investigation, first in the archives and then in Stockholm Harbor. In 1956 he began to probe likely locations in the harbor, which had changed markedly since the disaster. In the summer of that year, using a device of his own design—a punch attached to a weight— he brought up blackened oak plugs from a location that his research told him was the most likely resting place of the *Vasa*. That fall the Swedish navy, at the urging of Franzen, began to use the site for training, and by August the divers had found the *Vasa*. The following year a committee was organized for her salvage, and in 1958, with funding assured, the actual salvage effort began.

The ship lay in 110 feet of water, in a thick mud bottom that gripped the hull tenaciously. The problem was one of heavy salvage, with the added complication that the internal structural condition of the ship was not known. With great skill the Swedish divers stopped all the ports and repaired other breaks in the hull, Then, working in pitch-black darkness, the divers cut tunnels through the mud under the hull to receive the cables to be used in raising the ship. While the tunneling was being done, hundreds of carvings and other objects from the ship were found in the mud.

By September 1959 almost the entire hull had been raised and moved to water fifty feet deep. The ship had proved to be strong enough for complete salvage, and by May 1961, she was afloat and was towed into dry dock. Today she rests in the museum built over her. The study of the *Vasa* and the thousands of objects recovered from her is proceeding. The problem of preserving the hundreds of tons of wet timber is being solved through the application of the latest scientific techniques. The wood is preserved by replacing the water in the wood cells with polyethylene glycol—a wax-like, water-soluble substance that supports the cells and permits the wood to dry without distortion through shrinking. The metal objects are being preserved by chemical and physical methods that reduce the corrosion salts to their metallic form. This is accomplished with such devices as a hot blast of hydrogen on the iron to remove the oxygen and restore the metallic iron.

The *Vasa* is a unique find of the greatest historical importance, and it provides a perfect example of the value of archaeology in providing data that documents alone cannot. Here we have a complete package of maritime history in remarkable condition and containing everything that went into the maintenance and operation of a large warship of the early seventeenth century. The study of the ship and the collections from her will answer many questions that have been asked by maritime historians.

One year after the discovery of the *Vasa*, Danish archaeologists from the National Museum in Copenhagen began the investigation of an ancient ship site that had been located in Roskilde Fjord. Their purpose was to get experience in underwater excavation techniques. The site had been found by fishermen some twenty years earlier when they had broken a way through it to make a channel.

Working through three seasons, the divers found the site to be from the Viking Age and to consist of not one but five ships, all of different types. They apparently had been sunk together to make a defense barrier. Because of strong currents and poor visibility, the archaeologists concluded that an underwater excavation of the valuable site was not practical. In 1962, therefore, a coffer dam of locking steel-sheet piling was built around the site and the area pumped dry. By July of that year, excavation began. The five ships proved to be merchant ships of three different sizes, a warship, and a vessel used as a ferry on inland waters. This find has greatly increased our knowledge of shipbuilding during the Viking Age. Carbon-14 tests place the date of the ships at between A.D. 1000 and 1050. The ships had been stripped before being secured together, and virtually nothing was found in them except the stones used to sink them.

The studies of the *Vasa* and the Roskilde ships

represent an interesting combination of underwater and land archaeology. Underwater techniques were used to locate and raise or isolate the sites, but the final investigation was performed in the classical manner: by excavating both the interior of the *Vasa* and the muddy site in Roskilde provided by the coffer dam. These important finds are tantalizing samples of what can be expected to be found in Baltic waters, where the saline content and the temperature of the water prevent the teredo worm and fungus from living and doing their destructive work. Other sites of intact early ships are now known, and these waters promise to provide highly significant sites.

The increasing participation of Scandinavian representatives in international conferences on underwater archaeology is an indication of the beginning of a period of intensive exploration in the Baltic and the cold waters adjoining it. At the last two conferences of the Council for Underwater Archaeology, papers were delivered by representatives from Sweden, Denmark, Finland, and Germany.

In the English Channel and North Sea, recent discoveries—notably, the sites of the *Association* and the *Gerona*—have brought attention to the possibilities of underwater archaeology there. The *Association*, flagship of Sir Cloudesley Shovel, commander-in-chief of the British fleets, ran on the rocks near the Scilly Isles in October 1707. The *Gerona*, a Spanish Armada vessel wrecked on the coast of Northern Ireland on October 26, 1588, was found in June 1967 by the distinguished Belgian diver Robert Stenuit, who has recovered objects of great historical and intrinsic value from it. Other sites of Armada vessels, now being investigated, also are expected to yield important collections of precisely dated material that will be of great value to historians and archaeologists.

On the other side of the Channel, in the coastal waters of the Netherlands, the draining and reclamation program of the Dutch government is bringing to light vessels of all periods. Here the problem becomes one of land archaeology. Collections have been recovered and are to be found in the museums of the country.

Sunken Cities

The province of the underwater archaeologist includes the sites of ports and harbors that man has built, as well as the ships in which he has sailed.

The first detailed study of an ancient harbor was made by Gaston Jondet, Chief Engineer of Ports and Lights in Egypt, who surveyed the old port of Alexandria prior to construction of new breakwaters in the modern port in 1916. He discovered solid quays, more than 2,000 feet long, and a series of great breakwaters, more than a mile in length, that protected the anchorage. Most of the structures were no more than a few yards below the surface and could be seen easily, so Jondet had to use divers to carry out the mapping in only a few places.

Twenty years later, the first truly archaeological survey of a sunken harbor was made by another Frenchman, Père Antoine Poidebard, who explored the Phoenician port of Tyre in 1934–36. Working with local sponge divers and helmeted divers from the French navy, Poidebard developed new techniques for finding and accurately recording underwater structures. First, aerial photographs of the site were taken and the plans of the structures thus revealed were checked by free divers. Then, underwater photographs were taken and the points marked by buoys for surveying. Detailed examination of the structures was carried out by the helmeted divers, who measured depths, dictated notes from which sketch-plans and sectional drawings were prepared, and made small excavations. The plan developed in this manner by Poidebard revealed that Tyre had had two harbors, one to the north and one to the south of the island on which the ancient city stood. The general arrangement of the harbors, together with that of the outer roadstead, which was protected by long breakwaters, suggested Graeco-Roman construction.

Since World War II a number of other surveys of this sort have been carried out. In 1945 Poidebard initiated a survey of Tyre's sister-city, Sidon. A decade later, in 1955, a team of ten volunteer divers under John Leatham, of the British School of Archaeology at Athens, carried out a survey of sunken structures around the coast of Crete. They discovered the remains of ancient fish tanks at Chersonesos and Machlos, delineated the sunken harbor works at the former site, and explored Psarà and Ayia Galini. Their finds dated from Minoan to Roman times.

In 1958 a Cambridge University team under Nicholas Flemming began a two-year exploration of the Greek colonial city of Apollonia on the coast of the Libyan peninsula of Cyrenaica. Using the modern aqualung and standard surveying instruments, they delineated the harbor works and prepared the first plan of this important settlement. The next year, an expedition led by Edwin A. Link explored the sunken harbor of Caesarea Palestina and discovered the harbor works built by Herod the Great to be much larger than previously supposed. A startling find was a small pewter token with a representation

A 24-pound cannon is raised from the sunken city of Port Royal, Jamaica. The town, which thrived on buccaneer raids on the Spanish Main, slid into the sea during an earthquake in 1692. (Mendel Peterson)

of the harbor buildings—the only known rendering of this ancient city.

In the Americas, the most important sunken city to be explored so far is Port Royal, Jamaica. Unlike the Mediterranean sites, which seem to have been inundated as a result of gradual subsidence of the land (with the possible exception of Apollonia, whose submergence may have been due to an earthquake in A.D. 365), Port Royal was literally dumped into the sea within a matter of minutes. This was a tragedy for the town's inhabitants—2,000 of whom were drowned—but a boon to the modern archaeologist. Since the inhabitants had no chance to escape with their possessions, the site is now the world's richest deposit of seventeenth-century European antiquities.

Port Royal was established by the English after they wrested the island of Jamaica from the Spanish in 1655. The town quickly became a major port, initially for the buccaneers raiding the Spanish Main and then for merchants carrying on less violent, but also illegal, trade with Spanish settlers on the mainland. These activities made the town rich—its per capita wealth was greater than that of any other English-American city, and fine three-story buildings rose on its streets—and they gave it the reputation of being "the wickedest city in the world."

Moralists were not surprised, therefore, when on June 7, 1692, between 11:30 A.M. and noon, a tremendous earthquake, followed by tidal waves, caused two-thirds of the town to slide into the sea. The buildings in the remaining third of the town were tumbled down and in the following weeks more inhabitants fell victim to epidemic disease than had been drowned. The town lay under the water, disturbed through the centuries only by native divers and people in boats who grappled up from the water everything that was not covered by fallen structures.

In 1956 Link and the Smithsonian Institution made a brief reconnaissance of the sunken portion around Fort James, which marked one end of the site. Here were found iron twenty-four-pounders, shot, some remains of ships' lead sheathing, and rubble. The remains were covered by a thick layer of silt, which, when disturbed by flipper or airlift, rendered the surrounding water absolutely opaque. The conclusion was that a further exploration was in order, and Link returned north in his *Sea Diver* to proceed with plans to build a bigger and better salvage and research vessel. The result was *Sea Diver II*, a fine steel-hulled ship with every conceivable device to aid the progress of the underwater explorer.

With his new ship, Link returned to Port Royal in June 1959, in an expedition sponsored by himself, the National Geographic Society, and the Smithsonian Institution. In the ensuing two months an extensive probing of the sunken town from the area of Fort James westward to the area lying off the original governor's house produced a collection of some five hundred objects—cannon; a wrought iron, breech-loading swivel gun; pewter, copper, and brass utensils from a kitchen; pipes, bottles, and drinking vessels from a tavern; and many other objects of interest to the cultural historian. The probes provided information for a fine chart, relating the underwater ruins to the existing town, which was produced with the help of Captain P. V. M. Weems, the noted navigator and instrument designer of Annapolis, Maryland. The most dramatic discovery was a watch, whose hands had left an iron oxide deposit that showed up as a shadow on X-ray film, recording the time of the earthquake: 11:43.

After Link's explorations in 1956 and 1959, the site lay unexplored until 1965, when the Jamaica government retained Robert Marx to explore certain areas of the drowned city that were marked for a modern deep-water port development. Using native assistants and working with light equipment under horrible diving conditions, Marx recovered a remarkable collection of artifacts. Numbering in the thousands, his finds included precious gold jewelry, silver pieces of eight by the hundreds, and silver plate and pewter in many forms. Brass and copper ware and the common red pottery of everyday use came from the wrecked cook houses. The taverns were represented by thousands of clay pipes and large numbers of spirit bottles, many with the corks in them. Some of these objects will force a rethinking of our present ideas on the dating of utensils and bottles and other objects commonly found on historical land sites. The picture of life in a small but comparatively rich city is revealing indeed. This collection, while containing some materials dating from after the disaster, is the finest accumulation of English colonial material in existence. The study of the material will require years and will provide us with a new and more graphic idea of life in colonial times. Rich as this collection is, much more awaits the underwater explorer at Port Royal.

Underwater Archaeology Comes of Age

After the many exciting finds of the 1950's, underwater archaeology began to come of age in the 1960's. Appropriately enough, since the first important modern finds were made in the Mediterranean at Antikythēra, it was the Mediterranean that served as the crucible for the new development. Appropriately, too, the first major site to be studied in the new period proved to be the oldest shipwreck yet discovered.

In 1958 Peter Throckmorton, an American, began searching along the southern coast of Turkey near Bodrum, the nation's sponge center, for the site at which a bronze statue of Demeter had been hauled up in a sponge fisherman's net in 1954. Throckmorton lived and worked with the local sponge divers, particularly Captain Kemel Aras of the village, for two years. The rocky Turkish coast and the islets lying off it proved to be a veritable treasure trove of archaeological material. Throckmorton was either shown, or found himself, a number of wrecks in waters ranging from 100 to 150 feet in depth—most of them covered by mounds of amphorae and their shards. The captain also gave Throckmorton the location of an apparent wreck that presented quite another appearance. Situated between two islands off Cape Gelidonya in some eighty to eighty-five feet of water, the wreck was marked only by a conglomeration of bronze or copper grown into the calcareous seabed. Using Throckmorton's directions, a party of American divers, led by Drayton Cochran, managed to find the site and bring up several copper ingots. The ingots were of the so-called ox-hide shape, the form in which bulk copper was transported in ancient times. This find dated the wreck to the period prior to 1000 B.C.

In 1960 the University Museum of the University of Pennsylvania sent an expedition, led by George Bass, to join Throckmorton in excavating the site. Work began on June 14. After carefully photographing and mapping the area, the divers started in on the wreck itself. They soon realized that it would take them an extremely long time to cut away the

deposits of hard lime—sometimes several inches thick—that had formed over most of the wreck, so it was decided to raise the ingots in large lumps and perform the cutting operation on land. Even this simplified approach involved laborious underwater toil with hammers and chisels. In the case of some exceptionally large fragments, breaking lines were cut by chisels and then the pieces were freed with the aid of an automobile jack. Despite these apparently rough techniques, none of the objects were damaged, and, thanks to the careful plotting that had been done, it was possible after cleaning the cargo to reassemble it on land in the same arrangement in which it had been found.

Fragments of wood from the hull of the vessel and brushwood dunnage were found under the ingots. Subsequent study indicated the vessel to have been a small one, perhaps thirty-five feet long, carrying a ton of copper mainly in the form of hide-shaped ingots weighing about fifty-five pounds each, and a wide variety of other goods. The cargo of the ship included bronze hoes, axes, picks, one shovel, pins, knives, spearheads, and one spit. The site also produced some pottery, as well as scrap metal, unfinished cast tools, and unalloyed tin and copper. Whetstones and a large stone that may have served as an anvil were also found. Personal objects found in one area of the site included a cylinder seal, stone mace-heads, a razor, an oil lamp, balance weights, and five scarabs. From this evidence it was concluded that the crew not only traded in bronze implements but also made them during the course of the voyage. While the wreck contained no definite evidence of the ship's nationality, it is likely that the copper came from the mines of Cyprus, and that the ship was sailing in a westerly course when it sank, probably around 1200 B.C. In December 1967 the American Philosophical Society published Bass's report on the site—the first to completely describe an entire archaeological dig underwater.

The following year, 1961, Bass led an expedition, sponsored by the National Geographic Society and the University of Pennsylvania, to another of the sites that the sponge divers had shown Throckmorton. This site was in 120 feet of water off Yassi Ada, a small island lying off Bodrum. It was marked by a great number of the fat, round amphorae typical of the Byzantine period. During the next three years, excavation of the site proceeded in as precise a manner as it would have on land. Despite the hazards posed by the pressure of comparatively deep water, strong surface winds, and sea currents, the site was completely excavated and studied, and detailed drawings were made, using a frame grid system, stereo photography, and other techniques and devices. Layer by layer, the cargo of the vessel was revealed and plotted. The timber remains of the ship were uncovered and immobilized by pinning them together with sharpened bicycle spokes until they were measured and drawn. The group of specialists at the site—including archaeologists, artists, architects, and experts in photogrammetry and preservation techniques—had to work from a large barge during the first stages of the operation to avoid the rats on Yassi Ada. Later, with screens to protect them from the vermin, the crew worked from a camp on the island where there was adequate space for preservation and restoration activities, as well as rooms for preparing the drawings while the measuring and plotting of the site continued.

This first complete underwater excavation and study produced a wealth of information on seafaring in the Byzantine period. During the first season, the divers found a hoard of gold and copper coins in the cabin area aft. The coins dated from the reign of Heraclius, Byzantine emperor from A.D. 610 to 641. A great variety of pottery also was found in the cabin area, including cooking pots, cups, bowls, and lamps. A bronze censer, beam scale, and weights; a copper cooking pot; and a copper food tray came from the cabin. The provisions carried on the ship were represented by mussel shells and fish bones. The main cargo of the vessel was wine, contained in the hundreds of round amphorae that had revealed the location of the site to the sponge divers. Six iron anchors and numerous encrusted ship fittings were found on the site, along with the clay roof tiles that covered the cabin aft. A careful study of the wood fragments, encrusted fittings, and the position of the tiles when found enabled the architects to reconstruct the main features of the ship's structure, giving historians their first certain knowledge of the appearance of a large merchant ship of this period. The information gained from the dig has already produced significant essays on ship construction; trade; social customs; techniques of underwater excavation and measurement, and of preservation and reconstruction of materials recovered from underwater sites. The complete study of the data obtained will require years, and the final effect on the course of underwater archaeology will be profound.

North and Central America: New Finds

During the late 1950's and the 1960's, explorations continued in both the salt and fresh waters of North America.

In the United States, the Centennial of the Civil War provided impetus for the search and study of wreck sites associated with that conflict. Using records as a guide and a simple compass as a detection device, Edwin Bearss, a historian with the National Park Service, located the ironclad gunboat *Cairo* in the Yazoo River in 1956. This ship, which went down near Vicksburg in 1862, occupies a special niche in military history: It was the first vessel to be sunk by an electrically detonated mine. The wreck subsequently was cleaned out and large collections of equipment, ordnance, and supplies recovered from her, despite the difficulties of working in the muddy and at times turbulent river. The vessel broke up when an attempt was made to raise her in 1960, but enough of the *Cairo's* structure survived intact to enable a reconstruction to be made.

Off the coast of North Carolina, several Confederate blockade runners were discovered in 1962. The task of supervising the salvage and preservation of these ships was assigned by the state to Samuel Townsend, who organized a preservation laboratory near Fort Fisher. Hundreds of significant objects have been received at the laboratory since that time. The collections are being distributed to the museums of the state.

Another major discovery of the Civil War period was made in January 1967, when a party of engineers located the wreck of the *Tecumseh*, a single-turret monitor of 2,100-tons displacement that was sunk in the Battle of Mobile Bay on August 5, 1864, while trying to pass through a minefield (the torpedoes of Admiral David Farragut's famous quotation, "Damn the torpedoes! Full speed ahead"). The ship rolled over as she sank and now lies on her side in a deep mud bottom under thirty feet of water. Plans are being developed to raise the vessel and recover the equipment and other objects in her. The find is of particular importance, since the *Tecumseh* was one of the first of the ironclad vessels, and no other example of this significant evolution in ship design has been found.

The rapid rivers that flow in the midsection of the North American continent also have been searched for evidence of the great fur trade that flourished from the seventeenth to the nineteenth centuries. Reasoning that some of the traders' canoes must have overturned in the rapids, depositing goods in the cavities of the rough rock bottoms of the streams, E. W. Davis began an investigation in 1960 of the rapids at Grand Portage, a major rendezvous point for traders on the northwestern shore of Lake Superior. Davis and his group of scuba divers soon found a remarkable nest of ten brass and copper kettles. This discovery encouraged Robert Wheeler of the Minnesota Historical Society to begin systematic investigation. In 1962 the study was enlarged through the entry of R. Kenyon from the Royal Ontario Museum. These efforts have produced collections of tools, utensils, weapons, and other goods that were lost while being sent into the interior of the continent for trade with Indians. Coming from cold, fresh water, the objects are remarkably well preserved.

Archaeologists also have continued to explore the fresh waters of Central America for relics of pre-Columbian cultures. In 1957 Dr. Stephen F. de Borhegyi, an American teaching in Guatemala, heard of discoveries of ancient ceramics in Lake Amatitlán, Guatemala. Since 1954 a group of young Guatemalans had been scuba diving in the lake and had recovered a fine collection of pre-Columbian effigy pots and incense burners. De Borhegyi examined the finds, the lake, and the surrounding shores and reached the conclusion that the body of water had been a sacred place to the highland Maya, who had for centuries made sacrifices there. De Borhegyi was able to relate the finds on the bottom of the lake to certain sacred spots on the shoreline where the people gathered to worship. Leading several investigations of the lake, the young archaeologist uncovered a remarkable complex of water and land sites. Later, in similar investigations in Lake Petén Itzá, he continued his new approach to the study of the highland peoples and recovered significant collections of ceramics, some dating as early as 1500 B.C.

In Mexico, important new finds were made at the sacred *cenote* at Chichén Itzá—first explored by Edward Thompson in 1904—by a joint American-Mexican expedition in 1960–61. But the well has still not been emptied of its contents. Plans are now being made to drain the *cenote* in order not to miss anything. Another major discovery in the Yucatán was made in 1958 by Luis Marden and Bates Littlehales in the *cenote* at Dzibilchaltún, a Mayan city some seventy miles northwest of the smaller and newer, but better-known, Chichén Itzá. Diving in pitch-black water to depths of 144 feet, Marden and Littlehales recovered thousands of artifacts and human remains.

At about the same time that Marden and Littlehales were probing the *cenote* at Dzibilchaltún, divers commenced operations on the wreck of a Spanish merchantman that had been discovered off Yucatán opposite the island of Cozumel. Vast collections of merchandise destined for the markets of Mexico were recovered by Clay Blair, Robert Marx, and the Club de Exploraciones y Deportes Acuaticos de Mexico, under the direction of Pablo Bush

Romero. The cargo included manufactured goods from Germany, France, Italy, and probably England. Brass crucifixes and religious medals, probably made in Italy, were found in the thousands, along with base-metal and paste costume jewelry from France. Needles, made in Aachen, Germany, were still in their paper wrappers, the iron oxides having preserved the paper. The finds also included a gold encased watch, bearing English proof marks; a large number of brass spoons; the remains of brass-handled knives; and thousands of brass and pewter knee buckles of perhaps forty different designs. The ship, identified as *El Mantancero*, which sank in February 1741, took with it to the bottom the most varied cargo of merchandise ever found underwater.

The Pacific Ocean

Underwater archaeologists have barely begun to explore the Pacific Ocean basin. Nevertheless, some significant discoveries have already been made, especially along the coast of western Australia. Dutch ships rounding the Cape of Good Hope, sailing for the East Indies, frequently overran their destination and came to grief on the reefs and rocks of this coast. The sites of four of these ships have been found during recent years. The wrecks, each of which is well documented, have been identified as the *Batavia*, wrecked in the Abrolhos Archipelago in 1629; the *Zeewyk*, wrecked in the same area in 1727; the *Gilt Dragon*, wrecked in 1656 some seventy miles north of the present city of Perth; and the *Zuytdorp*, wrecked in 1712 on the coast north of the Abrolhos Archipelago at a point now known as the Zuytdorp Cliffs. The sites have already yielded significant collections of artifacts to the cursory explorations of the skin divers. Copper cooking vessels, a bronze cannon, and two astrolabes have been recovered from the *Batavia*, while the *Gilt Dragon* has given up Spanish pieces of eight and a fine Bellarmine jug. Dr. C. Jack-Hinton, Senior Curator of Human Studies of the Western Australian Museum at Perth, is organizing a systematic program for the archaeological investigation of the sites. Collections and information of great importance are expected to come from this study.

On Australia's eastern coast, a most dramatic find was made January 10, 1969, when an expedition from the Academy of Natural Sciences of Philadelphia located the six cannons that had been thrown overboard from Captain James Cook's *Endeavor* almost two hundred years before. Cook ditched the cannons during the night of June 10, 1770, in a desperate move to lighten his ship, which had struck on a jagged coral reef, now known as Endeavor Reef, some eight hundred miles up the coast from Brisbane. Cook's tactic was successful, and the *Endeavor*, after being patched, continued its circumnavigation of the world, limping back to England the following year.

Although the general location of the guns was well known, at least ten scientific expeditions and numerous other search parties had been unable to find them prior to the 1969 success. The academy's seven-man party, headed by Dr. James C. Tyler, had actually gone to Australia to collect specimens of fish. They were accompanied, however, by several laymen, led by Virgil Kauffman, who were determined to search for the cannons. While the academy men dived for fish, the laymen used a helicopter to drag a magnetometer through the area where the guns were believed to be resting. When this device indicated the presence of metal, divers checked out the location and found the coral-encrusted weapons in about sixty feet of water. After the cannons have been raised, they must be washed for months in fresh water and treated in an electrolytic bath in order to preserve them. It is expected that at least one of these cannons will be sent to the academy in Philadelphia; the others will go to Australian museums.

Another important Pacific discovery was that of an early eighteenth-century wreck on Ceylon's Great Basses Reef, a long and treacherous barrier some ten miles off the island's southern coast. This find was made in 1961 as the result of an expedition to the reef by a movie maker named Mike Wilson and two young Americans—Bobby Kriegel, aged fourteen, and Mark Smith, aged thirteen—to film a fantasy based on a boy's adventures underwater. All were excellent divers and well equipped for the assignment. Mike Wilson had been to the reef several times, filming sea life with Arthur C. Clarke, the noted expert on space and author of many works on space and on the sea. While filming in a spot near the Great Basses lighthouse, the boys discovered masses of coins protruding from the reef and a small bronze cannon lying at the base of the coral cliff. The coins proved to be rupees of the Mogul emperor Aurangzeb, dated 1113 in the Mohammedan Era—1702 in the Christian calendar. With one exception, the coins were all of the same year, and those undamaged by electrolysis were in mint condition, suggesting that the ship had sunk soon after they were struck. The small brass cannon bore a weight mark that proved to be English.

In the spring of 1962 Wilson and Clarke, accompanied by Peter Throckmorton, returned to the site and began as systematic an investigation as could be conducted in the turbulent waters around the reef. A collection of small arms parts, another bronze gun, some small artifacts, and more coins were recovered. Throckmorton prepared a master drawing of the site and related the finds to it. While the site has not been identified, it is believed to be the wreck of a ship going to China from India for the purchase of silk. This would account for the large amount of silver bullion on the vessel.

In the future the Pacific basin will see increasing activity in underwater archaeology, and that area, which has been so important to the history of Western European trade and exploration, can be expected to yield much knowledge of the activity of Europeans traversing it, as well as the no less important maritime history of the Pacific peoples themselves.

The Present State of the Art

The tools available to the underwater archaeologist are increasing so rapidly that techniques of

Light, slow-moving aircraft greatly expand the archaeologist's scope of vision over areas of clear water. Above, a helicopter sinks toward the landing pad of Jacques-Yves Cousteau's "floating island" laboratory in the Mediterranean. (French Embassy Press & Information Division)

search, survey, and recovery at underwater sites are in a constant state of change.

Visual search has been greatly facilitated by the use of aircraft. Small and slow-moving light planes; helicopters; and balloons towed behind slow-moving boats have greatly expanded the archaeologist's scope of vision and increased the rapidity with which he is able to search large areas where water conditions permit examination of the bottom. In addition, light submersible vehicles are now available for visual searches of the ocean bottom in both shallow and deep waters.

Progress in electronics, especially, has affected the course of underwater exploration. While the human eye will always be an indispensable means of search, new electronic devices permit the underwater archaeologist to make rapid and accurate surveys of comparatively large areas of ocean bottom where visual search would not be possible due to darkness or cloudiness of the water. Development of the flux-gate magnetometer, a device of great sensitivity and range, made possible the detection of sites otherwise completely invisible to the diver. The sensing head of this instrument contains two magnetized needles that are balanced against the earth's magnetic field and against one another. The presence of iron in the sea bottom interrupts the regular pattern of magnetic lines, and this imbalance is detected by the needles and indicated to the operator by means of a visual signal. A newer and more compact device, the proton magnetometer, has two containers of gas or liquid in its sensing head. When the regular pattern of protons traveling through the gas or liquid is disturbed by an irregular pattern in the earth's magnetic lines, this instrument, like the flux-gate magnetometer, indicates the presence of the iron. So sensitive is this device that a single planking nail can be detected in depths of water up to fifty feet. Larger masses of metal, of course, can be located in much greater depths. A recent example of the usefulness of this instrument was the discovery in international waters off Florida of the remains of one of the ships of the Spanish Plate Fleet of 1733. The wreck, explored by Tom Gurr, was covered by deep sand and a dense tangle of grass. Nothing was visible. It is inconceivable that the site could have been discovered without electronic equipment.

The work of Harold Edgerton, the pioneer electronics genius of Massachusetts Institute of Technology, has been of the greatest value to underwater exploration. His side-scanning sonar and mud pingers and boomers are capable of penetrating murky water and mud and sand bottoms to reveal the presence of

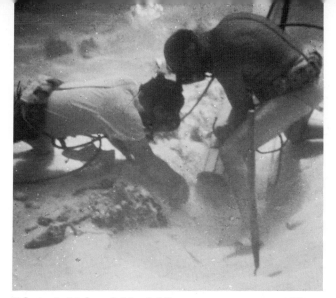

Edwin A. Link and Mendel Peterson operate an airlift on the site of the Spanish Plate Fleet off Florida. (Smithsonian Institution)

Vertical and horizontal distances are recorded on location with great accuracy. In this picture the author measures a wreck site with range and bearing equipment. (Smithsonian Institution)

229

materials of a different character than that of the surrounding natural sands and sediments. The equipment has proved capable of delineating the remains of a wrecked ship and its cargo when visible evidence is entirely lacking. Edgerton also has developed photographic equipment capable of working in the deepest areas of the oceans—recording the character of the bottom and the presence of any natural or manmade objects to be found there.

Once the site of a wreck is located, the present-day archaeologist has at his disposal a wide variety of tools with which to systematically survey and excavate the site according to the strictest rules of his profession. He can survey the site using stereo photography, recording the original surface of the site and each successive layer as it is excavated. Measuring and sighting devices such as the bearing circle, surveyor's tape, and sighting level may be used to record horizontal and vertical distances to a high degree of accuracy. The measuring frame, a metal grid set up in a level position above the site—developed by an Italian archaeologist, Professor Nino Lamboglia, and by Jacques-Yves Cousteau's associate, Frederic Dumas, and refined by the Bass group working at Bodrum—may be used to measure accurately all distances to individual objects and the elements of the ship structure. The problem of establishing an artificial horizon underwater from which to make vertical measurements underwater has thus been solved in several ways. Once the datum point is established and an optimum elevation determined, other points may be measured in relation to it by the use of an ingenious level developed by Donald Rosencrantz. This is a plastic tube, one end of which is closed and fastened to a measuring rod at the datum point, while the open end is held by the diver some distance away. When the tube is filled with air, it becomes a giant level, with the open end bleeding a bubble or two of air when it is held at the same level as that at the datum point.

For excavating the site, the underwater archaeologist can use an airlift in a variety of types and sizes. It may be a six-inch model, capable of moving large amounts of overburden—the shovel of underwater exploration; or a small one of three inches, capable of removing sand a grain at a time—equivalent, in effect, to the brush used by archaeologists in land excavations. When the underwater archaeologist has entered "pay dirt" on a site, he must exercise the same caution as his colleague on land, and much of this phase of the dig may be done simply with the hand "fanning the water" to set up currents that carefully uncover an object.

One of the great problems in working on an underwater site is the disposal of the waste being excavated. If the natural currents do not sweep it aside, the outflow of the airlift may be piped away from the site. Ideally, the discharge is flowed over a riffle board on a barge to recover any small objects that have been drawn into the airlift. In other cases, the discharge may be sent through a wire cage, which can be examined and dumped by a diver.

Sophisticated recovery vessels also are becoming available. The *Alcoa Seaprobe*, newest and most spectacular of these, will operate under Willard Bascom and Robert Marx to test the theory that ships may be preserved intact in the cold, dark oceanic depths. The vessel can electronically scan the ocean floor for wrecks and lift as much as 200 metric tons to the surface.

Building on the well-known techniques for preserving objects recovered from wet, sandy soil, rapid strides have been made in the preservation of objects taken from underwater sites. Organic materials, if covered with sediment or sand, survive very well in fresh and salt water—often better than they would in wet soil. These materials are preserved by replacing their water-filled cells with a solid substance, such as polyethylene glycol. Unlike organic materials, metals suffer varying degrees of damage in water, due to electrolytic currents. Damaged metals may be restored in a galvanic bath, by a direct electric current, and by other methods, such as the hot hydrogen blast. Missing parts may be restored by using new catalytic setting plastics. Ferrous metals submerged for long periods of time accumulate a crust of calcareous materials, iron oxide, and sometimes sand and pebbles. When an object in such a crust has been completely converted to amorphous iron oxide, the natural mold made by the disintegrated object may be filled with a rubber compound or soft metal and an exact reproduction of the lost object made. Another method of preserving an object is to embed it in a clear plastic shell, which prevents disintegration. A final resort is to leave the object in its calcareous encrustation and make X-ray profiles, which can be studied in lieu of the object itself.

The identification of a shipwreck site—i.e., the establishment of exactly what ship or ships went down, and the date and circumstances of the disaster—may proceed from one or two directions. In the first place, documentary evidence indicating the presence of a wreck in a certain area may be studied and a search made for a specific ship. Alternatively, a site may be discovered and then documentary evidence sought to make the identification. In the case of ancient inundated land sites, the name of the settlement is usually known. Shipwreck sites from

When submerged for a long time, ferrous metals collect a crust of calcareous materials and iron oxide. These encrusted cannon are from the *St. Joseph,* sunk in 1733 in the central Florida keys. (Smithsonian Institution, photo by Ed Reimard, courtesy T. Gurr)

ancient times, unless associated with some historical event of major importance, are usually identifiable only by period and nationality. Such sites, when dating from more recent times, are frequently identifiable when the wrecks are those of regular vessels of war but less often when they are of merchant ships. In most cases in recent times, shipwreck sites have been found by accident and only identified later by study of documentary evidence. Some outstanding exceptions are the discoveries of the *Philadelphia* in Lake Champlain, the *Vasa* in Stockholm harbor, and the *Cairo* in the Yazoo River.

The identification of a shipwreck site involves the study of every detail of the objects found on the site and of any remains of the ship structure that might be discovered. Dated objects such as coins, cannon, pewter, silver, and ceramics furnish important clues to the period of the wreck. Other objects marked with the arms, ciphers, or other insignia of princes, commanders, and governments may indicate the nationality of the ship. If the ship was a vessel of war, the number and size of cannon will indicate the size and type of the ship—valuable clues to its period and nationality. In addition, almost all ordnance materials bear government marks. In some cases, the family arms of officers and passengers on the lost ship will be found on personal objects, and these may furnish conclusive evidence. If the ship was a merchant vessel, exact identification generally is much more difficult, although period and nationality may be established. Names on charts are frequently clues, since reefs, bars, and cliffs frequently bear the names of vessels wrecked on or near them. Alligator Reef, Looe Key, and Carysfort Reef in the Florida Straits are examples. Other names such as Gun Key, Wreck Hill, and Double Headed Shot may indicate the presence of a wreck site but not the specific name of the ship that was wrecked. Finally, after the study of all internal evidence has been completed and, hopefully, the nationality and period of the ship established, the researcher must then turn to the appropriate archive to search for documents relating to the disaster. With perseverance and luck, he may find the name of the ship and documents relating

to its sinking. In this case the collections from the site take on added historical importance because they can then be connected with a specific ship, event, and period. These procedures may be used in identifying both ancient and more recent ships, although chances of positively identifying an ancient vessel—beyond its nationality and the approximate period of loss—are virtually nonexistent.

Publication of the results of an underwater exploration project, like those of land investigations, may take several forms. The basic publication is the site report, which gives a historical account of the operations, and of any research before or after the dig, and details of objects found and conclusions reached. Specialized papers discussing certain classes of objects, special aspects of research, and equipment and procedures of excavation will be a natural product of a successful dig. Popular articles for laymen will carry the account of the expedition to the public, as will film reports and lectures. And finally, if the site has yielded discoveries from which important historical conclusions may be drawn, the results will enter the general body of historical knowledge and have a lasting influence on the writing of history. Archaeological research may be said to take up where the documents leave off, and the knowledge flowing from underwater investigations can enrich our knowledge of sea commerce, warfare, and social history in general with a vital picture of how our ancestors traveled, traded, and fought on the world's waters.

It appears inevitable that in the years to come, as man goes deeper, stays longer, and travels farther under the ocean, important new discoveries will be made. New underwater vehicles can cruise at depths of 5,000 feet or more, enabling scientists and archaeologists to observe with care areas of the ocean bottom that previously have been hidden from them. The increasing use of sophisticated gas mixtures, using the rare inert gases, and other new diving techniques will enable men to work at great depths for long periods of time while they live underwater in pressurized shelters. Through these advances, an area of ocean bottom equivalent in extent to the land mass of the New World will be opened for exploration. Using improved techniques for deep-water salvage, which are now being perfected, explorers undoubtedly will bring to light undreamed-of archaeological riches, which will profoundly increase our knowledge of man's life on the sea. Marine archaeology, like other aspects of oceanography, will develop as a separate discipline but one closely related in technique to other underwater studies, and it will form an essential part of the study of the oceans that we call oceanography.

231

10 MARINE ECOLOGY
by Charles E. Lane

In the early spring of 1953 many of the village cats in Minamata, a fishing community in western Kyushu, the southernmost of Japan's four main islands, began to behave strangely. They developed muscular tremors, became extremely agitated, and lost their coordination as though they were drunk. Many jumped or fell into the sea and were drowned. This phenomenon would have caused little public concern, except that a similar affliction soon appeared among the village fishermen and their families. Among humans, the symptoms that characterized the illness included numbness of the feet, hands, and lips; a gradual narrowing of the field of vision; slurring of speech; and pronounced motor incoordination.

The mysterious affliction quickly reached epidemic proportions; the death rate was high. Of 116 cases officially recognized and recorded, 43 died. Most of the survivors suffered irreversible damage. Few recovered completely. Pregnant women who themselves showed no signs of the affliction bore defective children.

The ailment, termed "Minamata disease," was studied widely in Japan. By 1965, when another totally unconnected outbreak of the disease occurred at the mouth of the Agano River in the Niigata Prefecture, its cause was known to be mercury poisoning. Both the original episode and the second one, in which seven of the thirty people reporting the characteristic symptoms died, were shown to have resulted from eating fish or shellfish taken from polluted waters. The food organisms contained from 20 to 50 parts per million of methyl mercury. This was traced to wastes discharged in coastal streams by chemical plants using inorganic mercury compounds as catalysts in the synthesis of acetaldehyde and vinyl chloride.

Though industrial society increasingly has used the sea as a dumping ground for its waste materials, often containing large amounts of heavy metals and other toxic materials, the ocean volume is so vast—over 300 million cubic miles—that these pollutants are greatly diluted. The natural abundance of mercury in sea water—that is, in sea water unmodified by human agencies—is only about 0.03 parts per billion. Coastal waters near the continents on which man lives generally contain 0.5 to 2.0 ppb. However, this higher concentration is equivalent to just two drops of mercury evenly dispersed through the contents of a swimming pool measuring fifteen by thirty feet, filled to an average depth of five feet. No one would get mercury poisoning from drinking this water, but many marine organisms concentrate mercury and other trace materials in their bodies to many thousand times their level in the water. This characteristic of marine organisms, as illustrated by the deaths in Japan and by the more recent condemnation of canned tuna fish and frozen swordfish by the U.S. Department of Agriculture, poses a very real threat to mankind.

Contaminating the Sea

Generally, the trace materials are neither metabolized nor excreted, merely remaining with the organism until it dies or is eaten. Death from "natural causes" is probably unusual in marine communities; most organisms ultimately are eaten by predators. These acquire all the burden of trace elements accumulated during the life of the prey. The total amounts involved increase with the size of the animal until, in the case of swordfish or tuna, mercury may be present in amounts sufficient to poison man.

The principal sources of mercury pollution are the pulp and paper industries, chemical factories that make chlorine and caustic soda by electrolytic processes using mercury electrodes and the agricultural fungicides used for treating seed grains. Leached into the soil, mercury enters the drainage system and

AND POLLUTION

ultimately the ocean. Mercury in the atmosphere and in rainwater derives chiefly from burning fossil fuels.

Early in 1970 D. H. Klein and E. D. Goldberg of the Scripps Institution of Oceanography determined the concentration of mercury in the air over the San Francisco Bay area, finding between 0.5 and 50 nanograms (billionth of a gram; one pound is about 454 grams) of mercury per cubic meter, and reported that these levels increase in the summer. These investigators always found higher concentrations of mercury associated with smog. Smog is one manifestation of industrial and automotive pollution of the atmosphere, so human activity apparently governs the abundance of mercury in the atmosphere. Total world production of mercury in 1966 was estimated at 9,200 tons, of which 4,000 to 5,000 tons were discharged into the environment and ultimately reached the sea. There the metal is converted to the methyl derivative by bacteria in the bottom sediments and is concentrated in other marine organisms; the sea weeds *Fucus* and *Laminaria* contain 0.2 to 0.4 parts per million—nearly a thousandfold enrichment, and many fishes, including some tunas and swordfishes, contained four or five times this amount.

Lead, recently identified as a metallic pollutant, probably reaches the sea after being scrubbed from the air by rainwater. Estimates of the amount of lead annually emitted in automotive exhausts vary widely, but more than half the total lead production of the United States is used to improve the antiknock characteristics of gasoline. Lead contamination of the atmosphere, although global in extent, is proportional to automotive traffic density. It has been detected in air samples taken 150 miles off the California coast. Rainwater collected in San Diego contained 40 to 300 ppm of lead, depending on the rate of rainfall, compared with a national average in the United States of 0.36 ppm. T. J. Chow, a marine chemist from the Scripps Institution of Oceanography, has analyzed

livers of marine fish taken from coastal waters near metropolitan centers. Livers of sea bass caught off the southern California coast in the vicinity of Los Angeles had an average concentration of 22 ppm of lead. In fish caught 150 miles offshore the liver contained only 11 ppm of lead. Fish caught off the Peruvian coast had only about 9 ppm of lead in the liver. Sea water as far as 150 miles off the California coast contained up to 36 ppb of lead, while water from the Mediterranean Sea was found to contain only 2 ppb, and water from the mid-Atlantic contained only 0.7 ppb. The exact way in which lead circulates through the marine food web is not yet known, but it probably follows pathways similar to others that have been described; after being absorbed by microorganisms and unicellular plants, lead is transferred to herbivorous planktonic animals, thence to small carnivores, and ultimately it enters larger predatory fish in their food.

To understand how high concentrations of mercury, lead, copper, and other materials are achieved in living marine organisms when they are so dilute in the environment, it is necessary to examine the flow of energy through the community of marine organisms, describe feeding habits and other ways in which marine animals and plants affect each other, and determine how changes in the environment may modify the behavior of all the creatures living there. These are some of the principal problem areas comprising the division of biological science known as ecology.

Ecology Becomes a Science

Although predator-prey relationships between organisms have been recognized for as long as man has observed his environment, and the influence of climate on populations of animals and plants was well known to early nomadic herdsmen, the organization of these observations into a formal scientific discipline

The fruits of expanding technology in densely populated areas include such scenes as this: An estuary near Bay- side, New Jersey, that is being overwhelmed by garbage. (Ralph A. Schmidt)

is a relatively recent event. Carolus Linnaeus, the eighteenth-century botanist who founded modern systematic biology, was also among the first scientists to recognize the degree of interdependence between plants and animals. He emphasized the harmony in nature that results from the precise adaptation of each living creature to its way of life, correctly evaluating the influence of external conditions upon such aspects of plant life as size, flowering habit, and distribution. Linnaeus may quite properly be called the first ecologist, although this term was not employed until nearly a century after his death.

The word *oecology* was coined by the German biological philosopher Ernst Haeckel in 1869, but was apparently not used again for nearly thirty years. Thus ecology as a biological science is less than seventy-five years old. During this brief period it has evolved through several distinct stages. As in most other sciences, the early emphasis in ecology was descriptive. Multitudes of habitats, or niches, for animals and plants were described and classified; the physical and chemical peculiarities of each habitat recorded; and lists of organisms found in them drawn up. As these studies proliferated, it became clear that the communities of plants and animals in particular

niches were not static like the figures one might see in a museum display case, but alive, changing, influencing, and being influenced by, other members of the group—or, as Bostwick Ketchum of the Woods Hole Oceanographic Institute has expressed it, "eating and being eaten." Ecology, in short, passed from the purely descriptive phase of its development into a dynamic era when the full significance of the interdependence of organisms was recognized.

It also became clear that interactions involved not only the living members of a community of organisms but their environment as well. Not only do changes in the environment, such as temperature, illumination, and pressure, affect the organisms in a particular niche; but the organisms, in turn, may influence some aspects of their environment by exchanging materials with it. And, closing the circle, these environmental changes may have a rebound effect on the organisms that caused them. The result is a feedback system in which the biological community maintains itself in dynamic balance with its environment.

For example, dissolved oxygen and nutrient salts such as phosphates and nitrates are indispensable for the growth and survival of marine organisms, but in living and growing the organisms use up large quan-

234

tities of these compounds. The growth of phytoplankton depletes the supply of nutrient salts, limiting further growth of this plant population. As the phytoplankton "bloom" fades away, dead cells return nutrients to the water whence they came. Meanwhile animals have used the oxygen liberated during the phytoplankton bloom and, in their respiration, have made increased amounts of carbon dioxide available. Thus the cycle is completed: plants fix carbon dioxide during photosynthesis in the light and excrete oxygen as a waste product, and animals extract oxygen from the water and return carbon dioxide.

The Impact of Man

Alteration of any of the components of an ecological complex may upset the dynamic balance between a biological community and its environment, profoundly altering the destiny of the entire association. For marine communities occupying the relatively shallow waters of continental shelves, the environment is most frequently and most adversely modified by man and his works—in other words, by pollution.

An environment or niche is said to be polluted when it has been modified, generally adversely, by the addition of foreign, or "exotic," materials. These foreign substances usually are the waste products of a highly evolved, technical society: sewage, industrial wastes, detergents, garbage, and such solid rubbish as worn-out automobiles, metal cans, and imperishable bottles.

From earliest antiquity, man's communities have been erected on the banks of streams or protected bays. Disposal of refuse presented no serious problem since it could always be channeled into the river where it was washed away, diluted greatly, and rendered innocuous by various biological agencies. This simple disposal system was quite effective during much of recorded history, when the population of mankind was small and technology was primitive. In the last century, growth of population and of industry has been explosive. The world population now is doubling at least three times every hundred years and technology in most societies is expanding even more rapidly, with the resulting volume of waste materials threatening to overwhelm the capacity of nature to absorb them.

The magnitude of the disposal problem is illustrated by the volume of sewage sludge produced by a single large city. Each year in metropolitan New York about 5 million cubic yards of solids are as-

The pulp and paper industries have been among the principal sources of water pollution. *Above*, waste water being discharged from a paper mill in Oregon and, *below*, the resulting dead fish, together with a mat of sludge that has been lifted from the bottom by the gases of decomposition. In recent years, these industries have made great strides in cleaning up their activities. (C. P. Idyll)

235

The Gulf Oil Corp's huge *Universe Ireland,* more than four times the tonnage of the largest passenger liners afloat. The ever-increasing size of tankers means that when accidents occur, they will pose ever-greater ecological hazards. (Gulf Oil Corp.)

sembled from primary and secondary treatment plants and barged to a dumping area in New York Bight, within 25 miles of Ambrose Light. This mass of material is sufficient to make a layer 5 feet thick over an area of 4.6 square miles.

In 1968, after the dumping had continued for at least forty years, the U.S. Corps of Engineers, which controls and licenses this operation, commissioned a study by the U.S. Bureau of Sport Fisheries and Wildlife Laboratory at Sandy Hook to determine the biological consequences of this practice. Jack Pearce of that laboratory found that sewage sludge has had a disastrous effect on the fisheries resources of the area. Moreover, he recorded high concentrations of heavy metals: 200 ppm of chromium, 350 ppm of copper, and 250 ppm of lead. Persistent pesticides such as DDT, dieldrin, and some of their identifiable breakdown products were present in concentrations up to 200 ppm. The sewage sludge is rich in organic materials that are eagerly accepted by microorganisms as a food source. In this nutritious broth they grow vigorously, consuming and converting sewage material into bacterial substance. In the process, however, these organisms also require enhanced amounts of oxygen to support their explosive population growth. Dissolved oxygen in the sludge dumping area was reduced from a normal seawater level of 9.8 ppm to 1.5 to 0.9 ppm. This level is insufficient to support most aerobic (oxygen-requiring) organisms. In the dumping area, the normal population of microfauna has virtually disappeared, creating what Pearce described as a "desert."

Since population is most dense along the coasts, it is clear that conditions described for the metropolitan New York area must also prevail over other portions of the coastal United States where sewage and industrial wastes are disposed of by dumping in the sea. Indeed, the editors of the *Marine Pollution Bulletin* estimate that more than 48 million tons of such material was dumped into the waters adjoining the continental United States in 1969.

Of course, the problem is by no means limited to the United States. In Europe, for example, one-fifth of the volume discharged by the Rhine River into the North Sea consists of sewage and industrial wastes from communities, manufacturing plants, and factories that line the river banks along its 820-mile course from its origin in Switzerland. The Rhine, in-

Aftermath of a tanker accident: Dead crabs and fish on oil-soaked sand of Point Puntilla, Puerto Rico, following the wreck of the *Ocean Eagle*. (M. J. Cerame-Vivas)

cidentally, provides nearly two-thirds of the total fresh water supply of the Netherlands!

While the volume of variously treated sewage reaching the sea directly is considerable, and at least in localized areas may profoundly modify the composition of marine communities, by-products of industry may have even greater and more disastrous effects. Chief among these are the metals—copper, zinc, and lead—and the mixtures of chemicals excreted by such industries as paper manufacturing, plating establishments, and chemical manufacturing.

Oil-Troubled Waters

Crude oil from refinery wastes, accidental spills from tankers, and the various chemicals employed to disperse them also figure prominently in any list of pollutants of the marine environment. To satisfy industry's enormous appetite for energy, world petroleum production continues to skyrocket. Ever larger fleets of supertankers are required to transport oil from producer and processor to consumer. By 1970, some of the larger tankers in service displaced upward of 300,000 tons, making them more than four times the size of the great passenger liners the *Queen Elizabeth 2* and the *France*. The *Universe Ireland* and her five sister ships of the Gulf Oil Corporation fleet, for example, measure 1,132 feet in length by 175 feet in breadth by 105 feet in depth, with a dead

weight capacity of 327,000 tons and a draft of about 81 feet when fully loaded. Giant tankers of 500,000 tons displacement are expected to go into service shortly and plans already are being made for even larger vessels.

Huge tankers, such as those of the Gulf Oil Corporation fleet, traveling at 15 knots, lack maneuverability. They require more than two miles and more than 10 minutes to stop, and more than a minute just to alter course by 20 degrees. In narrow waterways with heavy traffic, such as the English Channel or the Straits of Dover, there have been numerous accidents, resulting in massive oil spills. For example, when the tanker *Torrey Canyon* stranded on Seven Stones Rocks off the southwest tip of Cornwall on March 18, 1967, her entire cargo of 118,000 long tons of Kuwait crude oil was released into the sea. The probability of collision or other damage to lumbering tankers in restricted waters subject to heavy traffic has weighed heavily on government agencies responsible for locating terminal facilities for petrochemicals.

Breakage of large-diameter pipe lines used to transport crude oil from producing areas to terminals or refineries also has caused extensive pollution of marine environments. Oil pollution of marine environments in cold latitudes may be even more disastrous than in the temperate zone. Toxic components of the oil will take far longer to evaporate at

237

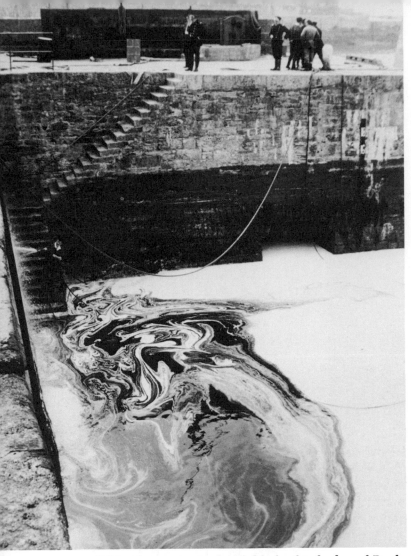

White detergent mixed with black oil in harbor of Porthleven, England, following the wreck of the *Torrey Canyon* in the English Channel in 1967. The detergent later was found to be more damaging to marine life than the oil that it was supposed to disperse. (G. W. Potts)

low temperatures and the increased viscosity of the oil will multiply the difficulty of cleanup of shores and beaches. Mishaps during drilling and subsequent pumping of offshore wells may release large volumes of crude oil to the sea. This occurred during 1969 in the Santa Barbara Channel and in 1970 in the Gulf of Mexico.

In addition to these man-made spills, geological disturbances of the sea floor may release crude oil from offshore pools. On November 10, 1793, the great English seaman Captain George Vancouver, visiting the California coast in the *Discovery*, made this entry in his log: ". . . the sea had the appearance of dissolved tar floating upon its surface, which covered the ocean in all directions within the limits of our view. The light breeze that came principally from the shore, brought with it a very strong smell of burning tar, or of some such resinous substance." On this date his ship was south of Point Conception on the California coast. Apparently the Santa Barbara Channel has experienced natural crude oil pollution for at least 200 years.

Oil spills at sea generally are treated in various ways to reduce the threat to communities of marine organisms and to the recreational use of the shore. The treatment methods have included detergents to subdivide the mass of oil into tiny droplets that are dispersed by wave action; straw or chalk to absorb the oil; and fire, which requires special pretreatment, since a crude-oil spill at sea is nearly fireproof.

The *Torrey Canyon* spill was treated with the best dispersant available at the time, BP 1002, and with more than 3,000 tons of chalk. The dispersant itself was quite toxic, killing test organisms in concentrations of only 10 ppm in later laboratory tests. The treatment, in this instance, was more damaging than the oil itself. Subsequently, better dispersants have been developed; some, such as COREXIT 7664, are essentially nontoxic to marine organisms.

For an oil-spill dispersant to be truly effective, it must subdivide each large oil drop into a myriad of submicron-sized droplets (1 micron equals $\frac{1}{25,000}$ of an inch). This enormously increases the area of contact between oil and water and greatly accelerates the liberation of water-soluble materials in the oil; thus, the toxicity of a mixture of oil and dispersant is generally greater than that of either component alone. Crude oil, since it will not mix with water and is lighter than water, floats on the surface, releasing its more volatile components by evaporation into the atmosphere. Soluble components are released to the water, diluted, and dispersed. The thick, tarlike residue may float harmlessly on the surface for months, affecting only sea birds that may be attracted to it. Winds and currents may drive such a slick onshore, where it kills many intertidal animals by smothering them.

Radioactivity and Heat

A particularly insidious pollutant may occur in the effluent of establishments processing radioactive chemicals. Organisms do not discriminate between radioactive and nonradioactive forms of the same elements. In the early years of the atomic age, highly diluted waste water from a plant located in the Savannah River drainage system was allowed to flow untreated into the river until commercial oystermen at the mouth of the river observed that many bivalves on the oyster bars actually glowed in the dark from absorbed radioactivity.

Elevated temperatures also affect marine organisms. The unwanted heat may come from water used to cool condensers of conventional or nuclear-powered generating plants, or other industrial sources, and may affect only a small area about the point of discharge on an open shore or influence a large enclosed bay. Effects of added heat to a marine ecosystem have been carefully studied in only a few temperate-zone localities, generally sites occupied by large electric generating plants. California, for example, has a total of twenty-two power stations, with a total generating capacity of 16,824 million watts (M.W.), all discharging into the sea or into estuaries. Included in this list is the largest fossil-fuel thermal plant in the United States (2,107 M.W.), whose cooling water is discharged into Monterey Bay. Studies in California by independent investigators and by scientists of the Pacific Gas and Electric Company suggest that the total heat input into coastal waters from existing electric generating plants is somewhat less than that added when the outgoing tide returns sun-warmed water from the coastal flats. Productivity, as measured by total mass of marine creatures collected from the warm water plume of effluent, was generally enhanced. Similar results were noted in studies conducted in Scandinavian fjords and in the north-temperate waters off New England.

In the Patuxent River estuary, a branch of the Chesapeake Bay system, R. L. Coney and J. W. Nau-man of the U.S. Geological Survey studied the river fauna before and for several years after the installation and operation of a steam electric generating plant on the river. Production of total dry tissue weight in the river fauna was nearly three times greater after the plant was put in operation than in the preceding period. These results suggest that the modest increase in water temperature occurring in the immediate vicinity of the outfalls of these installations in temperate to boreal waters may be beneficial to some members of the local fauna and flora.

Conditions in lower latitudes are vastly different. Tropical waters generally reach summer temperatures only a few degrees below the thermal death points of many members of the resident flora and fauna. Heated effluents of power plants and other industrial installations cause particularly serious environmental stress in shallow tropical waters. Corals, for example, die if they are exposed to temperatures as little as three degrees hotter than the maximum to which they are normally exposed on the reef.

A mounting problem in the tropics is the stress on marine communities resulting when hot salty brine from desalination plants is returned to the sea. In one of the methods of producing potable water from the oceans, seawater is used as a coolant for the condensers of steam electric generating plants. From the condensers the already heated effluent is pumped to the desalination department where it is exposed to greatly reduced pressures, causing the warm water to boil, "flash" distilling much of the water away from its burden of salts. The warm brine remaining may be

Light areas are warmest in this infrared aerial photograph of heat pollution in Biscayne Bay from Florida Light and Power Company plant at Turkey Point. Water taken into plant at 69° F by the diagonal canal at right emerges into inland canals at 81° and is discharged into bay by canals at far left at 75°. All plant and animal life in the small area opposite the two discharge canals has been wiped out; damage declines as the temperature returns to normal outward from this. (Bendix Corporation)

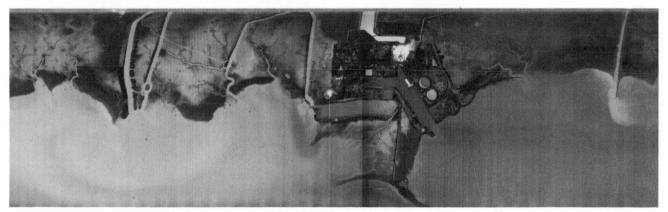

so dense that it sinks, despite its elevated temperature, doing great damage to the bottom community.

In Lindbergh Bay on St. Thomas, Virgin Islands, scuba divers R. P. Van Oepoel and D. I. Grigg were studying the corals and other invertebrates in the vicinity of the outfall from a desalination plant. On one occasion the plant effluent was so hot that they were unable to approach the outfall closely enough to complete their measurements.

Heat pollution also has had a serious impact on life in southern Biscayne Bay, Florida, a shallow (about ten feet), partially enclosed body of water in which summer temperatures approach 95° F. Some bottom-dwelling organisms are found in the outer bay throughout the year but disappear from the warmer inner bay during the summer months. Heated water from a Florida Power and Light Company generator plant on Turkey Point killed many marine organisms in the bay even when the plant operated far below its designed nuclear capacity, burning fossil fuels only.

The sensitive corals, as might have been expected, were killed at a greater distance from the outfall than most other creatures. Damage also was particularly severe among fixed plants and sedentary animals like bivalves, barnacles, and bottom-dwelling marine worms. Other animals simply moved out of the area of heated water or never entered it. As this plant became operational as a nuclear generator, cooling water requirements increased to 1,152 acre-feet each day. This is the volume in a twelve-inch layer of water spread over 1.73 square miles.

Persistent Pesticides

The chlorinated hydrocarbon pesticide DDT was developed and widely used as an insecticide during World War II. Since that time other related compounds have been developed that are even more effective. All the compounds in this family, which includes dieldrin, endrin, aldrin, and others, are highly toxic and very stable. They have been used to reduce crop damage by insects, in public health programs designed to eliminate mosquito carriers of malaria and yellow fever, and to control such domestic insect pests as houseflies and roaches. Pesticide consumption in the United States increased from a few million pounds in 1945 to over a billion pounds in 1965. Pesticides are presently applied to at least 90 million acres—more than one quarter of the total area of land devoted to agriculture in the United States, excluding Alaska.

These compounds are washed into the soil in rainwater and ultimately permeate watersheds to reach the sea. Within the decade of the nineteen-sixties numerous investigators reported finding chlorinated pesticides in concentrations of from 5 to 800 ppm in an alarming array of marine animals ranging from Arctic polar bears through oceanic birds of the middle latitudes to Antarctic penguins. Since these animals live in remote areas, far from the primary sources of pesticide pollution, they must have acquired the insecticide either from the seawater or from their food.

Analyses of seawater distant from known sources of pollution have shown only the faintest traces of DDT and its chemical relatives, generally only fractional parts per trillion. Examination of marine sediments and a variety of marine organisms, however, has revealed the ubiquity and magnitude of pesticide contamination. Marine microorganisms, the tiny shrimplike copepods and other planktonic animals, mid-Pacific skipjack tuna, anchovies, English sole, ocean perch, and many others contained chlorinated pesticide concentrations up to 10 ppm—twenty times the allowable levels for fishery products shipped in interstate commerce. A king crab taken five miles offshore in the north Pacific contained nearly 3 ppm of DDT. Mature porpoises in Pensacola Bay were found to carry over 800 ppm of DDT. This distribution suggests that chlorinated pesticides—like trace metals in extreme dilution—may be concentrated many thousand times by marine organisms, with each trophic level in the food web absorbing increased amounts of pesticide until, in the larger predators, the load may be so great as to be incompatible with life of the individual or of the species.

In my laboratory at the University of Miami, we have been investigating the absorption and distribution of the chlorinated hydrocarbon pesticide dieldrin in marine fishes, pink shrimp, and other organisms. Although the sensitivity of different animals to this pesticide varies somewhat, all show a remarkable ability to extract the material from seawater. When shrimp are exposed to only 0.25 ppb of dieldrin the gills contain nearly 1 ppm by the end of one hour. This is more than a thousandfold concentration. We have followed the pesticide through the body of the shrimp, seeing it pass from the gills into the blood and from there to tissue depots such as the brain, the abdominal muscles, and the hepatopancreas. Dieldrin concentration in these organs continues to increase as long as the animal survives and remains in contaminated water. Few shrimp survive longer than a week in water containing as little as 2 ppb of dieldrin. The ecological significance of these observations is clear: when shrimp, in their migrations, enter contaminated estuaries they acquire a pesticide load that increases in proportion to the amount of time they spend there. If they survive and return to the open sea they may be captured by man or some other predator, passing their pesticide on to the next higher level in the food web. This, when added to what the predator may himself have absorbed from the water, could be lethal. If the shrimp were eaten by man, its load of pesticides would be added to that already present in his body fat and other tissue reservoirs.

The exact mechanisms by which pesticides kill fishes and invertebrates are so far unknown. Several lines of evidence suggest that tissue enzyme systems may be among the earliest victims of pesticides in animals. These protein molecules regulate such fundamental life activities as the rate of energy flow, growth, distribution of ions, permeability, and secretion.

Declines in the populations of marine as well as terrestrial birds have been attributed to pesticide-induced aberrations in calcium metabolism resulting in the formation of unusually thin, fragile eggshells. Such defective eggs are easily broken during incubation. Penguin eggs collected prior to World War II from museums around the world were measured, weighed, and compared with eggs collected during the decade of the sixties. Recent eggshells were thinner than pre-DDT eggshells. Modern penguins collected in Antarctica contain several parts per million of DDT; gulls, terns, and albatross bear similar amounts of pesticides.

The Web of Life

The conversion of inorganic to organic material—that is, the rate of generation of living substance—in the sea is potentially several times greater than the productivity of most land areas. It is measured by the rate of incorporation of carbon into complex molecules of starch, protein, and lipid in the process of photosynthesis, a function characteristic of the teeming planktonic plant life of the sea. These organisms all contain chlorophyll and all are able to perform photosynthesis. Since photosynthesis depends on sunlight, and since this may penetrate as far as 150 feet in most areas of the earth's oceans, at least in summer,

Plants and animals of the plankton photographed alive. Two principal groups of plants shown at left are the diatoms (*Coscinodiscus, Biddulphia,* and *Stephanopyxis*) and the dinoflagellates (anchor-shaped *Ceratium tripos*). Among the animals at right are the transparent arrow-worm, *Sagitta setosa;* zoea (early larva) of a crab; round gadoid egg with a developing fish inside; and the crustaceans, *Centropages* and *Acartia*. In this photograph, the diatoms appear as small chains and squares of cells. (Douglas P. Wilson)

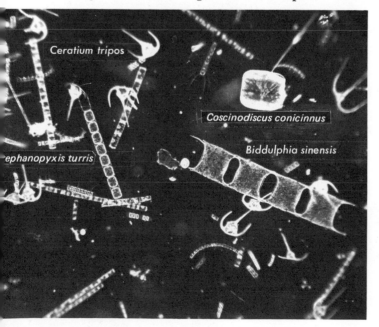

Ceratium tripos

Coscinodiscus conicinnus

ephanopyxis turris

Biddulphia sinensis

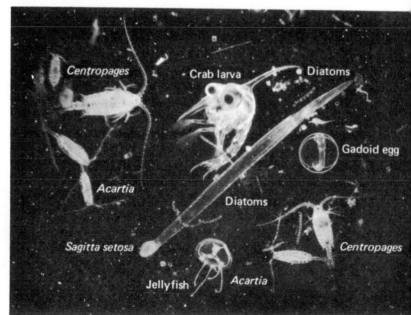

Centropages — Crab larva — Diatoms

Acartia

Gadoid egg

Diatoms

Sagitta setosa

Jellyfish *Acartia*

Centropages

photosynthesis is not limited to a flat surface, as on land, but goes on in a three-dimensional environment. Oxygen, a by-product of photosynthesis, is released into the atmosphere by terrestrial plants or into the water by aquatic green plants.

Among phytoplanktonic organisms particularly important in photosynthesis are diatoms, microscopic plants consisting of a single cell, the wall of which is organized as a pair of valves, or plates, similar to the shells of a clam or mussel, consisting of a pectin-like substance covered by a layer of hydrated silica that is chemically similar to opal. The siliceous layer of the shell is generally highly patterned by regularly arranged grooves and perforations. In temperate and boreal waters diatoms are the most abundant vegetable plankton and form the lowest nutritive level in the food web of the sea. These are the organisms that are browsed by small herbivorous planktonic animals.

The siliceous portion of the cell wall is not affected when the cell dies or is eaten. In areas of the sea where diatoms are particularly abundant, the bottom sediment may consist chiefly of diatom "shells." Fossil deposits of "diatomaceous earth" consist principally of these siliceous remains. Wells in the Santa Maria oil fields of California pass through a layer of marine diatomaceous earth about 3,000 feet thick. Near Lompoc, California, surface beds of marine diatomaceous earth several miles long and over 700 feet thick are presently being quarried to be used in insulation, as industrial filter aids, and as a mild abrasive agent, as in silver polish.

Another important component of the phytoplankton includes the motile, microscopic plants consisting of a single cell that are known as dinoflagellates. Some of these are encased in a cell wall of overlapping cellulose plates that may be intricately ornamented;

The upper portion of a diatomaceous quarry near Lompoc, California. The surface beds of fossil diatoms in this area are more than seven hundred feet thick. The diatomaceous earth is used for insulation, industrial filters, and as a mild abrasive. (Johns Manville Corporation)

others are unarmored. Dinoflagellates are generally not preserved in bottom sediments, nor do they appear in fossil beds.

Phytoplankton are the true foundation of the nutritional pyramid of the sea. Their numbers determine the abundance of microscopic browsing animals such as crustacean copepods, euphausiids, and the larval stages of many different kinds of marine animals, on which a smaller number of larger forms feed. This process is repeated until, finally, at the top of the pyramid, are the relatively few large predatory invertebrates, such as the octopus and giant squids, large fishes and marine mammals, and birds. An example, first given by N. J. Berrill in *You and the Universe*, clarifies the complexity of a single food chain: "A hump-back whale . . . needs a ton of herring in its stomach to feel comfortably full—as many as five thousand individual fish. Each herring, in turn, may well have six or seven thousand small crustaceans in its own stomach, each of which contains as many as one hundred and thirty thousand diatoms. In other words, some four hundred billion yellow-green diatoms sustain a single medium-sized whale for only a few hours."

When physical and nutritional conditions are optimal, diatoms and dinoflagellates reproduce explosively by simple cell division, each cell dividing as often as twice each day. Under these conditions phytoplankton may become so numerous as to discolor the surface of the sea over wide areas. In such "plankton blooms" the number of cells may be greater than 40 million per quart of seawater. Because of the abundant red plant pigments contained in these cells, the patches of discolored water are often called "red tides." The first written record of such a planktonic bloom occurs in the Bible, Exodus, chapter 7, "and all the waters that were in the river turned to blood, and the fish that was in the river died, and the river stank and the Egyptians could not drink of the water of the river."

When dinoflagellate blooms occur in the relatively shallow waters of the continental shelf they sometimes result in massive kills of fish and other marine animals. These mass mortalities were once attributed to a severe oxygen deficiency created by the life activities of the phytoplankton and by the oxygen consumed by the bacteria that thrive on dead phytoplankton cells. Modern analyses have shown some decrease in dissolved oxygen in the immediate vicinity of a red tide, but they have also revealed that many marine dinoflagellates produce virulent poisons that probably are the immediate cause of extensive fish kills. Toxins also account for most of the discomfort experienced by human residents of affected shores. A particularly potent example is saxitoxin, secreted by the dinoflagellate *Gonyaulax catenella*. Saxitoxin was named for the Alaska butter clam, *Saxidomus giganteus*, the source of one of the first pure preparations.

The poisonous dinoflagellates also present a danger to humans who eat shellfish taken from areas in which blooms are occurring. Bivalve mollusks and other filter feeders circulate considerable volumes of water over their gills in respiration. The common mussel, for example, pumps more than five gallons of water per day; an average Alaska butter clam may pump twice this much. As this is forced over and through the gill membrane, much of its particulate matter is trapped on the layer of mucus covering the gills. Specialized areas on the gills are clothed with cells bearing minute hairlike cilia that move the mucous sheet with its trapped particles into the mouth. This is the normal feeding method of bivalve mollusks. In areas of red tides this feeding mechanism traps large quantities of dinoflagellates, rendering the bivalve flesh poisonous. Numerous human deaths have resulted from eating shellfish taken from such areas. To prevent localized epidemics of "paralytic shellfish poisoning," it is customary to close affected regions to fishermen. Interestingly, the clams, mussels, and oysters do not appear to be affected by toxins of the red tide organisms.

Historically, most red tides have been phenomena of the continental shelves, generally occurring in areas fertilized by major river systems with nitrates, phosphates, and other organic materials from the land. Those blooms occurring at some distance from the shore generally mark areas of upwelling where nutrient-rich bottom waters approach the surface.

The Essential Nutrients

The major plant nutrients in seawater (see chapter 5) are sulfate, magnesium, silicates, and potassium. Such essential elements as nitrogen, phosphorus, and iron are present only in minute amounts. In the ecology of the sea these latter elements often limit the growth rate and abundance of phytoplankton. In the English Channel there is a regular seasonal variation in the abundance of phosphate. The maximum occurs during the winter and phosphate may almost disappear from the surface layers during the summer. Decline in phosphate in the surface water coincides with the springtime bloom of phytoplankton. During the summer, phosphate values remain low, rising again in the autumn to the winter maximum. Phytoplankton populations fluctuate in response to many environmental variables, including light and temperature, as well as nutritional factors, but there remains a strong correlation between the availability of phosphate in the water and the rate of phytoplankton growth.

Nitrate is the most abundant source of inorganic nitrogen for phytoplankton growth in the sea. Variations in nitrate concentration generally parallel those of phosphate. Thus, in temperate zones surface waters may be largely depleted of nitrate by the vernal diatom outburst. Generally the concentration of both nitrate and phosphate is low in the surface waters in subtropical and tropical regions. For this reason phytoplankton populations are usually more sparse over these regions of the oceans.

Phytoplankton blooms may be rapidly destroyed by grazing, because most zooplankton herbivores feed voraciously. For example, the common Atlantic copepod *Calanus finmarchicus* so important to the herring and mackerel fishery, was reported by H. W. Harvey to discharge a green fecal pellet every twenty minutes. Others have observed that some copepods placed in a culture containing phytoplankton in concentrations similar to those in the open sea will fill the gut within one hour. If such a gorged animal is removed to a dish without food, the gut empties itself within one hour.

The efficiency of conversion of phytoplankton substance to zooplankton substance is usually estimated to be not more than 10 percent. This means that each pound of animal plankton must have consumed at least 10 pounds of phytoplankton. This same level of energy conversion efficiency seems to prevail in all subsequent levels of the feeding pyramid, so each pound of small fish that prey on the copepods must represent at least 100 pounds of phytoplankton.

Nitrogen compounds and phosphates are particularly abundant in sewage. Many other nutrients abound in industrial wastes. Domestic detergents are a particularly rich source of phosphates. When fresh water rivers and lakes are fertilized with these pollutants contained in raw, inadequately treated sewage, resident phytoplankton are stimulated to extravagant reproduction, often covering the surface with unpleasant viscous green mats of algae. When such plant blooms exhaust the nutrient supply, the plants die. Their degradation reduces dissolved oxygen below the tolerance levels of higher organisms, often producing massive fish kills. Plant debris falling to the bottom creates a thick layer of organic debris from which dissolved oxygen may be lacking. With time the layer of ooze thickens, reducing the depth of the water and further accelerating the deterioration of the lake or stream. This general process of destruction of bodies of water by overfertilization is called eutrophication. It has advanced so far in the western basin of Lake Erie that there is serious doubt in many quarters that that portion of the lake can be salvaged.

Eutrophication has not been as widely studied in the oceans as in fresh water, but recently A. H. Banner and his colleagues at the University of Hawaii have described conditions in Kaneohe Bay on the eastern side of the Hawaiian island of Oahu. From neighboring communities the bay receives daily more than three and a half million gallons of sewage that has received only primary and secondary treatment. For more than a mile around the sewage outfalls, most of the corals have been killed. Reef sediments, nearly white in nonpolluted areas, are medium to dark gray within the area fertilized by sewage. In the middle reaches of the bay the reefs are being overwhelmed by the "green bubble alga," *Dictyosphaeria cavernosa*, that blankets entire areas of reef. In 1970 solid masses of algae a foot or more thick and many yards in extent covered reef communities that were unaffected only a year earlier. Denied food, light, and free water circulation, the enveloped corals die, as do many of the plants and animals that normally occupy niches in the reef. The overgrowth of this alga has paralleled increased sewage input and the accompanying increase in nutrient levels of the bay.

Maxwell Doty, a plant ecologist of the University of Hawaii, has described the effects produced by seepage from a single cesspool servicing a public restroom in Honaunau Bay on the western coast of the island of Hawaii. A nearby coral community has been largely destroyed, and bottom-dwelling algae are more abundant than in unfertilized waters. Algae are also affecting the coral reefs at Waikiki Beach in Honolulu because of increased nutrients in the water from the 45 million gallons of raw domestic sewage that flow into the ocean each day within four miles of Waikiki Beach.

The Crown-of-Thorns

Coral reefs of the Red Sea and Pacific Ocean have lately suffered extensively from animal predation as well as from pollution.

In 1963, while investigating the ecology of the coral reef community in the Red Sea, the late Thomas

Eutrophication in Hawaii: Sewage pollution in Kaneohe Bay has led to overgrowth of "green bubble alga," *Dictyosphaeria cavernosa,* which smothers coral reefs. *Above,* normal growth of the coral *Porites compressa* in nonpolluted area; *center,* a colony of *Porites* being engulfed by the darker alga; and *below,* a close-up of the destructive alga about the base of another coral, *Montipora verrucosa.* (A. H. Banner)

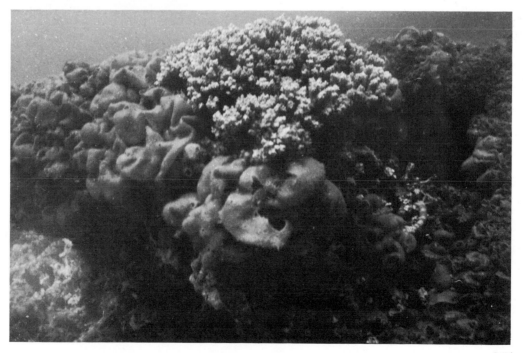

The crown-of-thorns starfish, *Acanthaster planci*, destroying living coral on the Great Barrier Reef, Australia. (Robert Endean)

Goreau, a physiologist from the University College of the West Indies, suggested that the crown-of-thorns starfish, *Acanthaster planci*, might be a predator of reef corals. This starfish was named by the Swedish taxonomist Carolus Linnaeus in the early eighteenth century, but until recently it was a rare species, many museums having no specimens. It is a sixteen-armed animal that grows to a total diameter exceeding twenty-four inches. Its life span is about eight years from the time it terminates its planktonic larval stage and settles on a solid surface to change into its adult form. Normally active only at night, the starfish conceals itself in reef crevices by day.

Dr. Jack H. Barnes, a physician in Cairns, Australia, reported in 1966 that extensive areas of the Great Barrier Reef off the Queensland coast of Australia were being destroyed by *Acanthaster planci*, which was present in large numbers. An experienced biologist as well as a busy medical practitioner, Dr. Barnes has identified the jellyfish causing a characteristic affliction of sea bathers on the northeast coast of Australia, and has been prominent among those studying the biology and toxicity of the sea wasp *Chironex fleckeri*, a jellyfish capable of inflicting stings that may be fatal to man within minutes. His report on the depredation due to the crown-of-thorns starfish was instrumental in the decision of various governmental agencies in the Indo-Pacific region to investigate this new menace, to describe its general biology, and to seek methods for its eradication, or at least control.

Richard Chesher, from the University of Guam, was commissioned by the Guam legislature to investigate the causes of the destruction of coral reefs off that island. He documented the population explosion of *Acanthaster planci* on the western shore of this island. Before 1967 they were not common in Guam, but early in that year they became abundant on localized reefs and were observed feeding actively, even in daylight, in from ten to thirty feet of water. Both Barnes and Chesher have reported that crowding may cause the animals to abandon their usual nocturnal existence and feed continuously. Within one year the starfish had destroyed all the corals on the reefs where they were first observed and had moved on to deeper water. During the next few months the animals moved out from their original dispersal point, until by April 1969, for twenty-two miles, "over 90 percent of the reef corals of the northwestern coast of Guam were dead between low spring tide level and the lower limit of coral growth." Marked individuals traveled as fast as 300 yards a week. The starfish were observed to arrange themselves into rows from fifteen to fifty feet wide, parallel with the coastline, and to move inexorably as an advancing wave, destroying all the coral they swept over.

Acanthaster planci feeds like other starfish, everting its stomach through its mouth, then spreading the membranes over the coral. Powerful digestive enzymes secreted by the wall of the stomach digest the soft tissues of the prey. After the living coral polyps have been digested and absorbed the predator moves on to a new location on the reef. The coral skeleton of the browsed patch stands out sharply in pure white relief against the darker background of the living reef. The daily feeding rate was observed to destroy all the coral polyps in an area twice the diameter of the central disklike part of the body. Corals are killed, therefore, over an average area of one square yard per animal per month. This means that when the population measures only one animal per square yard of reef, all living coral animals will be destroyed in one month. In regions where the infestation is severe, as many as twelve individuals have been found on a single coral head measuring only a few feet in diameter.

The reasons for the population explosion of the crown-of-thorns starfish in the Pacific and in the Red Sea are obscure. A large marine snail, *Tritonia*, is said to be a natural predator of the starfish. Shell collectors prize *Tritonia* specimens for their size and relative rarity. Depletion of the population of this snail by shell collectors has been mentioned as one of the factors permitting relatively unrestricted growth of the starfish population. In other remote regions of the Indo-Pacific area, such as Palau and Rota, however, that support large populations of the crown-of-thorns starfish, the snail population has not been disturbed by shell collectors.

Chesher suggested that the destruction of reefs by blasting and dredging to create new airfields or to enlarge existing facilities may have removed normal encrusting organisms from wide reaches of the coral reefs. Such bare expanses would provide a suitable substratum for the settlement and metamorphosis of the planktonic larval stages of *A. planci.* However, during World War II, much of the reef area of the Indo-Pacific was heavily blasted by aerial bombardment and naval gunfire. There was no population explosion of crown-of-thorns starfish immediately following cessation of bombing. Dredging and blasting are not new in the Pacific, but the widespread use of modern persistent insecticides is. These substances introduced into the food web at a low level and concentrated by passage through successively higher trophic levels may limit the reproductive rate of other normal predators. It is possible that the crown-of-thorns starfish, like some other echinoderms, may be particularly insensitive to these pesticides, giving it an additional—perhaps critical—advantage in the unending struggle for position and nourishment that characterizes life in general, and the life of the coral reef community in particular.

When Natural Barriers Fall

In addition to such ecological factors as salinity, temperature, light, and nutrients, animal communities in the sea are limited as well by geographic barriers. The narrow isthmus of Central America, for example, separates the Pacific and Atlantic tropical faunal communities. The Panama Canal might provide ready communication between these two life zones except for Gatun Lake, a 23-mile segment of the canal that consists of fresh water. This barrier effectively prevents migration of marine organisms between the east and the west.

On August 15, 1914, when the Panama Canal was first opened to ship traffic between the Pacific Ocean and the Caribbean Sea, its limiting depth of 40 feet, lock width of 110 feet, and minimum length between gates of 1,000 feet comfortably accommodated the largest ships. In less than sixty years the canal has become obsolete, many warships, passenger vessels, and all supertankers being too big for transit. A new and larger sea-level canal is being considered that would eliminate the necessity of raising ships by locks to Gatun Lake, 85 feet above sea level, and would lessen some of the inconveniences attending transit through the existing canal.

The biological consequences of opening a direct channel between the Atlantic and Pacific—or of bypassing any other natural barrier—deserve most serious study in advance of the action. Indeed, at least one American fishery disaster may be attributed directly to the construction of a sea-level migration pathway for marine creatures. The Erie and Oswego canals to Lake Erie and Lake Ontario were completed by 1828 and the Welland Canal around Niagara Falls in 1829. These waterways provided a migration route for the sea lamprey, *Petromyzon marinus,* a voracious predator that established itself in the deep lakes in less than one hundred years and destroyed the thriving commercial fishery for lake trout. In Lake Michigan, for example, 6,860,000 pounds of lake trout were taken in 1943; in 1966 less than 1,000 pounds were landed. There have been numerous examples of the disastrous effects of introducing exotic organisms into established ecosystems. The European rabbit became a scourge after its introduction into Australia. The English sparrow and the starling have spread throughout the United States since their introduction, and the mongoose has upset the balance of nature in the Caribbean islands where it was imported to control the rat.

The eastern Pacific Ocean fauna includes at least two dangerous animals that do not occur in the Caribbean: the poisonous sea snake *Pellamis platurus,* and the crown-of-thorns starfish, *Acanthaster ellisi.* If the latter form gained access to the abundant coral reefs of the Caribbean, it might cause as much havoc as its western Pacific relative, *A. planci.* There are presently no poisonous marine snakes in the Caribbean. Any plan for a continuous sea level connection across the Isthmus of Panama must incorporate an efficient barrier to the migration of animals from one environment to the other.

The examples of the crown-of-thorns starfish in the Pacific and the sea lamprey in the Great Lakes illustrate a paradoxical truth about the oceans: Despite their enormous volume and the variety and complexity of the life forms they contain, the oceanic ecosystem is surprisingly fragile. The reduction in the population of a natural predator, or the elimination of a natural barrier, may have profound consequences.

Clearly, too, the capacity of the oceanic ecosystem to recover from man-made physical and chemical insult also is limited. Established practices of dumping waste products of all kinds into the sea or in tributary streams should be reviewed most critically. Since the oceans bathe the shores of almost all nations, their protection and preservation are international responsibilities that merit consideration by scientists of every nation as well as by diplomats. There is no dearth of problems for scientists who seek to expand the boundaries of our understanding of the dynamics of the oceanic ecosystem.

11 MAN BENEATH

by Captain Edward L. Beach

Although all life probably originated in the sea, for thousands of years man has intuitively dreaded the sea and peopled it in his mythologies with monsters, gods, and malevolent spirits. This fear is perhaps an atavistic remnant of the fear that drove his remote ancestors from the sea to the land, probably to escape their enemies. Only in relatively recent times has man begun to return to the sea, conquering both the environment and his fear of it.

Not all primitive men dreaded the sea, of course. Some even built their villages on stakes in shallow water as a defense against marauding bands of other men, and some, from earliest days, made their living on the water. Few went far, or often, beneath the surface in ancient times, but as history began to be recorded it became clear that there were some who did. As always, there were those exceptional individuals who were not awed by the unknown. These were rare specimens indeed, but in the human experience it has been these, always, who have pushed back the frontiers.

The first historical account of undersea operations is the reputed exploit of a Greek named Scyllis, who, with his daughter, Cyane, contributed to the victory at Salamis in 480 B.C. by cutting the anchor cables of some of Xerxes' ships the night before the battle. The account that has come down to us is rather hard to believe, for it has Scyllis and Cyane swimming underwater from anchor cable to anchor cable for several hours; but if one allows them great physical endurance, hollow reeds for rudimentary snorkels, unusual swimming ability, and an enemy who could not comprehend the idea of underwater operations, it is a plausible exploit, and it is likewise understandable why none of the Persian lookouts ever saw them.

One hundred and fifty years later, according to Aristotle, Alexander the Great exercised some of his military ingenuity by devising—or having devised

for him—some sort of a watertight vessel with a viewing port (most artists' illustrations—none contemporary and only one really ancient—show his "diving bell" to resemble a nautical hogshead) in which, so the story goes, he had himself lowered to the bottom of the sea to proclaim his sovereignty over this part of the world too. Little is known for sure about this near-mythological incident, nor the apparatus devised for it. Probably the first artist, from whom most later illustrators drew their details, merely pictured what seemed logical in his own experience. Apparently the hogshead was about six feet in diameter and the same in length; obviously it could not have been lowered more than a few feet beneath the surface, and one wonders what sort of viewing ports there were and of what use the vehicle was. Aristotle, to be sure, refers to Alexander's employment of "diving bells"—open at the bottom—at the siege of Tyre (332 B.C.). It may be that the whole incident is merely an enlargement of his sabotage operations before the walled island of Tyre.

With the exception of these few true or fictional occasions, however, the occasional evidence of primitive salvage efforts, and the activities of pearl divers (of which more later), ancient man looked at the sea as a great, trackless, watery desert through which his only guidance was by the stars and the influence of deities whom he propitiated regularly—and below whose surface there was certain death and destruction. Such has been our heritage, until the last thousand years or so.

The Chinese had the first oceangoing ships, and the Polynesians first learned to navigate offshore, but to Europe belongs the honor of the first detailed investigation of the sea. The man responsible was Prince Henry the Navigator (d. 1460), the first recorded person to have channeled money, political power, and a lifetime of single-minded interest into a systematic investigation of winds and currents in

THE SEA

the Atlantic Ocean. His object was to make contact, for military alliance and mercantile advantage, with the legendary Christian kingdom of Prester John (most likely ancient Abyssinia) and thus establish a route to India. In this he never succeeded; but the discoveries of his captains and the gold, ivory, and slave trade that followed upon his studies made Portugal wealthy. And he knew before he died that a passage to India, discovered by the Portuguese and entirely under their control, was only a question of time. Long after Henry's time, the results of his research were jealously guarded by the Portuguese crown. Portugal used them to develop a seaborne commerce with the Far East, and, to protect her monopoly, her war caravels attacked the ships of all other nations attempting similar voyages.

The Pope, in an effort to terminate this war on the distant seas, divided the world by fiat into two parts, assigning to Portugal undisputed domain over the Atlantic waters surrounding Africa and all the Indian Ocean. To Spain he assigned the other half of the world, well knowing that it was round but totally unaware of the existence of the linked American continents barring the way to the Orient and the source of trade.

Thus it was that Spain gained the New World and, ironically, her two greatest navigators were products of Prince Henry's tradition. Columbus, an Italian by birth, learned not to fear the Atlantic while in the service of Portugal; and Magellan—a veteran of Portuguese voyages to India, Indonesia, and the Philippines—obtained Spanish backing for his expedition to find a passage through the American continents after being cashiered from the Portuguese navy for making the same proposal.

It is not, however, our purpose here to retell history—at least the history of the surface of the sea. The foregoing summarization is intended only to illustrate the past impact of investigation and research into a totally new area that touches every part of the world, and to suggest an analogy. The nations of the West and the East stand today, regarding the undersea, where Spain and Portugal stood at the beginning of the fifteenth century. We believe that we are wiser than they were, that we have a better understanding of the subtle forces with which we shall be dealing. It is also true that both our destructive and constructive capabilities are far greater than those of our forebears. The struggle for control of the sea may be far more crucial to mankind in the future than it has ever been in the past.

The Impelling Forces

Man does have a noble, altruistic side—at least, so are we assured. For his own corporate and individual sanity he must strongly hold to this belief. Nevertheless, it is also accepted that basically man carries on his investigations, does his research, makes his inventions, and develops new expertise in the hope of practical rewards or solutions to pressing problems. Often, his incentives are military ones.

So far as future exploitation of undersea resources is concerned, it will not be many years before certain areas of the ocean will be known for their oil production, others for their food production, still others for their mineral deposits and mining. The Persian Gulf, for example, if closed by nets at its mouth and transformed into a gigantic fish farm, might be able to produce all the proteins required by a doubled or even tripled world population. By the year 2000, it can be predicted, there will be a determined effort for exploiting, yet conserving, the resources of the sea. Hopefully, we now know how to do a better job both ways.

Treasure will still be lost in the sea, but much that has already been lost will be recovered. Some will be in gold, silver, and other precious cargoes, but

The year after Columbus discovered the New World, the Pope mediated the rivalry between Spain and Portugal by decreeing a line of demarcation 100 leagues west of the Azores and Cape Verde Islands. Spain was given the rights of discovery and conquest to the west of the meridian; Portugal, to the east. A treaty between the two countries in 1494 moved the line farther west and gave Portugal the rights to what turned out to be Brazil.

there will be far more of value in the artifacts of antiquity. In some cases the actual substance of these artifacts will be long gone, but their form will have been saved by the preserving action of encrustation, for study by antiquarians and historians. The entire early history of mankind is liable to revision as a consequence.

There will be unlimited production of cheap, inexhaustible power if a feasible method of harnessing the tides is developed. The world's need for sweet water to restore its parched lands will be satisfied as soon as there is an abundance of power, either electric or atomic, and the land of history, the Middle East, may ultimately grow green again.

As a rule, however, military exploitation of any new development comes before its peaceful uses. Around the middle of the sixteenth century, written descriptions of a submersible boat, supposedly proposed to Charles V of Spain, excited interest in Europe and worry in England. It is probable that this boat was entirely fictional, or at least was never built; but in 1620 the well-authenticated boat of Cornelius van Drebbel was built in London. Little is known of this submarine except that King James I submerged in it, according to some accounts, and that Van Drebbel is said to have discovered the "essence of air" for renewal of its atmosphere—a claim which indicates that he must at least have considered one of the principal problems of submergence. In 1653 a "near-submarine" seventy-two feet long was built in Rotterdam. Not intended to submerge fully but to operate awash, it was wisely distrusted by the sagacious Dutch and never put to sea. Another pioneer submarine, built by an English inventor named Day in 1772, was designed to submerge, but Day and his boat were lost when it failed to surface after his second dive.

Less than two hundred years ago an American, David Bushnell of Saybrook, Connecticut, built the first operational submarine, the *Turtle*. In 1776 it attacked the British 64-gun ship-of-the-line *Eagle*, moored off Governors Island in New York harbor. Although the *Turtle* failed to sink the *Eagle,* the operation—the first of its kind in the annals of undersea warfare—was successful in the sense that the submarine reached its target, attacked, evaded counterattack, and returned safely to its base.

The submarine resembled two huge turtle shells joined together, with the long dimension—around nine feet—being the vertical one. Its crew consisted of a single man, who sat on a bicycle-type seat, worked a ballast pump with foot pedals, and cranked a propeller for propulsion with one hand while he steered with the other. A vertical shaft operated a second propeller for ascent or descent, and another vertical shaft operated a large screw by which an explosive charge, detachable from the *Turtle*, could be attached to the bottom of another vessel.

The little submarine's intrepid "crew"—an army sergeant named Ezra Lee—actually succeeded in getting under the *Eagle*, but there he ran out of luck; he could not get the screw to bite into the warship's bottom. Whether this was because she had been copper-sheathed (unlikely) or because Lee had unluckily struck upon some iron on the *Eagle's* bottom (he did believe he was in the vicinity of the rudder) cannot be stated for sure. What is known is that Lee, never able to get the screw started into the wooden hull above him, finally had to give up and return with the tide to his starting point near the present Battery Park on Manhattan Island. Once clear of the *Eagle* he surfaced for air, which by this time he must have needed badly, and was spotted by an alert British sentry on Governors Island. The alarm given, he cranked madly to escape and finally was forced to detach his "torpedo," set on short fuse, as a diversion. In this the resourceful sergeant was fully successful: The pursuing longboat was doused with water from the detonation of 150 pounds of black powder, and in the confusion he made his escape.

Twenty-four years later, the dominance of the British navy on the surface of the seas was again threatened by submarine attack, when Robert Fulton received funds from Napoleon to develop his *Nautilus* for France. Far more advanced than the *Turtle*, Fulton's vessel could sail on the surface, strike down its sail and hinged mast, and proceed beneath the surface, powered by a hand-driven propeller. Fulton demonstrated the practicality of his design by making several dives at Le Havre and Brest and blowing a forty-foot sloop to smithereens; then, apparently in fear that someone would try to steal his "plunging boat," he destroyed her immediately after completion of his public trials. Napoleon had evidently wished to inspect the *Nautilus,* and when Fulton's impetuous destruction of his craft made that impossible, he may have made up his mind that the inventor was merely another of those cranks with great ideas and no ability to put them into operation—even though the Prefet Maritime at Brest witnessed the tests and reluctantly reported their success. Napo-

leon, in fact, is reported to have characterized Fulton as "a charlatan and a swindler, intent only upon extorting money."

The British Secret Service had heard of Fulton's success, however, and in the naval archives there is a letter in which commanders of British fleets were warned to be alert for possible submarine attack. In 1803 His Majesty's commissioners made contact with Fulton, and in April 1804 he arrived in London, ready to do business. For two and a half years he remained in England, initially buoyed by hope of bringing his plans to reality, and finally in increasing petulance as he came to the realization that England had no intention of developing a submarine. The First Lord of the Admiralty, Earl St. Vincent, is reported to have advised Pitt, the Prime Minister, "Don't look at it and don't touch it. If we take it up, other nations will; and it will be the greatest blow at our supremacy on the sea that can be imagined." Embittered, Fulton returned to America in 1806, and finally won fame and a modest fortune with his steamship, the *Clermont,* in 1807.

Submarine development languished during the first half of the nineteenth century but received impetus again from the American Civil War. This war was the first modern war in so many respects that it is not surprising to find that it included the first sinking of a full-fledged man-of-war by submarine attack. This feat was achieved by the Confederates, who built three submarines: the *Pioneer,* one which was never named, and the *H. L. Hunley.* These were boats that could totally submerge and carry out submerged attacks patterned after Fulton's scheme of towing a floating powder charge astern on a long wire as they dived under the enemy and detonating it upon contact with her bottom. Not to be confused with the "real" submarines were a number of semisubmersibles called "Davids," also built by the Confederates. These were small, armored, steam-driven launches that attacked by ramming a "torpedo," rigged on a spar that projected over their bows, into the enemy's side. The David itself presented a minimum target, since it attacked while in an awash condition.

The most notable of the Confederate submarines was the *H. L. Hunley,* named after its inventor, Horace L. Hunley, a wealthy sugar broker, who perished in the sub's fifth and last accident. This boat, between thirty-six and forty feet long, was constructed by adding bow and stern to a four-foot-diameter ship's boiler that had been cut in half lengthwise and deepened. When completed, its cross-section was an ellipse with its long dimension, six feet, in the vertical plane. Propulsion was by a large propeller turned

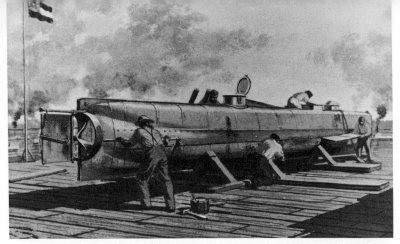

The first destruction of a man-of-war by a submarine took place during the Civil War, when the Confederate *Hunley* (shown here under construction) sank the Federal sloop-of-war *Housatonic* in Charleston Harbor. It was not a total victory, however, as the 36-foot submarine was also sunk. (U.S. Navy)

by hand by eight men sitting on the port side of the boat; and a ninth member of the crew, the commander, handled all details of submergence, steering, and attack (some accounts give *Hunley* a ten-man crew, with a separate helmsman).

After five crews had been lost in training, General Beauregard, who had sponsored the craft, forbade further submergence. Lieutenant George E. Dixon, of the 21st Alabama Infantry Regiment, now *Hunley*'s commander, requested permission to rig a spar torpedo similar to that already in use on the Davids and to operate the boat, trimmed down to the awash condition, as though she were indeed a David. In this manner the daring crew made a night attack February 17, 1864, on the new Federal sloop-of-war, *Housatonic*, then blockading Charleston. The outcome was the exact opposite of that in the Revolutionary War, when both participants survived; both attacker, *Hunley*, and victim, *Housatonic*, were sunk.

The lesson of submarine warfare was already clear to those who considered themselves analysts of war at sea. The usefulness of an effective submarine—if one could be devised—to a weak navy trying to contest the sea with a far stronger adversary was apparent. Possession of even a few submarines would have a massive effect in terms of restrictions on enemy tactics. Radical changes, perhaps even lifting of the blockades by the British in the Revolution and the Union in the Civil War, might have resulted had the Colonies had more *Turtles* and the Confederacy more *Hunleys*.

The final decades of the nineteenth century saw a spate of submarine invention. Two factors were primarily responsible for this, one technological and the other psychological. First was the invention of the electric motor and the electric storage battery, which provided a source of power, albeit a severely limited one, for submerged operations. Until then,

the only successful submerged propulsion was by means of the hand crank, for all other power sources required combustion of air in quantities that made their employment totally impracticable. To be sure, Thorsten Nordenfeldt of Sweden attained considerable success with a submarine having a steam-charged boiler. The steam engine, however, was unsatisfactory for submarine use for several reasons. On the one hand, it was difficult to keep the interior temperature down to bearable levels, and on the other hand, the furnace had to be closed down when the boat submerged, with the result that the gradual dissipation of heat from the boiler severely limited underwater running time. The second and perhaps determining factor in the new wave of submarine development was the announcement in 1888 by the United States of a large prize and a construction contract to be awarded to the designer of the first practicable military submarine that could fulfill certain specifications. Heretofore it had been the inventors who had tried to interest governments in their boats; now the shoe was on the other foot, and wealth and fame awaited the lucky man.

Some of the competitors, together with the dates of their first submarines, were: Claude Goubet (France), 1885; Gustave Zédé (France), 1886; Thorsten Nordenfeldt (Sweden), 1881; Isaac Peral (Spain), 1889; George C. Baker (United States), 1892; and John Philip Holland (United States), 1875. Because of this contest the United States is generally accorded the credit for developing the submarine, despite the earlier work of numerous inventors of all nations—this and the fact that the winner of the contest, Holland, had been in the field for nearly twenty years before the decision was announced in 1893, and was himself a naturalized citizen of the United States. There were some, of course, who claimed that this sequence of circumstances could not have been entirely accidental or coincidental, but the truth of the matter is that Holland deserved the prize as much as anyone else of his time simply on the basis of the number of submarine boats he had built and the soundness of his designs.

Holland won the prize with the design he proposed, but this was not the submarine which the United States Navy commissioned him to build. More requirements were added, in the time-honored manner by which genius is stultified by ignorance, until Holland protested that the resulting submarine would never work. Nevertheless he was directed to construct it in accordance with the terms of the contest, and he made the effort. As predicted, the boat, steam-propelled, was a disaster and could not be tested at sea. Holland requested and was granted the privilege of building another boat to his own design without interference, at his own financial risk. This boat, which he always called "Holland No. 6," was a complete success. For surface running it used a gasoline engine and for submerged operations an electric motor and storage battery. On the surface, with the engine running and turning both motor and propeller, the electric leads to the motor could be so switched that the motor became a dynamo to recharge the battery. The sub had a torpedo tube and a "pneumatic cannon," but no periscope, which was yet to be invented. Sighting for submerged navigation or firing of the torpedo was through glass eye-ports set into the side of the entry hatch, requiring that the diminutive craft broach the surface momentarily.

On April 11, 1900, the U.S. Navy formally took title to the boat, paid her inventor, commissioned her into the Navy, and named her *Holland*. The United States Submarine Service consequently calls this date its "birthday." The *Holland* was followed by a number of improved versions of the same design, all lumped together as the "A-class," and every few years a new and better design, meriting a new letter designation, came forth. Thus A-boats, B-boats, and C-boats came out in quick succession, and by World War I there were the K's and L's. The entire alphabet, in fact, with certain gaps (Q, U, X, Y, Z) had been employed before World War II. The impetus of the two World Wars was, of course, responsible for a tremendous growth in submarine design, but even more extraordinary strides have been taken since the Second World War, with the introduction of nuclear power, which has opened horizons to submarine development several orders of magnitude greater than were previously conceivable. Military submarines today must rightly be considered, with their weapons and all their other accouterments, as among the most remarkable things ever built by man. As a measure of the change wrought in an extraordinarily short time, a single nuclear-attack submarine in the hands of either side at the battle of Jutland would have turned that famous naval conflict into a shambles. The same submarine deployed against the attacking Japanese force at the battle of Midway would have had an equivalent effect. And today, the huge, silent Polaris submarine, cruising deep on her secret station, unseen from the moment of departure from home base until return, is not an attacker of ships but a destroyer of nations.

In the meantime, development of undersea vehicles for other than military purposes could only be characterized as insignificant until very recent years. In the important turn-of-the-century period there was but one inventor, the United States's Simon Lake,

who constantly emphasized peaceful exploitation of his submarines' capabilities. Lake designed boat after boat—and some of them he built—to explore the ocean bottom, to salvage cargoes of sunken ships, and to assist in underwater construction operations. He pioneered the idea of bringing divers to and from the scene of work in the safety and relative comfort of his submersible. By and large, his imaginative proposals were looked upon with amused disdain. Lake's was a solitary and unheeded voice, for there were three things he needed and did not have. In ascending order of importance, they were: (1) adequate technology, particularly in metallurgy; (2) a source of power to operate the complex mechanisms his systems should have had (but many of which, themselves, had yet to be invented); and (3) most important of all, strong public and commercial interest in the seabed. Simon Lake's contemporary critics believed that man was at the end of the age of exploration, that the Arctic and Antarctic were the only frontiers left to conquer, and that Lake was a crusader born too late. A more recent evaluation is that he was, on the contrary, far ahead of his time.

Dating the submarine from the *Turtle* of 1776, its nearly two centuries of history and development have been almost entirely military in nature. The concept of man going beneath the sea for any other purpose began with Lake, but its serious development actually began after World War I; as the result of a number of submarine disasters, there was a need for finding ways to rescue personnel trapped in submerged submarines. During the same period there were the explorations of Dr. William Beebe, who developed a tethered diving sphere made of steel—about eight feet in diameter and capable of being lowered to great depths—that he called a "bathysphere." In 1931 he reached a record depth of more than 3,000 feet, near Bermuda. Since World War II skin diving has emerged as a new form of water sport, and, of course, overtones of serious research and exploration were present from its inception. Finally, the past decade has seen development of stout little "work" submarines, over which the spirit of Simon Lake must be hovering with unalloyed paternal delight. Already these have proliferated greatly in numbers and designed purpose, from exploration to heavy commercial work to tourism, and the field is only beginning to be exploited.

Control of the Sea

The German navy in World War I provided the first dramatic demonstration of the capability of the

253

submarine to contest a stronger navy's control of the sea. Starting with relatively few submarines, which today would be considered of the most primitive design, Germany achieved astounding results. It may be asked why it was that England's submarine force, approximately the size of Germany's, did not have the same massive impact. The reason is neither the method of employment nor the quality of the equipment (essentially equal) but quite clearly the size and nature of the target.

After the war started there was no German merchant marine, for it had been swept off the seas by far-ranging British cruisers. British submarines consequently had nothing to shoot at except in special situations, like those in the Sea of Marmara—a German "lake" until British submarines entered it—and except upon the very rare occasions when they came upon a German ship of war. German submarines, by contrast, were continually met by ships of all types at all times. Their only problems were to identify the neutrals and, until declaration by Germany of "unrestricted submarine warfare," to obey restrictive rules of sea-war that had been established prior to the invention of the submarine. Under these circumstances, had the German submarines been only half as many or half as effective they still would have had tremendous effect upon England. Had there been twice as many they would either have won the war for Germany or brought the United States into it at a much earlier date—most likely the former.

The breakdown of the restrictive "cruiser rules" with the advent of the submarine provides a good example of what can happen when a totally new element is introduced into a heretofore static situation. These rules permitted a ship of war accosting an enemy merchant ship to capture it, or even sink it, but not before providing for the safety of its crew and passengers. The customary procedure was to take them aboard the warship until they could be landed somewhere, or (after several captures had swelled the prisoner group) to give them a small ship of little value as a "cartel" to make their own way to port. Lifeboats, unless very close to shore in ideal weather, were expressly disallowed as "a place of safety." If a merchant ship refused to halt, the warship would fire a shot across her bow to indicate she meant business, had the power, and was in position to use it. If the merchantman still persisted, the warship was justified in shooting to hit but even so was obligated to stop shooting as soon as resistance ceased, and from that point on to render all possible assistance to save life. Neutral ships might be taken into custody in case of doubt as to actual neutrality,

or suspected contraband cargo, but they had to be sent into harbor for court action. The fundamental point was that bona fide neutral ships and owners thus could obtain redress in a court of law.

Manifestly, the submarine was far too small to take extra passengers on board, nor had it extra personnel to place aboard captured ships as prize crew. While a submarine running surfaced on its diesels might be able to overtake a slow merchantman, it could never hope to do so submerged, where its speed and endurance were much reduced. Being a very small ship herself, the submarine carried only a small gun, very low to the water, with the result that gunnery from a submarine plowing surfaced through even small seas was generally ineffective; thus the shot across the merchantman's bow was no longer meaningful, for the threat of its being sunk by a surfaced sub was not a credible one. Furthermore, on the surface, the submarine's low buoyancy made it extremely vulnerable to being rammed and sunk; a single lucky shot from a gun of even very small caliber could disable or sink it, or make further dives impossible. Submergence was the submarine's only weapon, and its only defense. In effect, the cruiser rules required this to be given up.

Germany's position was that she would obey the cruiser rules if in return any merchant ship accosted by a submarine would be obliged to surrender, would not try to ram, not radio for help, and not try to escape. International law in fact did permit a fast merchant ship to escape from a warship by speed alone if she could, and a "weatherly" sailing ship might sail closer to the wind and thus avoid a less handy sailer attempting to accost her; but no other counteraction was permitted on pain of losing noncombatant status—hence the original significance of the shot across the bow. England responded with orders to all British skippers to surrender under no circumstances whatsoever, but instead to call for help by wireless, giving the position of the submarine. They were directed to maneuver always to keep the submarine astern, or, if the opportunity arose, to ram it. The decoy ship was invented—an innocent-looking steamer made as unsinkable as possible, armed with several large, concealed guns, and commissioned in the British navy as a warship. The tactic of these "Q-ships" was to appear to carry out the merchant ship's role of halting when accosted, and when the submarine approached in accordance with the rules, to suddenly unmask the concealed guns and open fire at murderously close range.

The two sides of England's campaign are clear today. If Germany acquiesced to the cruiser rules, her submarine campaign would lose a great percentage of its potential effectiveness. If Germany reacted as England expected she would—with "unrestricted" submarine warfare—England's campaign to get the United States into the war on her side would be greatly aided. The latter, of course, is what happened.

The importance of the submarine to sea warfare was proved by the fact that it changed the rules of war (incidentally making it less humane). It would take another generation and another war to realize the truth of this, and for twenty-one years more the legal fiction that submarines would henceforth use the cruiser rules of warfare remained in the statutes of international law. The change, however, was a fact by November 1918.

The Two World Wars

Persons accustomed to thinking that advance of years must always bring change will find it something of a surprise that submarining—at least in its operational practice—changed so little between the two wars. Possibly this opinion might not be held by an envious 1918 submariner, admiring the superior welded construction, more efficient equipment, and better living quarters of the boats twenty years later; but the World War II submariner would see little difference, for the improvement in his equipment was more than matched by improvements in antisubmarine warfare, and with the exception of the "cruiser rules" there was little difference either in operations or in results.

Thus, in World War I, Kapitänleutnant Otto Weddigen, in the tiny, 450-ton, prewar, kerosene-burning U-9, sank three British cruisers within the space of an hour and a half. The three had been cruising in formation, and when the first was torpedoed the other two rallied to its support. But the submarine, though detected and attacked, was armored with two dozen feet of seawater over its hull and was totally immune to counteraction. Fourteen hundred men went down with the three ships. Later the U-21, a newer, larger submarine, torpedoed two old British battleships near the Dardanelles, actually diving beneath one of the sinking ships to evade pursuing destroyers, which by this time had rudimentary sonar equipment and carried ammunition that theoretically could shoot into the water and hit a submerged submarine. And a German commander bearing the improbable name of Lothar von Arnauld de la Periere became the outstanding submarine commander of all time in terms of enemy tonnage de-

stroyed, mostly in strict conformity with the cruiser rules. The ship he surrendered in 1918 was a veritable submersible cruiser, the biggest ship of her kind in the world—nearly 400 feet long, displacing more than 4,000 tons while submerged, able to stay at sea for months and to remain totally submerged, without replenishment of air, for three days. She was not rivaled in size or power until the between-wars British *X-1*, American *Argonaut*, and French *Surcouf*.

The British also had their submarine aces in World War I. With submarines but little superior to the U-9 they penetrated into the enclosed Sea of Marmara to wreak havoc among the German and Turkish shipping supporting the Turkish troops on the Gallipoli Peninsula. To enter the sea the tiny submarines had to pass through nets and minefields laid by the Turks across the Dardanelles Channel, but once free in the Sea of Marmara they could operate with impunity, concerned only with the expenditure of torpedoes and provisions and the ever-worrisome need of running the same gauntlet on the return trip.

One British skipper, Lieutenant Commander Martin Nasmith, gained special fame by his imaginative exploits. For example, he rigged his torpedoes so they would remain afloat, instead of sinking, if they missed their targets. Nasmith then pursued his stray torpedoes, brought them aboard through a torpedo tube, overhauled them, and fired them a second time. He also shelled beaches and roads at strategic points, and once put his executive officer ashore to place explosive charges under a railroad track that carried supplies to Gallipoli.

In World War II, except for a few halfhearted incidents on both sides in which their impracticability was again demonstrated, no one worried about the cruiser rules of submarine warfare. Both sides simply ignored them, as they also ignored concern about civilians killed in bombing raids from the air. Technology had gone forward, could no longer be denied. Henceforth, civilians in the path of war were simply unfortunate. Industries, most of which were located inside cities, were legitimate targets; by extension, so were the cities themselves, and there was no way to separate them from the people living and working there. The same was true of ships and their passengers.

The exploits of German submariners in World War II paralleled those of their predecessors a generation before. Lieutenant Gunther Prien daringly penetrated into the British anchorage of Scapa Flow to torpedo and sink the battleship *Royal Oak*. The magnitude of his feat is intensified when one appreciates that he had to make several repeated attacks before achieving success. Kapitänleutnant Otto Kretschmer, warmhearted personally but with a mind of steel, perfected the convoy-attack technique, shooting torpedoes from the surface between the convoy lines.

England also had its heros. Lieutenant Commander David Wanklyn's daring exploits in the Mediterranean (97,000 tons of Axis ships, plus a destroyer and three submarines sunk) won him the Victoria Cross, and Lieutenant Commander James S. Launders in 1945 accomplished a feat hitherto considered impossible: He detected, tracked, and successfully attacked a German submarine—the unprecedented aspect of this attack being that both submarines were totally submerged during the entire period. He knew the other submarine was German because of its characteristic noises—and the fact that no British submarine was authorized to transit his area.

United States submarines, which did not figure in World War I, played an important role in World War II against Japan. The American submarines were similar to those of other nations but larger (because they normally had to travel greater distances to reach their combat zones) and more uniform in design. While German and British subs were of several classes, even the largest of which was smaller than United States subs, American boats were generally of a single class—300 feet long and 1,600 tons in displacement. The American subs carried crews of seventy-five men and eight officers and could remain at sea for up to ninety days. Commander Eugene Fluckey—a cool number, like Nasmith—engaged in convoy actions in the South China Sea, entered a Chinese harbor to destroy shipping at anchor, landed a commando party to blow up a train, and bombarded shore points. Commander Lawson P. Ramage cut into the middle of a Japanese convoy and, like Otto Kretschmer, remained on the surface and shot his way out, leaving behind him a trail of sinking transports and tankers. Commander George Street slipped along the shallow side of a Japanese minefield at night, on the surface, to reconnoiter a small harbor. The harbor had two entrances; finding nothing at the first, he ghosted around to the second, where he found a tanker and three escorts. He promptly attacked, sinking the large ship and two of the three small ones, and retired unscathed. The author of this chapter stood beside him on the bridge, had the privilege of actually aiming and firing the torpedoes, and saw the results.

The differences between the exploits of the two world wars were in degree, not kind, and the similarities were not caused by the flattery of imitation but by the similarity of circumstances in both cases. There are strong parallels between Nasmith and Fluckey, Prien and Street, Ramage and Kretschmer—and, indeed, between Weddigen, who sank the three British cruisers in 1914, and Commander Sam Dealey, who sank five destroyers in five days, thirty years later. Once entered into the lists of combat, the submarine proved to be a magnificent tool for perceptive persons who intuitively knew how to exploit a new dimension in the art of sea-war. It gave free play to ingenuity and creativity—if the reader will accept anything connected with war as creative—and redemonstrated a fundamental aspect of man's nature: As his capabilities increase, he invariably will probe them until he has satisfied himself as to his new natural boundaries. Thus the combat submarine gave the sea warrior a tremendous new spectrum of battle, into which he leaped with verve and originality.

The Nuclear Age

It can be said that the two world wars brought submarine capability to a plateau, where it might still be had nuclear power not come on the scene. At the end of the war some submariners became concerned as to the possible effect of an atomic depth charge, should one be developed, as appeared likely. It did not take them long, however, to appreciate what nuclear propulsion could do for the submarine. All one had to consider was the prospect of fantastically increased submerged speed and endurance. A nuclear-powered sub could cruise submerged at twenty knots for one thousand hours, compared to eight knots for one hour or four knots for twelve hours, which had been the approximate previous limits before surfacing—or later, snorkeling—became necessary. Nuclear power freed the submarine from the need to return constantly to the surface to recharge its electric storage batteries, for it is the only mobile source of power that does not require combustion, and thus the consumption of oxygen, somewhere in the cycle (the battery, of course, was not a true "source," for it required frequent recharging of its comparatively small storage capacity). Freeing submarine engines from the need for air converted the sub from a surface ship that could occasionally submerge for short periods to a submerged ship that could stay down indefinitely; thus, the use of nuclear power produced the first "true submarine."

The crew still needed oxygen, but this much smaller requirement could be satisfied by large steel bottles containing the vital gas compressed to high pressure. Later, mechanisms using some of the abundant new power to produce oxygen from seawater and dump carbon dioxide waste into the sea became standard. As a consequence of these far-reaching developments, the entire appearance of the submarine changed. Designers optimized its shape for submerged performance and paid no further attention to surface speed, which had hitherto been an important criterion. It seems only natural, today, that the United States should have selected the new, tremendously improved, submarine for the most important mission placed upon any of the military forces of any country: that of preventing nuclear war through constituting itself an invulnerable retaliatory system—therefore a deterrent—of such awesome destructive power and implacable surety that a nuclear attack on the United States would amount to national suicide for the attacker.

At the same time as the new submarines called forth for themselves a new mission, they also required far more support in areas that had previously been almost totally ignored. It was necessary, for example, that the new submarines (more than 400 feet long, with an 8,000-ton displacement submerged) remain totally submerged at all times in order never to give away their operational methods, locations, strength, or procedures. This required new navigation systems. The necessity of being able to fire missiles accurately from any location required much larger crews (140 men) than submarines had previously had (with some exceptions, such as the 177-man U.S.S. *Triton*); much more training of a very precise sort for the crews; far more accurate charts than had heretofore existed; and far more specific knowledge of the gravity of the earth, its force fields, and the amount and direction of the slow, precise movement of its magnetic lines of variation. For most efficient operation of the huge, cigar-shaped submarines, with their sixteen intercontinental ballistic missiles, and the highly secret, complicated gear needed to run them, far greater knowledge was required of the characteristics of sound in the sea. Since bottom conditions affect sonar conditions, the bottoms of all the seas, shallow and deep, had to be studied, and their effects noted. In short, military interest in submarines gave impetus to research into all facets of the sea.

Nor was this all, for the United States recognized that the tremendous improvements in its submarines required corresponding improvements in antisubmarine warfare (ASW), not only because Soviet nuclear-missile submarines were bound to appear (as they since have) but also to render all possible protection to American submarines.

The result of both the pro-submarine and antisubmarine processes was the same: a world-wide increase in interest in the undersea. Among the first nations, along with the United States, to be so interested were France and, predictably enough, the Soviet Union. As always happens, this additional interest developed offshoots unrelated to the original objective of improving national combat potential. For all nations the undersea became, in effect, the new frontier of research that Simon Lake had predicted. Today—as these words are written—investigations in all sorts of diverse fields are now reaching into the sea. As our interest increases we find new ways to employ our new technological capabilities, and these new employments in themselves increase our efforts, our interests, and our research. The involvement of man in the undersea is only beginning.

The Little Research Submarines

Ocean explorers in the past conducted their investigations from surface vessels, using dredges, nets, corers, and other tools lowered to the bottom on long wires. The work was slow, frustrating, and generally only approximate, for the oceanographer was working blind from a great distance and could only guess about what might have barely missed his probe. In effect, oceanographers were "sampling" the seas, not "exploring" them. There obviously was a need, in addition to increased and improved sampling techniques, for direct personal observation and survey—in other words, for a manned undersea vehicle.

Neither military submarines nor diving bells were adequate to the job. Submarines were once equipped with ports for undersea viewing, but these have been frowned on since World War II because of the danger that they will be shattered by exploding depth charges; periscopes, meanwhile, are good only for viewing above the surface of the water. Even if a military submarine did have a "window" of some sort, it would be much too risky to operate so large a vessel near the ocean bottom. The conventional diving bell, on the other hand, although more suitable than the submarine for observational purposes, did not permit explorations since it was suspended by wire or cable from the surface. First, interest turned to diving bells that could be maneuvered within the scope of their tether; then came the proposal to remove the tether entirely, releasing the bell from

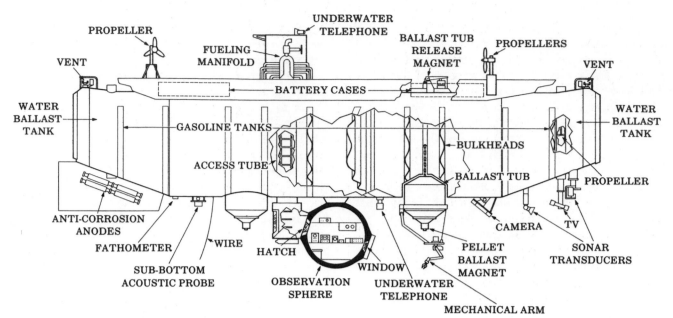

PROPELLER · VENT · FUELING MANIFOLD · UNDERWATER TELEPHONE · BALLAST TUB RELEASE MAGNET · PROPELLERS · VENT · BATTERY CASES · WATER BALLAST TANK · GASOLINE TANKS · BULKHEADS · WATER BALLAST TANK · ACCESS TUBE · BALLAST TUB · PROPELLER · ANTI-CORROSION ANODES · FATHOMETER · WIRE · HATCH · WINDOW · CAMERA · TV · PELLET BALLAST MAGNET · SONAR TRANSDUCERS · SUB-BOTTOM ACOUSTIC PROBE · OBSERVATION SPHERE · UNDERWATER TELEPHONE · MECHANICAL ARM

Having soared to unprecedented heights in a balloon, Professor Auguste Piccard built with his son Jacques a gondola to explore the deepest regions of the oceans. Their invention, the bathyscaphe *Trieste,* consisted of a sphere seven feet in diameter supported from a sixty-foot "hull."

the heavy drag of its wire, greatly increasing its radius of observation—and in effect converting it into a tiny submarine. The result was the bathyscaphe, which was so named to distinguish it from Beebe's bathysphere of a quarter of a century before.

First honors for development of the bathyscaphe belong to the Swiss Piccard family. Auguste Piccard and his brother Jean had experimented with stratospheric balloons and ascended to unprecedented heights in a special balloon fitted with a pressurized gondola. The balloon with gondola carried the initials F.N.R.S. (for Fonds National de la Reserche Scientifique) and became known as FNRS I. Auguste then applied the ballooning principles to the ocean and was instrumental in building the first bathyscaphe (FNRS II), which made an unmanned dive to 759 meters off Dakar in 1948.

The French Navy began construction, with Piccard's assistance, of FNRS III, the second generation of deep-sea balloons. Relations between Piccard and the French deteriorated, however, with the result that Piccard left the Navy and built the *Trieste* on his own, with the help of his son, Jacques. In the meantime, the French continued construction of FNRS III and made a record manned dive off Dakar to 4,050 meters in 1954. Their success has continued, and in 1960 FNRS III was replaced with a newer model named *Archimede.*

Trieste, like FNRS III and *Archimede,* is an underwater rigid-framed dirigible. The original *Trieste,* fitted with a sphere built by Krupp in Germany which was suspended from a 60-foot, boat-shaped "hull" containing gasoline for flotation, made the deepest dive on record, 35,600 feet, in 1960. *Trieste II* has a new sphere and a new 75-foot hull, and is designed for 20,000 feet depth. Everything is made to be operable at deepest depth, and everything is exposed to pressure except the interior of the personnel sphere. Her motors are low powered, giving a speed of two to three knots and endurance up to ten hours. The Krupp sphere has not been used since the deep dive of 1960; the original hull may be seen at the U.S. Navy Memorial Museum at the Navy Yard in Washington, D.C.

About 1959 Jacques-Ives Cousteau, a retired French naval officer and one of the inventors of the scuba, designed and built his first version of the diving saucer, a highly maneuverable little submarine propelled by water jets. The sub is small enough to be transported in a plane, yet it can take two men down several hundred feet for several hours and pick up small specimens from the bottom by means of a remote-control, hydraulically operated claw.

In the United States, interest in submersibles grew quickly, reflecting the nation's growing commitment to deep-sea research and exploration. One result was that the *Trieste's* record-setting dive in January 1960 to the bottom of the Challenger Deep was made under the auspices of the U.S. Navy, which had purchased the little submersible two years before.

Shown shortly after it was built in 1954, the *Trieste* hangs from a crane before being lowered into the Mediterranean for tests at Castellammare di Stabia, Italy. (Wide World Photos)

The Navy is continuing to use the *Trieste*, now based in San Diego, California, in a systematic program of deep-sea research.

One of the impelling reasons for this research was the fact, clear to all submariners, that the new military submarines were already capable of diving to depths far beyond those from which the crews might be rescued. Heretofore, rescue equipment had been competent to go without collapsing to any depth to which a submarine could conceivably go; thus if a submarine were to be damaged and bottomed with some of her crew still alive, there was a fighting chance that they could be rescued. Even the early rescue equipment was a response to disaster. Deep in the soul of the U.S. Navy was the 1927 tragedy of the *S-4*, in which seven men were slowly suffocated off Provincetown, Massachusetts, inside the hook of Cape Cod, at a depth of only one hundred feet, while rescue ships impotently hovered above them. The system this trauma produced was proved in 1939, when most of the crew of the bottomed *Squalus* were saved by a just-developed rescue chamber, a tethered diving bell able to attach itself to the hatch of a damaged sub. But the new submarines could go much deeper than the *Squalus*, much deeper than the rescue chamber. Very conceivably,

a sub could lie on the bottom at a depth below reach of the rescue chamber, with survivors doomed to an agonizing repetition of what had happened to the men of the *S-4*. A rescue capability able to match the deepest noncollapse depth of any submarine was an obvious necessity.

Added to the need for rescue was the dawning realization that an object lost in the sea need not actually be lost forever. The capability was at hand, needing only development and application, to provide the means of finding and salvaging small objects, say the size of a basketball, from the bottom of the ocean.

The beginning of this concern, and the associated thought processes, could be dated from the mid-1950's. Some developmental progress had been made in solving some of the problems—although by no means was everything resolved, or even heading in a clear direction—when in April 1963 the new U.S. nuclear submarine *Thresher* went down off the coast of Maine. She had been on a post-overhaul test dive, and was escorted by the submarine rescue vessel *Skylark*, a converted Fleet Tug, with whom she could converse through a short-range, acoustic, "underwater telephone." Something went wrong on *Thresher's* deep dive—what it was will never precisely be known—and she sank in water a mile and a half deep, less than a day's cruise from her home base. *Skylark* heard the entire tragedy through her sonar and the underwater telephone.

There were no survivors, nothing found to indicate why she had failed to surface. There was only an emotion-wracked last message from her captain—garbled, distorted by the watery distance between his undersea telephone transmitter and the receiver on the escort vessel above. The memory of this doomed voice, endeavoring in its final report to explain for the benefit of others what it was that had gone wrong, will undoubtedly live forever in the minds of the few people who heard it. Unfortunately, reproduction of what it said or any recollection of the captain's message, other than a few phrases, as "now exceeding test depth," was impossible. Vainly did those in charge ask themselves why there had been no tape recorder to make a permanent record for later dissection, vibration by vibration if necessary. To this there can be no answer. It was a situation that had never happened in exactly the same way before. Needless to say, such a tape recorder will be available and ready from now on, even though another occasion so clearly requiring its use will most likely not take place very soon, if at all.

The circumstances of the loss added an emotional need to the statutory investigative requirement of finding out, as nearly as it could be found out, what had happened to the ill-fated ship. Men must learn from experience, and this means that disaster can never be accepted with resignation. The U.S. Navy intensely felt its collective responsibility. The *Thresher* was down at 8,500 feet, far below the depth at which the pressure of the seawater would collapse her sturdy hull. Nothing could be done for her crew. Death had been instantaneous; no one could possibly have suffered, except in the realization of impending dissolution. But the living had to learn why their friends had died.

A tremendous search was organized; shore beacons for triangulation purposes were set up; promising locations of possible wreckage on the bottom were carefully plotted. *Trieste* was loaded into a Dock Landing Ship at San Diego and started around through the Panama Canal to the scene. Months of careful effort were rewarded at last, and the smashed and scattered remains of the lost submarine were viewed from the *Trieste* and by means of a camera that was towed from the research ship *Mizar*. The wreckage was surveyed and photographed, and a piece of it was recovered. The distribution of parts of the submarine over a wide area, and their con-

dition, made possible some authoritative evaluation of what had occurred in the last moments of the *Thresher*.

The Lost H-Bomb

Early in 1966 a U.S. Air Force bomber collided with its refueling tanker over the coast of Spain. Both planes were destroyed and the four unarmed hydrogen bombs carried by the bomber fell to earth. Three bombs landed in fields near the town of Palomares, and they were recovered by the search teams immediately sent to the area, but the fourth, nowhere to be found, had evidently landed in the sea. Thus began the most massive concentrated underwater search in history, with every resource of the United States committed to the task. The prediction that it might be necessary to recover small objects had come true with a vengeance, and far sooner than might have been expected. As usual, there had not been time to get ready, even though the need to develop this capability had been foreseen.

But the experience of the *Thresher* disaster did provide many dividends. Some of the rescue personnel involved were available, and so were some of the surface ships, their special equipment still operational. The new techniques that had been developed had not been forgotten. New deep-diving research submarines, nonexistent in 1963, had been built. The needs of the operational organization, so recently experienced, were remembered; the reports and recommendations were quickly reviewed. A large naval task force was hastily assembled on the spot, the ocean floor was divided into small squares, and a program of carefully searching every square was instituted. Instantaneously it was apparent that there were many new problems, too, for the depth of water varied from a few feet, close offshore, to several thousand feet; and the bottom, far from being relatively smooth, was a great jumbled mass of rocky outcroppings, ravines, the debris of ages, and mud. An incredible effort was made, and eighty days after the collision the missing hydrogen bomb was picked up from the ocean floor six miles offshore and half a mile down.

Honor for the recovery of the bomb must go primarily to the two Deep Submergence Vehicles (DSV's) *Alvin* and *Aluminaut*. Both had only recently been completed and tested, and as matters turned out, these two carried the brunt of the labor of the search in deep water. Despite the willingness to help of every member of the 3,000-man and 25-ship

Alvin's pressure hull is barely seven feet in diameter, with a viewing port straight ahead and three more at various angles. Her small size and maneuverability helped her locate an unarmed hydrogen bomb in the murky water more than half a mile below the surface at Palomares, Spain. (U.S. Navy)

Alvin moves into her "hangar," the well of a U.S. Navy Landing Ship Dock, during the search for the missing bomb. She found it after 34 dives. (U.S. Navy)

Fifteen feet of *Alvin's* 22-foot length consist of a conical attachment, which terminates in a swiveling propeller. The DSV, lost in deep water in 1968 while being readied for a dive, was recovered the following year and is now back in service. (U.S. Navy)

The research submarine *Aluminaut* begins a descent off Bimini Island in the Bahamas. (Reynolds Metals Company)

Navy task force that had been assembled, the small size of the two research submarines (the pressure hull of *Alvin* is a 7-foot sphere; that of *Aluminaut,* an 8-foot cylinder, 44 feet long) and their unusual technology made it necessary for their two- or three-man crews literally to work around the clock day after day. More than once operations had to be slowed, or the excuse of bad weather gratefully accepted, to provide some much-needed rest for the played-out technicians and scientists, who were the only ones qualified to work on or operate the unique pressurized, oil-encased, electric propulsion motors, and the pressure-proof articulated arms and claws for picking up small objects.

In the end, it was the *Alvin* that found the bomb. Her crew, from observation of the extremely rugged bottom topography, recognized that there was a strong probability the bomb might have landed on one of the steep slopes which were everywhere in the area of interest. If so, the heavy weapon might very possibly have slid for some distance downhill and left a recognizable track. So *Alvin* asked for a hypothetical description of what kind of tracks might be left by a hydrogen bomb rolling or sliding sideways, or sliding end-on, down one of the ooze-covered slopes, perhaps into a narrowing crevasse. With this description firmly in mind, *Alvin* began to search along the jagged contours of the bottom, choosing courses perpendicular to the most probable track of the weapon, venturing into close quarters between outcroppings of the earth's mantle, which only she could explore. The area to be searched was so large, the visibility (only five or six feet) so poor, and the object searched for so small, that in retrospect it is difficult to conceive of any other method of search which could not have been called mere "sampling." Even the terrain contour search could hardly be better classified. But it did have the advantage of maximizing the chances of success under the existing conditions.

A sophisticated explorer, *Aluminaut* carries a pair of manipulator arms, underwater lights, television and still cameras, and complex electronic equipment. She moves along the ocean floor either on skids or on detachable wheels. (Reynolds Metals Company)

Some description of the *Alvin* is necessary. The pressure hull of the little submarine is a sphere just seven feet in diameter, with one viewing port straight ahead and three more ahead and down, at various angles. External propulsion machinery and considerable other equipment is aft, blocking any possible viewports in the astern direction. The vehicle is twenty-two feet long, and some fifteen feet of its length—consisting essentially of a conical attachment on the seven-foot sphere—terminates in a swiveling propeller used for both propulsion and steering. This unexceptionable arrangement became of crucial importance in the discovery of the lost weapon and perfectly illustrates why the research submarine must be of such small size.

On *Alvin's* tenth dive a furrow on the bottom was sighted that approximated the best estimate of what the lost bomb's track might resemble if the weapon had slid downhill end foremost. Promptly, *Alvin* began to follow the track, but she was near the end of

her limited battery endurance, and difficulties resulting from the steepness of the slope—she could not go straight downhill without hitting the mud-covered basalt with her propeller—caused so much expenditure of time and energy that her battery became completely exhausted before she reached the bottom of the incline. She was forced to surface, her contact broken. For a week the crew of the *Alvin* strove without success to find the furrow again, carrying this time a sonar beacon to plant alongside it. Then they were further frustrated by being temporarily shifted away for four more days in pursuit of another lead. The day after her return *Alvin* saw the furrow finally, but again she had to surface from exhaustion of the battery. This time, however, she had planted the beacon. Two days later she saw the track for the third time, and now she backed down the slope, her propeller well clear of the bottom. Her two-man crew, following the trail through her bow ports, were dramatically rewarded by the sight of a parachute enshrouding a cylindrical object of the right size and shape.

But this was not the end of difficulty, for the first attempt to lift the weapon failed when the hoisting line parted. Nine days of near-microscopic search of the bottom were required before the bomb was again located, 350 feet down-slope and 300 feet deeper than before—this time on the lip of a narrow crevasse into which even tiny *Alvin* could not have followed.

During the entire period of search *Alvin* was ably seconded by her much larger sister DSV, *Aluminaut,* which has an overall length of more than fifty feet. Significantly, and not surprisingly, it was the smaller and more maneuverable *Alvin* that several times ventured into places where the larger craft could not go, three times found the telltale track, and twice located the bomb on the bottom, more than half a mile below the surface of the Mediterranean Sea.

It must be admitted that a not-inconsiderable share of good luck, helped out in large measure by brilliant deduction and performance, was required for the successful recovery operation. The hard fact is that just as in the case of the *Thresher,* the necessity arose before the U.S. Navy was completely and comfortably ready. In both cases failure might easily have resulted, and in both cases the all-out effort required was much greater and more expensive by every measurement than it need have been. Considering only those operations that directly and specifically contributed to the objective, much effort in both cases was wasted.

There was, in short, still very much to learn when, in May 1968, the nuclear submarine *Scorpion,* homeward bound from the Mediterranean, failed to arrive at Norfolk. She had last been heard from in the vicinity of the Azores, and it was here that the search was concentrated. The problem was immensely more difficult than the search for the *Thresher* or the lost H-bomb, since in both previous cases the areas of search had been small. Again the full resources of the U.S. Navy were galvanized for the task, and again improvisation was the order of the day—but with the body of accumulated experience to draw on there was greater economy of effort. Some of the techniques used are still classified, and full disclosure of the findings of the investigation has not yet been—and may never be—made. Significantly, however, the *Scorpion* was discovered and photographed, wrecked on the deep seabed nearly two miles down, somewhere south of the Azores. Significant, too, is the fact that it was the specially

designed ocean research ship *Mizar* that finally located and photographed the wreckage with her towed camera apparatus. *Mizar* had also found the *Thresher,* and she had proved immensely valuable in the search for the lost bomb.

The Deep Divers

But the outstanding successes in finding the *Thresher,* the H-bomb, and the *Scorpion* have not deluded the Navy into thinking that it has solved the problems of locating objects lost in the depths of the sea. Only those who were intimately involved have any conception of the arduous effort required and the extraordinarily small margin by which success was achieved in each case. They were a remarkable training program, and they produced a capability that did not even remotely exist a few years previously; but three successes in a row only make future failure in any fourth comparable task totally unacceptable. Yet the certainty of success with present equipment and techniques is all too small, for the world is only at the threshold of the science of the ocean floor.

The United States Navy, charged by Congress with responsibility for salvage in the deep sea, has fortunately seized the bit in its teeth. Salvage of small or even large objects lost in the depths is not far removed from the rescue of personnel trapped at the same depths, whether they be members of a scientific expedition, workers on a submerged oil well, or the crew of a military submarine. The capability to perform such salvage-rescue missions can be developed and maintained only by the government, as opposed to private interests, because of the fantastic cost of the equipment, the uncertainty of how and where it will be required, and the necessity that, though infrequently needed, it be ready to respond instantly to emergencies in which lives are at stake. The conclusion was that such an effort had to be established at the expense of all the people and maintained as a national resource in the event of emergency. This having been recognized, the Deep Submergence Systems Project has been established with the combined goals of exploration, salvage, and rescue from the depths of the sea. Divided into five phases, the effort has already been continuing for a number of years, and one of its first accomplishments has been the construction of small submarines known as Deep Submergence Rescue Vehicles, or DSRV's. These rescue vehicles are small enough to be transportable by airplane, rugged enough to descend far below the test depth of any submarine yet de-

Navy personnel lower into the Mediterranean the Controlled Underwater Recovery Vehicle (CURV) that picked off the bottom the missing H-bomb located by the *Alvin*. An unmanned, tethered vehicle, CURV transmits a television picture of what it "sees" to an operator on the surface tender. (U.S. Navy)

After the nuclear submarine *Scorpion* failed to return home on schedule to Norfolk in May 1968, the Navy galvanized a massive search in the vicinity of the Azores. Five months later the hull was located and photographed in more than 10,000 feet of water, some 400 miles southwest of the islands. In this photograph of the after section, a mooring line can be seen protruding from the stowage locker into the messenger-buoy cavity. (U.S. Department of Defense)

The first of the Deep Submergence Rescue Vehicles (DSRV's) constructed by the Navy is small enough to be transported by aircraft. The checkered transfer skirt at the bottom will attach to the escape hatch of a stricken submarine. (U.S. Navy)

signed or contemplated, able to find a stricken submarine, and able to attach themselves to it for rescue of the survivors.

The first two DSRV's have been constructed, and rescue will be their primary mission. But in recognition of the need for research and the further fact that a capability not exercised tends to become stagnant, it is intended that the DSRV's will be sufficiently versatile to participate also in many other underwater operations, at deep or shallow depth. What these may be it is difficult at this time and distance to predict; that there will be many needs for these little submarines is a certainty.

The governments of the world are not, however, the only organizations interested in the deep sea. Industry is also looking to the rich resources of the sea bed—oil, manganese, and phosphorite, for instance. Abandoned riches, salvageable metal if nothing else, lie entombed in the *Titanic, Lusitania, Bismarck, Prince of Wales,* and *Andrea Doria*—to name only a few of the great lost ships. In response to the demand for vehicles with which to exploit this oceanic wealth, there has been in recent years a proliferation of small, commercially-built submarines, most of them designed for specific tasks in connection with industrial needs, yet all of them able also to make a contribution to exploration and research.

Basically they are of two types: tethered and untethered; and of the tethered kind there are manned and unmanned models.

Movement of the tethered Deep Submergence Vehicle is restricted to the length of its tether, though its working area may be increased if its surface tender is a ship able to follow where the DSV leads. Its advantage is that most power and instrumental functions can be supplied directly by the tender, with consequent economies of weight and space in the vehicle. For specific purposes, working around a deep oil well for example, it may ultimately be the most popular as well as the least expensive vehicle to use. The manned tethered vehicle presents some special safety problems, one of the main ones being the need for a sure method of clearing its tether from bottom entanglements. Examples of such tethered vehicles are the *Kuroshio* of Japan and the Personnel Transfer Capsules (PTC's) used by the U.S. Navy to transport divers under pressure from the surface to their working depth. Unmanned tethered vehicles include the CURV (Controlled Underwater Recovery Vehicle), which recovered the H-bomb off Spain; the towed camera apparatus of the *Mizar*; and various types of towed sonar equipment for bottom-topography investigation. Some DSV's, meanwhile, are built so that they

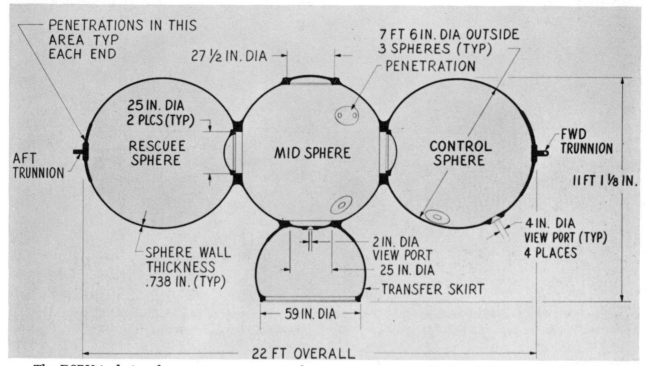

The DSRV is designed to rescue up to twenty-four men at a time at depths of 3,500 feet. (U.S. Navy)

can operate either with or without a tether, thus increasing their observation or instrumentation capability at the expense of mobility when required. The Westinghouse *Deep Star* embodies this feature.

As for the far more numerous untethered DSV's, *Trieste, Alvin,* and *Aluminaut* already have been mentioned. Other DSV's include the "tourist" submarine *Auguste Piccard* in Lake Geneva and Captain Cousteau's *Diving Saucer,* copies of which have been employed at a variety of jobs in different parts of the world. The U.S. Navy has, in addition to *Trieste II* (now designated DSV–1) and *Alvin* (DSV–2), *Turtle* (DSV–3), *Sea Cliff* (DSV–4), *Nemo* (DSV–5), and DSRV's 1 and 2. One of the most active commercial builders of work submarines is the Perry Submarine Company of Florida, which has produced several versions of its *Cubmarine* with various depth capabilities, and a special diving work boat called *Deep Diver,* which transports divers to a work area in safety and comfort under the necessary special atmosphere conditions (pressure, helium, reduced oxygen) and subsequently returns them to their pressure chamber on the surface.

Costs at the moment are high (charter of a DSV and its trained operators runs several thousand dollars a day, plus transportation costs of the heavy equip-

ment). Launch and recovery of the boats in any seaway is always fraught with danger and difficulty, but this will become less as new devices and techniques come forward. It was the failure of one such new device, the submersible platform by which *Alvin* was raised out of and returned to the water between dives (one of the lifting cables broke), which caused the famous little sub to be dumped into the sea 120 miles south of Cape Cod in October 1968. Fortunately there were no injuries to any of the operators, and the DSV was recovered a year later with the assistance of *Aluminaut* and the Naval Research Laboratory ship *Mizar.*

Another versatile research submarine is the *Ben Franklin,* designed by Jacques Piccard for the Grumman Aircraft Engineering Corporation. This 50-foot-long, 130-ton submarine is able to hover for long periods of time at depths ranging from 300 to 2,000 feet, while its 6-man crew makes extended observations and experiments beneath the sea. NASA, meanwhile, is interested in studying the *Ben Franklin's* crew, since the medical and psychological factors associated with its long undersea voyages are similar to those associated with space flights of similar duration. The *Ben Franklin's* first mission was a month-long study of the Gulf Stream. It submerged on July 14, 1969, off West Palm Beach, Fla., drifted northward in

Named for its ability to operate at depths of 1,350 feet, the *Deep Diver* is the first submarine to feature a "lockout" chamber that can be pressurized to match the surrounding water. (Ocean Systems, Inc.)

A simplified engineering drawing of *Deep Diver* illustrates the personnel compartments, gas sphere, battery pod, and power units. (Ocean Systems, Inc.)

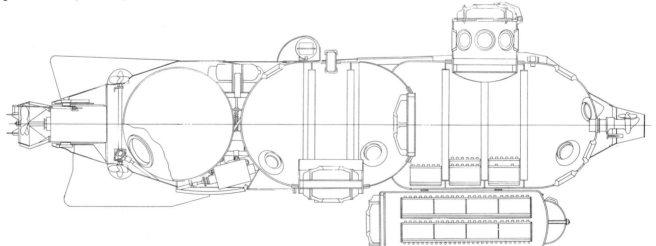

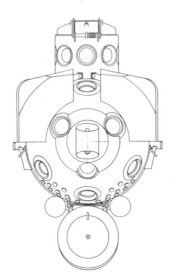

the stream's 3.3 knot current, and surfaced on August 14 three hundred miles off the coast of Halifax, N.S.

The entire problem of work and safety under the sea and under pressure is receiving exhaustive study at this time. A great deal more is known than was the case five years ago, and the state of the art is advancing weekly. Simon Lake predicted it all, half a century ago.

Pressure and the Environment—Sealab

Pressure is probably the greatest single factor in the environment under the sea, to which all else is subordinate. The average person can ascend eight to ten thousand feet in the air without feeling any discomfort, but in the water many people cannot dive deeper than fifteen to twenty feet without acute pain and in some cases incapacitation. But not everyone is so vulnerable, and the ability to dive without such pain—essentially requiring only open, and probably tougher, Eustachian tubes—can be developed through doing it. The ancients, like the moderns, had those daring souls who dived deep below the surface in hope of great reward. Long before Christ, pearl divers in the Pacific found that by developing their lung capacities and taking a large number of deep breaths (hyperventilating) before each dive, they could stay at remarkable depths for absolutely incredible periods. Men and women both (in Japan and Korea to this day there are communities in which diving for the cultured pearls is done exclusively by women) since earliest times have dived below 100 feet, without equipment of any sort except a large stone to speed their descent, and they have remained there for periods of several minutes, harvesting the giant pearl-producing shells from their encrusted beds. Some incredible tales of endurance up to fifteen minutes are recorded, but five minutes appears to be a reasonable maximum. In the Mediterranean Sea, and along the shores of what are now India, Iran, and Pakistan, strong men have always dived in the shallows for the valuable cargoes of wrecked trading ships. But even the most rugged modern diver, with the latest and most modern equipment, including "full 'hard hat' diving gear," cannot dive below about 400 feet, breathing what we normally refer to as "air under pressure," and his ability to do useful work at that depth is almost nil. Yet work must today be done, and much of it will have to be at depths below 400 feet. This has led to a great deal of investigation of the physical effects of pressure upon human beings.

It has long been known that a diver breathing air under pressure for a prolonged period—the greater the pressure the shorter the period—was in danger of getting the dread "bends" when the pressure was reduced. Bends is not a disease but a condition, named from its most obvious effect upon the human body. It is caused by nitrogen, one of the "benign gases" making up the atmosphere of earth. Air is roughly 21 percent oxygen, 78 percent nitrogen, and 1 percent other gases. The pressures at which gases enter into solution with human blood—and come out again—vary with each gas. So does the time and manner of the process.

When a diver goes under pressure, some of the nitrogen in the air passes through his body tissues and dissolves in his bloodstream. Here it has several effects: the most obvious is that when pressure is reduced it comes out of solution as minute bubbles in joints, muscles, and nerve-control centers. Permanent, severe crippling (the bends), brain damage, or death may result if the pressure is not instantly increased again to redissolve the gas. It is to avoid the bends that the lengthy decompression schedules which have so restricted underwater work were developed (the deeper a diver goes the less time he may stay there, and the longer time he must spend "decompressing"). A diver might, for example, work for half an hour and spend half a day decompressing.

Another undesirable effect of nitrogen is "nitrogen narcosis," or "rapture of the deep." Below about 300 feet—about nine times normal atmospheric pressure—the amount of dissolved nitrogen in the blood may cause mental unbalance. Some persons are more susceptible than others and show it more; the longer one stays at depth the more apt he is to show the symptoms and the more serious they may be. The common effect is to make a diver's mind fuzzy; he reacts almost as though he were drunk, his state of mind is euphoric, and he becomes a hazard to himself and others. Many a veteran diver can eloquently describe the tremendous concentration required for the simple operation of turning a valve under such conditions, even despite full prior awareness of the difficulty and the reason!

As a result, breathing mixtures containing a substitute for nitrogen have been tried. The most common, which has met with considerable success despite certain disadvantages, is a helium-oxygen mixture (actually, a small percentage of nitrogen is retained). Neon-oxygen and even hydrogen-oxygen have also been tried. Helium-oxygen does not have the adverse effects of nitrogen-oxygen (greatly reduced chance of the bends, no narcosis), but its own side-effects include changing the human voice so that speakers sound like Donald Duck, making it practically impossible to pass information or instructions without the still-experimental "helium speech converter." In addition, special protection against chilling is required for health and efficiency since body heat is lost at a much faster rate in the helium-oxygen atmosphere. Under the greater pressures possible with helium it is necessary to greatly reduce the oxygen content of the atmosphere so that too much oxygen is not taken in at each breath. At a depth of 400 feet, for example, the pressure is twelve times that of the normal atmosphere and a given quantity of oxygen occupies only one-twelfth the volume that it would on the surface. As a result—all other factors being equal—a man would take in twelve times the oxygen by weight, or molecular count, with each breath as he does at sea level. Oxygen poisoning (sudden convulsions with no prior warning) may result, depending on individual tolerance and length of exposure to the "elevated partial pressure" (the technical term for the condition) of oxygen.

It was found also that different individuals react radically differently to changes in pressure. Some cannot ever become "accustomed" to it, but to most, once the change in pressure has been experienced and adjusted to, the new conditions are essentially unnoticeable until it is time to return to normal atmospheric pressure. Yet the evidence suggests that certain complex body processes may be affected by changes in pressure. It is possible that there may be some finite biological limitation to the pressure man may subject himself to. The fear of coming on it accidentally and catastrophically is one of the basic constraints upon all programs. There is a great deal yet to be discovered about the human body.

From the years of accumulated experience there has developed the theory of "saturated diving." An individual exposed to an increase in pressure will gradually absorb into the tissues of his body more of the gases of which the atmosphere is composed until he reaches a point of physical equilibrium with the new pressure. This is known as "saturation" (the body, of course, normally is saturated at atmospheric pressure, 14.7 pounds per square inch at sea level). He may remain in this condition for a long period, perhaps indefinitely. No ill effects of this procedure have been observed, but concern about safety increased following the death in February 1969 of one of the Sealab divers, Barry L. Cannon. This accident later was determined to have resulted solely from human error, but if there are limitations to the procedure, they must be discovered. Barring a med-

ical finding to the contrary, however, it clearly seems to be far easier physically on the divers if they are subjected to pressure just once—from the beginning of the work period through rest, sleep, and food breaks, until completion of the period—and then slowly released from the pressure, under stringent safeguards. This contrasts very favorably to the older system of pressurizing and depressurizing divers for each dive, with long, wasted hours for decompression and the danger of the bends each time.

Thus began the idea of creating living habitats under the sea, where men could sleep and rest between work periods, always subject to the same pressure. In the U.S. Navy this was called the Man-in-the-Sea project, and it became one of the five parts of the Navy's Deep Submergence Systems Project. The U.S. Navy has carried out two submerged-living experiments, known as Sealab I and Sealab II, in which groups of "aquanauts"—a term copied from "astronauts"—lived for ten days at 192 feet in the warm water off Bermuda (Sealab I) and from fifteen to thirty days at 205 feet in the much colder water off San Diego, California (Sealab II). A third Sealab experiment, scheduled to be conducted at a depth of 600 feet, was postponed following Cannon's death in 1969 and subsequently canceled for fiscal reasons. The Sealab habitat was a steel cylindrical tank much like a submarine, containing living quarters and basic laboratory equipment, able to withstand considerable pressure, internally or externally, on its own. In an emergency it could be closed and hoisted

The Navy's Man-in-the-Sea program began in 1964, when four aquanauts lived for ten days in 192 feet of water off the coast of Bermuda. (U.S. Navy)

The Sealab I habitat rests on its support vessel. The forty-foot cylindrical tank contains living quarters and labora-tory equipment. (U.S. Navy)

Sealab II spent forty-five days 205 feet down in the cool waters off San Diego, California. It was manned by three teams of ten men, who alternated for fifteen days each. (U.S. Navy)

out by the support craft always in attendance, while the vital pressure within would be maintained for lengthy and careful decompression to normal sea-level pressure. Sealab I was terminated at the threat of an approaching storm on August 31, 1964. Sealab II remained on the ocean floor for the entire test period, with divers transferring between it and the surface in a small bell, officially called a PTC, or Personnel Transfer Capsule, but commonly referred to by the divers themselves as "the elevator."

Sealab II experiments, in the late summer of 1965, involved three teams of ten men each for fifteen days per team. One man, Commander Scott Carpenter, remained continuously for thirty days and might have stayed longer, but for a painful and near-incapacitating swelling resulting from the sting of a scorpion fish. The teams performed work in the vicinity of their quarters and found, as expected, that they were gradually able to increase their exposure to the 48-degree water, but never to the extent they desired. This limitation was attributable partly to the very rapid loss of body heat caused by the helium atmosphere. Psychologically there was little or no difficulty, although some suspicion exists that helium or prolonged pressurization may have some effect, psychologically or in some other way, upon judgment factors. This is difficult to evaluate because the crews were all carefully chosen and highly motivated. One environmental factor was a constant annoyance: The habitat had not been located on level ground, and everything in it was by consequence some five degrees off the horizontal or vertical. If there was a psychological problem, this was it!

The U.S. Navy now has two deep diving systems capable of supporting saturation diving to depths of 1,000 feet: the Deep Dive System (DDS) Mark I Mod O and DDS MK II Mod O. Improved designs of these systems are scheduled to be provided to operational Navy vessels in the near future.

At present there seems to be no reason why man cannot safely go to considerably greater depths than 1,500 feet. The French have successfully pressurized men to a depth in excess of 1,700 feet in an experimental chamber. However, wisdom and experience dictate that extensions in the known parameters of endurance of men and their materials be explored cautiously. There are indications that the Soviet Union recently lost a team of divers at a depth of more than 1,000 feet when structural failure caused catastrophic pressure reduction in a laboratory test chamber. Cousteau, financed by America's National Geographic Society, has maintained six French "oceanauts" for three weeks at 325 feet, and in one

experiment even had a pressurized undersea "garage" in which his diving saucer could be surfaced for repairs, upkeep, and exchange of crews. One difference in Cousteau's technique is that his undersea habitats are inflatable fabric instead of steel, and are therefore not capable of supporting pressure unless equalized on both sides of the light membrane. This equalization, of course, is the standard condition, but Cousteau lacks the safety factor inherent in the ability to lift the entire habitat as a pressure chamber. On the other hand, his habitats are far more versatile, offer more possibility as to size and complexity, and may indeed be the shape of things to come—provided only that the rubber and fabric membrane can be made proof against a maddened swordfish or killer whale.

A Sealab II aquanaut swims by a California sea lion while filming an underwater documentary. (U.S. Navy)

The idea has been popularly advanced that entire communities could be built underwater, their populations permanently subjected to pressure (the idea takes little account of the many problems introduced by families and children, and may have resulting difficulty in getting general acceptance). There seems to be no reason, however, why it would not be possible to build permanent floating cities in sheltered areas of the sea, nor why permanent, submerged, pressurized complexes of workers, technicians, and researchers could not be established, in which highly paid and highly trained men and women might live and work for long periods. An experiment of the latter sort was conducted in early 1969: Tektite I, in which four scientist-aquanauts remained for sixty days at a depth of fifty feet off the island of St. John in the

Virgin Islands. They spent a portion of each day swimming about in scuba gear, performing oceanographic studies. Though the pressure was relatively low—only one and a half atmospheres—the period is the longest anyone has spent under pressure.

Tektite II began at the same site in April 1970 and was completed in October. Ten different teams, each composed of five scientist-aquanauts, stayed on the ocean floor for periods of up to twenty days. One of the teams was made up entirely of women; they worked out of the habitat for two weeks. The Soviets have been conducting similar tests of undersea living. In September 1971 an experiment in the Black Sea had to be terminated after fifty-two days upon the onset of a storm.

One item of possibly great potential: Fish do not

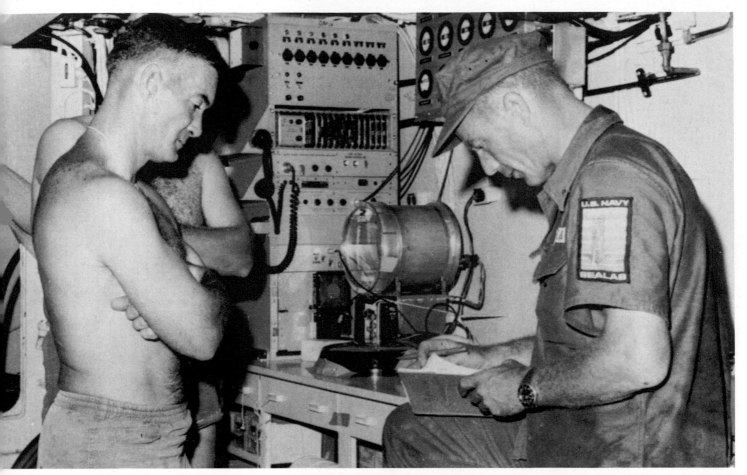

Astronaut-turned-aquanaut M. Scott Carpenter (right) checks the experiment schedule before his descent in the Sealab II habitat. (Wide World Photos)

breathe water; they subsist on the oxygen dissolved *in* the water and will "drown" in nonaerated water, as many home aquarium keepers have discovered. Fish gills contain membranes through which dissolved oxygen can be absorbed and carbon dioxide given off. If man could invent a similar device, permitting him to extract oxygen directly from water, he could "breathe" water exactly as fish do and free himself from all dependence on air. Experimenters recently have done precisely this, and a United States patent has been granted for the process. The equipment required is too large to compete with a scuba diver's outfit, but the principle has been proved. Someday, man-in-the-sea may be able to strap on a pair of gills and "breathe" water—extracting the oxygen—exactly as fishes do, directly from the sea!

The problems of pressure, of course, do not only affect human organisms. There are also a tremendous number of problems connected with the mechanisms men need to use. Submersibles, for instance, must stand pressure at all the depths at which they are expected to operate, and any mechanical device extending through their hulls must also be able to stand this pressure. It is for this reason that it has been so difficult to design an efficient mechanical grab for the retrieval of objects at great depths. Running an electric control cable through the pressure hull is an engineering problem of the first magnitude because of the essential fluidity of the cable sheathing. At extremely great depths even a miniscule leak can be serious because of the tremendous pressure behind it (a man carelessly walking in front of such a leak might be seriously injured). The pressure hull of the submarine itself, even well within the safety limits, will undergo a certain amount of compression with increase of depth and thus lose some inherent buoyancy. For greatest strength and least squeeze a completely spherical pressure hull is obviously the best. Second best is a reinforced cylinder with spherical ends. In any case there is a weakness at any point of discontinuity, such as the entrance hatches and viewing ports.

In the Navy's Deep Dive System MK I and its sister system, the MK II, diving teams remain under pressure for lengthy periods, transferring between the habitation aboard ship and the work site on the seafloor by means of the spherical capsule at top of their living chamber. (U.S. Navy)

All these pitfalls, of course, can be avoided if pressure inside the submersible can be made equal to that existing outside. Thus the Sealab structure in which the men live under pressure could be very lightly built. It has been proposed that future habitations be made of some light, leakproof material similar to balloon cloth, its only requirement being not to leak under any circumstances whatever. Such a habitat probably would be similar to the fabric structure already developed by Cousteau, and, while being more versatile than the steel chambers in many ways, would suffer from the limitation that it could not be hauled quickly to the surface in an emergency and still maintain its internal pressure.

Where Do We Go from Here?

Naval officers are fond of citing the fact that seventy percent of the earth's surface is covered by water, that it is important to the survival of the United States that control of the sea remain in friendly hands. By and large, however, control of the sea has until now meant only power to control the surface of the sea because of its importance to the land. Man has prided himself on having explored—and controlled—virtually the entire surface of the earth. But of recent years he has begun to think of the depths of the sea, and hence of exploration and utilization of the bottom of the sea. This still leaves the water column between the surface and the bottom in the old category of "open range," but if investment is to be made on the bottom, a degree of control must go along with it. What this may mean for the future of mankind can only be very generally described, but it will certainly have a profound impact.

If man must consume the resources of the earth in order to live, then it is only logical that he must seek these resources wherever they actually exist and that therefore he will ultimately explore the depths of the sea, as well as the depths of the land, to find them. If one can make the assumption that the earth's

natural resources are generally randomly spaced throughout, one would expect that about seventy percent of these resources (with certain exceptions because of the manner of their creation) lie under the sea, yet to be discovered. It is thus possible that full exploitation of all the resources under the sea could double or triple the period man may expect to survive on earth. Certainly we must probe much deeper under the surface, and one may imagine, therefore, that all sorts of legal as well as technical problems will arise when we begin to carve up the hitherto free range of the open seas. The distinction between the continental shelf (and its definition) and the deep sea will come in for inspection. Claims to sea resources by noncoastal nations may have to be given some consideration. Lawyers of the sea may happily look forward to generations of lucrative business.

To take a fanciful tack, consider the normally accepted theory of the development of man with which this chapter began—the idea that man is descended from some sort of fishlike creature that crawled out on land. One might also theorize that some cousin of man, subject to slightly different environmental conditions, did not stay on land but instead returned to the deep sea. One branch of man's family might then have developed on land into *Homo sapiens*, and the other, constrained by a different environment, might today be an air-breathing sea-creature, for example the porpoise.

There are those who say that the porpoise, or bottle-nosed dolphin, is highly intelligent, with a brain equal to that of man. It is conceivable that the more difficult environment in the sea prevented our intelligent aquatic cousin from starting on the developing accumulation of knowledge that man has enjoyed for the past few hundred centuries (only an instant in the Darwinian theory of evolution). Perhaps this thinking is not so fanciful as it sounds. Some very interesting experiments are already being made. If it can be shown that the dolphin reacts by the power of reason, and not through intuition or reflex, this will be something of which great account will be taken.

One thought cannot help but arise: If porpoises do have a basic power of reason, if they too are a high order of life and have a creative mind, what does this presage for the world of the future? Considering that scientists say that from among the millions of planets which must exist there are undoubtedly some populated by intelligent beings, how can we know what these intelligent beings look like? How can we assume a distant planet to be like earth, its population to resemble ours, when the total environment must be different in significant details? Returning to consideration of our own planet, can we visualize what intelligent life might have been had there been only water, and no land? In a few thousand more centuries, perhaps after man is gone, is it possible that the survival of intelligent life on earth will be in the sea? In other words, is it possible that the heirs of man are already waiting in the wings, already started upon their own slow climb upward to civilization, research, organized war, heightened capacity for destruction, and in their turn (perhaps), ultimate oblivion?

A porpoise, trained to help divers in their work, carries out an exercise with one of its human colleagues. (U.S. Navy)

Oceanographic Date List

1500 B.C.	Egyptian Queen Hatshepsut goes by sea to the "Land of Punt," probably the Somali Republic.
1000 B.C.	Homer regards the world as a flat disk, encircling the Mediterranean.
1000 B.C.	Hesiod refers to "lands beyond the Sea," the Isles of the Blessed and the Hesperides.
600 B.C.	From Tyre and Sidon in the Mediterranean the Phoenicians circle Africa from east to west.
6th Century B.C.	Miletus and Pythagoras declare the earth to be a sphere.
4th Century B.C.	Aristotle describes 180 species of marine animals, noting patterns of structure and habit and showing how they can be grouped by body form.
330 B.C.	Alexander the Great is said to have been lowered into the sea in a barrel.
3rd Century B.C.	Plato, repeating an old Egyptian tale, describes the lost land of Atlantis, destroyed by a sudden cataclysm.
250 B.C.	Eratosthenes calculates the circumference of earth to be about 25,000 miles—remarkably close to the fact.
1st Century B.C.	Hippalus notes behavior of monsoon winds in the Indian Ocean, blowing one way during half the year, then reversing their direction.
Ca. 100 B.C.	Poseidonius measures the depth of the ocean near Sardinia, recording about 1,000 fathoms.
1st Century A.D.	Pliny the Elder lists 176 species of marine animals, believing he has included all that exist.
2nd Century A.D.	Ptolemy produces a map of the world depicting the earth as a sphere—but erroneously stating its circumference as 18,000 miles.
A.D. 1000	Vikings reach Iceland, Greenland, North America.
1416	Prince Henry the Navigator of Portugal founds a school of navigation at Sagres, and launches the Golden Age of Exploration.
1487–1488	Bartholomeu Diaz sails around the Cape of Good Hope and back to Portugal.
1492	Christopher Columbus rediscovers America. Even after four voyages and mountains of evidence, he persists until his death in regarding the new lands as parts of Asia.
1497–1499	Vasco da Gama sails around the Cape of Good Hope, opening up a new trade route to India.
1519–1521	Ferdinand Magellan sails to the Pacific. His ship, under the command of Juan del Cano, completes the first circumnavigation of the globe after Magellan is killed at Cebu in the Philippines.
1537–1590	Flemish geographer Gerhardus Mercator produces new kinds of maps of immense value and stimulation to ocean explorers. He postulates (erroneously) that land and sea must occupy about equal areas of the globe.
1576, 1577, 1578	Englishman Martin Frobisher, during three voyages, seeks "a northwest passage" from Europe to India.
1585, 1586, 1587	Englishman John Davis makes three voyages in attempts to find "a northwest passage."

1609–1611	Henry Hudson, an Englishman in the employ of the Dutch, seeking "a northwest passage," sails up the river which later bears his name. A year later, on a fourth voyage, he sails into Hudson Bay, and in 1611 is cast adrift there by mutineers.
1616	William Baffin discovers Baffin Bay, but becomes convinced that an ice-free "northwest passage" does not exist.
1642	Dutchman Abel Janzoon Tasman circumnavigates Australia, and shows that the land masses of the southern ocean are far smaller than in Mercator's concept.
1725	Count Luigi Ferdinando Marsigli publishes *The Physical History of the Sea*. Most famous as a soldier, he was also a pioneer oceanographer, apparently the first to use a naturalist's dredge. His observations of the current system of the Bosporus are remarkable in terms of the knowledge and techniques of the time.
1736	Carolus Linnaeus provides the orderliness of a rigid classification of plants and animals, stimulating collections.
1768–1779	James Cook undertakes the first scientific ocean expeditions—primarily to observe the transit of Venus rather than to pursue oceanography. But he also makes soundings to depths of 200 fathoms and accumulates an impressive amount of data about the sea. He is killed in 1779 in the Sandwich (Hawaiian) Islands.
1769	Ben Franklin, as postmaster general for the American colonies, discovers that delay in the delivery of mail from Britain is a consequence of the ships' bucking the east-flowing Gulf Stream. A map of the surface currents, prepared with the help of American whalermen, helps mail packet skippers to reduce their times of passage.
1776	American pioneer *Turtle* submarine unsuccessfully attacks HMS *Eagle*.
1802–1804	Alexander Baron von Humboldt makes observations of the ocean and its inhabitants off the west coast of South America, describing the great north-sweeping current that now bears his name.
1818	Sir John Ross invents a "deep-sea clamm" to collect animals from the bottom of the deep sea. From 1,050 fathoms he collects worms and a starfish, proving that animals live at great depths.
1831–1836	Charles Darwin sails on the English survey ship *Beagle* as naturalist. He makes a major contribution to marine science in his theory of the formation of coral reefs. His observations during this voyage contribute significantly to his culture-shaking theory of evolution.
1839–1842	Charles Wilkes, an American, uses copper wire for the first time for soundings.
1841–1854	Edward Forbes, an Englishman, improves the naturalist's dredge and makes extensive collections off the Isle of Man and in the Aegean Sea. He believes there are eight zones of abundance of animals in the sea, and theorizes (incorrectly) that life ceases at 300 fathoms. His major contributions lead him to be regarded as the cofounder, with Matthew Maury, of the science of oceanography.
1841–1861	Matthew Fontaine Maury, U.S. Navy, Officer in Charge of the Depot of Charts and Instruments, correlates immense amounts of data on wind and currents from naval vessels, and persuades other ships to collect and report information. From these data, ships' navigation is greatly improved. In 1855, Maury publishes the first classic book in oceanography, *The Physical Geography of the Sea*. He and Forbes are credited as the "Founding Fathers of Oceanography."
1850	Robert McClure finally makes a "northwest passage," but travels part way over ice.
1850	Michael and G. O. Sars, Norwegian father and son, collect animals from the ocean between one and two miles deep.
1864	Norwegian Svend Foyn invents the explosive harpoon, and the modern whaling industry begins.
1868–1870	Dr. Wyville Thomson and Dr. W. B. Carpenter lead expeditions on the British ships *Lightning* and *Porcupine*. Temperature observations and deep-sea dredgings excite scientists greatly, and lead to the momentous *Challenger* expedition.
1872	Anton Dohrn founds the first marine biological station at Naples, Italy.

1872–1876	The British *Challenger* expedition under Wyville Thomson marks the beginning of scientific oceanography, during a 68,890-mile world-girdling voyage.
1873	Louis Agassiz founds the first American marine biological station at Penikese Island, near Cape Cod.
1875	The Woods Hole Laboratory of the U.S. Commission of Fish and Fisheries is founded by Spencer Baird.
1877–1905	Alexander Agassiz conducts highly important voyages on the *Blake* in the Caribbean Sea and the Gulf of Mexico, and on the *Albatross* in these same areas and in the Pacific. *Albatross* was the first ship designed specifically for oceanographic research.
1879	Laboratory of the Marine Biological Association is founded at Plymouth, England.
1882	T. W. Fulton of the Fishery Board for Scotland begins studies of the changes of paterns of North Sea currents in relation to supplies of commercial fish and of the biology and migrations of fish.
1886–1906	Prince Albert of Monaco outfits four yachts as oceanographic research ships. He founds the oceanographic museum and laboratory at Monaco in 1906.
1886–1889	The Russian ship *Vitiaz* cruises round the world, making pioneer observations of sea temperature and density.
1889	Victor Hensen leads the German "Plankton expedition," during which the first quantitative investigations of the biological economy of the sea are made.
1892	Hopkins Marine Station of Stanford University is founded.
1892	E. W. L. Holt of England establishes the precedent of combining a fundamental knowledge of the biology of fish and the pattern of their catches. He is sometimes called the "Father of Fishery Science."
1893–1896	Norwegian explorer Fridtjof Nansen allows his ship *Fram* to be frozen in the arctic ice to study the polar icecap.
1894	C. G. J. Petersen, a Dane, develops a pioneer theory of overfishing and invents a fish tag still widely used.
1900	The U.S. Navy Submarine Service is born with the commissioning of the USS *Holland*, named after her inventor, John Philip Holland.
1902	The International Council for the Exploration of the Sea, coordinating body for European fishery research, is founded.
1905	Scripps Institution of Oceanography is founded.
1906	Roald Amundsen fulfills the old dream of "a northwest passage," landing in Nome.
1908	Bingham Oceanographic Laboratory is founded at Yale University.
1910	Norwegian Johan Hjort leads *Michael Sars* expedition, resulting in greatly improved knowledge of the North Atlantic and its inhabitants.
1912	Alfred Wegener proposes the theory of continental drift, challenging the view that the present forms of the continents and ocean basins were defined in the earliest geologic times; most scientists reject Wegener's theory.
1919	College of Fisheries of the University of Washington is founded.
1920–1922, 1928–1930	Cruises of the *Dana*, sponsored by the Danish Carlsberg Foundation, result in numerous important discoveries about deep-sea animals, including the astonishing story of the life history of the eel.
1925	British scientists on the *Discovery* begin a study of antarctic whales.
1925	Chesapeake Biological Laboratory is founded.
1925	Fritz Haber launches the *Meteor* expedition to extract gold from the sea for Germany's World War I debt to the Allies. He fails in this task, but the expedition contributes important information on the chemistry of the sea.
1930	Woods Hole Oceanographic Institution is founded.
1934	William Beebe and Otis Barton dive to 3,028 feet off Bermuda in a bathysphere, a small spherical steel chamber suspended from a surface ship by cable.
1937	Narragansett Marine Laboratory of the University of Rhode Island is founded.
1937	Duke University Marine Laboratory is founded at Beaufort, N.C.

1943	Marine Laboratory (now Rosenstiel School of Marine and Atmospheric Science) of the University of Miami is founded.
1943	Scuba (self-contained underwater breathing apparatus) is invented by Jacques-Yves Cousteau and Emile Gagnan, enabling divers to descend freely as deep as 300 feet.
1945	Swedish geologist Börje Kullenberg develops a piston corer that takes samples of bottom sediments up to seventy feet long.
1947	Dr. Harold Urey develops technique for "aging" ocean sediments by measuring the ratios of oxygen isotopes.
1947–1948	Round-the-world cruise of the Swedish *Albatross,* headed by Hans Pettersson, emphasizes marine geology.
1950–1951	Round-the-world voyage of the Danish *Galathea* emphasizes deep-sea dredging. Fishes are trawled from 23,000 feet and other animals from 33,000 feet.
1950–1952	A new British *Challenger* makes precise soundings of the ocean depths. The Challenger Deep, now thought to measure 35,600 feet, is found to be the ocean's deepest point.
1951	An immense subsurface current is discovered, stretching at least halfway across the Atlantic; it is named for Townsend Cromwell, American oceanographer who studied it.
1954	FRNS III dives to 4,050 meters off Dakar.
1957	A south-drifting subsurface current is found beneath the Gulf Stream by British and American scientists employing the Swallow float.
1957–1958	International Geophysical Year; oceanographic studies play an important role.
1958	U.S. nuclear submarines *Nautilus* and *Skate* sail under the North Pole, collecting information on the Arctic Sea and the icecap.
1959	First International Oceanographic Congress is held in New York.
1959–1965	International Indian Ocean Expedition employs scientists of twenty-two nations in a coordinated study of a neglected part of the ocean.
1960	Jacques Piccard and Donald Walsh take the bathyscaphe *Trieste* to the bottom of the deepest part of the sea, the Challenger Deep, nearly seven miles down.
1962–1963	Jacques-Yves Cousteau occupies *Conshelf I* and *II.*
1963–1964	International Cooperative Investigation of the Tropical Atlantic conducts coordinated surveys of the tropical Atlantic by seven nations.
1964–1968	U.S. Navy Sealab I is occupied by four men for eleven days, 192 feet deep, near Bermuda. Sealab II, at 205 feet near La Jolla, is home for Commander Scott Carpenter for thirty days, and other men for lesser periods. Sealab III is built but the program is then canceled for fiscal reasons.
1965	Cooperative study of the Kuroshio Current and adjacent regions of the Pacific employing oceanographers of several nations is begun.
1966	Second International Oceanographic Congress is held in Moscow.
1967	The Torrey Canyon is wrecked, dumping oil on beaches of England and France. This, and the blow-out of a well in the Santa Barbara Channel in 1969, spreading oil on California beaches, jolt the world into realization that ocean pollution has become a major problem.
1967	Deep Sea Drilling Project begins, with cores obtained by *Glomar Challenger* helping to confirm theories of sea floor spreading and continental drift.
1968–1971	International cooperation in ocean research gains momentum; multinational investigations begin in the Caribbean, Mediterranean, and Atlantic.
1969	*Alvin* is recovered from a depth of 5,000 feet; the largest deep recovery to date.
1970	Concept of "global tectonics," holding that earth's crust is divided into about eight moving and interacting rigid plates is now widely accepted as a result of findings during the preceding decade.
1970	The General Assembly of the United Nations asks that a Conference on the Law of the Sea be held in 1973.
1971	DSRV-1 makes the first submerged mating with its mother submarine, simulating a rescue operation.

Bibliography

1 The Science of the Sea

Bruun, Anton F., ed. *The Galathea Deep Sea Expedition 1950–1952*. New York: Macmillan Company, 1956.

Coker, Robert E. *This Great and Wide Sea*. New York: Harper & Row Publishers, Torchbooks, 1962.

Cutting, Charles L. *Fish Saving: A History of Fish Processing from Ancient to Modern Times*. London: Leonard Hill, Ltd., 1955.

Darwin, Charles. *The Voyage of the Beagle*. New York: Doubleday & Company, Inc., Anchor Books, 1962.

Deacon, G. E. R., ed. *Seas, Maps, and Men: An Atlas-History of Man's Exploration of the Oceans*. New York: Doubleday & Company, Inc., 1962.

Gaskell, T. F. *The World Beneath the Oceans: The Story of Oceanography*. Garden City, N.Y.: The Natural History Press, 1964.

Hardy, Sir Alister. *Great Waters*. New York: Harper & Row Publishers, 1967.

Herdman, W. A. *Founders of Oceanography and Their Work*. London: Edward Arnold & Co., 1923.

Idyll, C. P. *Abyss: The Deep Sea and the Creatures That Live in It*. Rev. ed. New York: Thomas Y. Crowell Company, 1971.

Maury, Matthew F. *The Physical Geography of the Sea, and Its Meteorology*. Cambridge, Mass.: Harvard University Press, 1963.

Murray, Sir John, and Hjort, Johan. *The Depths of the Ocean*. London: Macmillan & Co., Ltd., 1912.

Radcliffe, William. *Fishing from the Earliest Times*. London: John Murray Publishers, Ltd., 1921.

Thomson, C. Wyville. *The Depths of the Sea*. London: Macmillan & Co., Ltd., 1873.

2 The Underwater Landscape

Beebe, William. *Half Mile Down*. New York: Harcourt, Brace & Company, Inc., 1934.

Carson, Rachel. *The Sea Around Us*. Rev. ed. New York: Oxford University Press, 1961.

Dietz, R. S., and Holden, J. C. "The Breakup of Pangaea." *Scientific American*, September, 1970.

Emery, K. O. *The Sea off Southern California*. New York: John Wiley and Sons, 1960.

Fisher, Robert L., and Revelle, Roger. "The Trenches of the Pacific." *Scientific American*, November 1955.

Heezen, Bruce C. "The Origin of Submarine Canyons." *Scientific American*, August 1956.

——— "The Rift in the Ocean Floor." *Scientific American*, October, 1960.

Hurley, P. M. "The Confirmation of Continental Drift." *Scientific American*, April 1968.

Kuenen, P. H. *Marine Geology*. New York: John Wiley and Sons, 1950.

Maury, Matthew F. *The Physical Geography of the Sea, and Its Meteorology*. Cambridge, Mass.: Harvard University Press, 1963.

Menard, H. W. *Anatomy of an Expedition*, New York: McGraw-Hill, 1969.

Menard, H. W. *Marine Geology of the Pacific*. New York: McGraw-Hill Book Co., Inc., 1964.

——— "The East Pacific Rise." *Scientific American*, December 1961.

Murray, Sir John, and Hjort, Johan. *The Depths of the Ocean*. London: Macmillan & Co., Ltd., 1912.

Piccard, J., and Dietz, R. S. *Seven Miles Down: The Story of the Bathyscaph* Trieste. New York: G. P. Putnam's Sons, 1961.

Shepard, F. P. *The Earth Beneath the Sea*. Baltimore: Johns Hopkins University Press, 1959.

—— and Dill, R. F. *Submarine Canyons and Other Sea Valleys*. New York: Rand McNally & Company, 1965.

Thomson, C. Wyville. *The Voyage of the* Challenger. New York: Harper & Brothers, 1878.

Time, Inc., Book Division. *The Sea*. New York, 1961.

Wilson, J. T. "Continental Drift." *Scientific American*, April 1963.

3 Biology of the Sea

Carson, Rachel. *The Edge of the Sea*. Boston: Houghton Mifflin Co., 1955.

—— *The Sea Around Us*. Rev. ed. New York: Oxford University Press, 1961.

—— *Under the Sea Wind*. New York: Simon & Schuster, 1941.

Hardy, Sir Alister. *Great Waters*. New York: Harper & Row Publishers, 1967.

—— *The Open Sea, Its Natural History*. Boston: Houghton Mifflin Co., 1965.

Nordenskiöld, Erik. *The History of Biology*. New York: Tudor Publishing Co., 1946.

Russell, Frederick S., and Yonge, C. M. *The Seas: Our Knowledge of Life in the Sea and How It Is Gained*. 2nd ed. London and New York: F. Warne & Co., 1963.

Tressler, Donald K., and Lemon, James McW. *Marine Products of Commerce: Their Acquisition, Handling, Biological Aspects, and the Science and Technology of their Preparation and Preservation*. 2nd ed. New York: Reinhold Publishing Corp., 1951.

4 Physics of the Sea

Bascom, Willard. *Waves and Beaches: The Dynamics of the Ocean Surface*. New York: Doubleday & Company, Inc., 1964.

Cotter, Charles H. *The Physical Geography of the Oceans*. New York: American Elsevier Publishing Co., 1966.

Darwin, Charles. *The Voyage of the Beagle*. New York: Doubleday & Company, Inc., Anchor Books, 1962.

Daugherty, Charles M. *Searchers of the Sea: Pioneers in Oceanography*. New York: The Viking Press, Inc., 1961.

Davis, Kenneth S., and Day, John A. *Water, the Mirror of Science*. New York: Doubleday & Company, Inc., Anchor Books, 1961.

Deacon, G. E. R. *Seas, Maps, and Men: An Atlas-History of Man's Exploration of the Oceans*. New York: Doubleday & Company, Inc., 1962.

Defant, Albert. *Ebb and Flow: The Tides of Earth, Air, and Water*. Ann Arbor: University of Michigan Press, 1958.

Fairbridge, Rhodes W., ed. *The Encyclopedia of Oceanography*. New York: Reinhold Publishing Corp., 1966.

Guberlet, Muriel L. *Explorers of the Sea*. New York: Ronald Press, 1964.

Long, Edward J. *New Worlds of Oceanography*. New York: Pyramid Publications, Inc., 1965.

Maury, Matthew F. *The Physical Geography of the Sea, and its Meteorology*. Cambridge, Mass: Harvard University Press, 1963.

Neumann, Gerhard, and Pierson, Willard J. *Principles of Physical Oceanography*. Englewood Cliffs, N.J.: Prentice-Hall, Inc., 1966.

Ritchie, George S. Challenger: *The Life of a Survey Ship*. New York: Abelard-Schuman, Ltd., 1958.

Sverdrup, H. U., Johnson, M. W., and Fleming, R. H. *The Oceans: Their Physics, Chemistry, and General Biology*. Englewood Cliffs, N.J.: Prentice-Hall, 1942.

Williams, Frances L. *Matthew Fontaine Maury, Scientist of the Sea*. New Brunswick, N.J.: Rutgers University Press, 1963.

5 Chemistry of the Sea

Coker, Robert E. *This Great and Wide Sea*. New York: Harper & Row Publishers, Torchbooks, 1962.

Colman, J. S. *The Sea and Its Mysteries*. London: G. Bell & Sons, Ltd., 1950.

Cowen, R. C. *Frontiers of the Sea*. New York: Doubleday & Company, Inc., 1960.

Dietrich, G. *General Oceanography, an Introduction*. New York: John Wiley & Sons, Inc. (Interscience Publishers), 1963.

Harvey, H. W. *The Chemistry and Fertility of Sea Water*. New York: Cambridge University Press, 1960.

Hill, M. N., ed. *The Sea*. Vol. 2. New York: John Wiley & Sons, Inc. (Interscience Publishers), 1963.

Mason, B. *Principles of Geochemistry*. New York: John Wiley and Sons, 1960.

Murray, J., ed. *The Voyage of H.M.S.* Challenger. London: H. M. Stationery Office, 1884.

Riley, J. P., and Skirrow, G., eds. *Chemical Oceanography*. Vols. 1–2. New York: Academic Press, 1965.

Sears, M., ed. *Oceanography*. Washington, D.C.: American Association for the Advancement of Science, 1961.

Stumm, W., and Morgan, J. J. *Aquatic Chemistry*. New York: John Wiley & Sons, Inc. (Interscience Publishers), 1970.

Sverdrup, H. U., Johnson, M. W., and Fleming, R. H. *The Oceans: Their Physics, Chemistry, and General Biology*. Englewood Cliffs, N.J.: Prentice-Hall, 1942.

Williams, J. *Oceanography*. Boston: Little, Brown & Co., 1962.

6   Food from the Sea

Bardach, John. *Harvest of the Sea*. New York: Harper & Row Publishers, 1968.

Borgstrom, Georg, ed. *Fish as Food*. Vols. 1–4. New York: Academic Press, 1961–65.

Cushing, D. H. *The Arctic Cod*. Oxford: Pergamon Press, 1966.

Gilbert, De Witt, ed. *The Future of the Fishing Industry of the United States*. University of Washington Publications in Fisheries, New Series, Vol. 4. Seattle: 1968.

Graham, Michael. *The Fish Gate*. London: Faber & Faber, Ltd., 1946.

———— ed. *Sea Fisheries, Their Investigations in the United Kingdom*. London: Edward Arnold & Co., 1956.

Grant, Leonard, ed. *Wondrous World of Fishes*. Washington, D.C.: National Geographic Society, 1965.

Hardy, Sir Alister. *The Open Sea: Fish and Fisheries*, pt. 2. London: William Collins Sons & Co., Ltd., 1959.

Idyll, C. P. *The Sea Against Hunger*. New York: Thomas Y. Crowell Company, 1971.

Jones, J. W. *The Salmon*. London: William Collins Sons & Co., Ltd. (New Naturalist Series), 1959.

Smith, F. G. W., and Chapin, Henry. *The Sun, the Sea, and Tomorrow*. New York: Charles Scribner's Sons, 1954.

Stansby, Maurice E., ed. *Industrial Fishery Technology*. New York: Reinhold Publishing Corp., 1963.

Walford, Lionel A. *Living Resources of the Sea*. New York: Ronald Press, 1958.

7   Farming the Sea

American Association for the Advancement of Science. Publication 83, *Estuaries*, ed. George H. Lauff. Washington, D.C.: 1967.

American Fisheries Society. *A Symposium on Estuarine Fisheries*. In *Transactions of the American Fisheries Society*. Supplement to Vol. 95, No. 4, 1966.

Cooper, L. H. N., and Steven, G. A. "A Further Experiment in Marine Fish Cultivation." *Nature*. Vol. 167, March 31, 1951.

Gross, F. "A Fish Cultivation Experiment in an Arm of a Sea-Loch." *Proceedings of the Royal Society of Edinburgh*. Series B, Vol. 64. Edinburgh: 1950.

Hall, D. N. F. *Some Observations on the Taxonomy and Biology of Some Indo-West Pacific Penaeidae*. Colonial Office Fishery Publications, No. 17. London: Her Majesty's Stationery Office, 1962.

Hickling, C. F. *The Farming of Fish*. London: Pergamon Press, 1968.

———— *Fish Culture*. London: Faber & Faber, Ltd., 1962.

———— *The Recovery of a Deep-Sea Fishery*. Ministry of Agriculture and Fisheries, Fishery Publications Series II, Vol. XVII, No. 1. London: Her Majesty's Stationery Office, 1946.

———— *Estuarine Fish Farming*. In *Advances in Marine Biology*, Vol. 8, pp. 119–213. London and New York: Academic Press, 1970.

Hudinaga, M., and Miyamura, M. "Breeding of the 'Kuruma' Prawn." *Journal of the Oceanographical Society of Japan*. 20th Anniversary Volume, pp. 694–706. 1962. In Japanese with English abstract.

Iversen, Edwin S. *Farming the Edge of the Sea*. Fishing News (Books) Ltd. London: 1968.

Kalle, K. "Der Einfluss des englischen Küstenwassers auf den Chemismus der Wasserkörper in der südlichen Nordsee." *Berichte der Deutschen Wissenschaftlichen Kommission für Meeresforschung*. Vol. XIII, No. 2, pp. 130–135. Stuttgart: October 1953.

Lin, S. Y. *Milkfish Farming in Taiwan*. Taiwan Fisheries Research Institute, Fish Culture Report No. 3. February 1968.

Schuster, W. H. *Fish Culture in the Brackish-Water Ponds of Java*. Indo-Pacific Fisheries Council, United Nations Food and Agriculture Organization, Special Publications No. 1. Rome: 1952.

Shelbourne, J. E. *Advances in Marine Biology*. Vol. 2, *The Artificial Propagation of Marine Fish*. London and New York: Academic Press, 1964.

Vatova, A. "The Salt-Water Fish Farms of the North

Adriatic and Their Fauna." *Journal du Conseil International pour l'Exploration de la Mer.* Vol. 27, No. 1, pp. 109–115. Copenhagen: 1962.

8   Mineral Resources and Power

Bascom, Willard. *A Hole in the Bottom of the Sea.* New York: Doubleday & Company, Inc., 1960.

Coker, Robert E. *This Great and Wide Sea.* New York: Harper & Row Publishers, Torchbooks, 1962.

Cowen, R. C. *Frontiers of the Sea.* New York: Doubleday & Company, Inc., 1960.

Deacon, G. E. R., ed. *Seas, Maps, and Men.* New York: Doubleday & Company, Inc., 1962.

Dugan, James. *Man Under the Sea.* New York: Harper & Brothers, 1956.

——, et al. *World Beneath the Sea.* Washington, D.C.: National Geographic Society, 1967.

International Oceanographic Foundation. *Sea Frontiers.* Periodical. Miami, Fla.

Mero, John. "Minerals on the Ocean Floor." *Scientific American,* December 1960.

—— *The Mineral Resources of the Sea.* New York: American Elsevier Publishing Co., 1965.

Shepard, F. P., *et al. Submarine Geology.* New York: Harper & Row Publishers, 1963.

Woods Hole Oceanographic Institution. *Oceanus.* Periodical. Woods Hole, Mass.

9   Underwater Archaeology

Bascom, Willard. "Deep Water Archaeology," *Science,* October 15, 1971.

Bass, George F. *Archaeology Under Water.* New York: Frederick A. Praeger, 1966.

Bearss, Edwin C. *Hardluck Ironclad.* Baton Rouge: Louisiana University Press, 1966.

Borhegyi, Suzanne de. *Ships, Shoals, and Amphoras.* New York: Holt, Rinehart, and Winston, Inc., 1961.

Bush-Romero, Pablo. *Under the Waters of Mexico,* New York: Carleton Press, 1964.

Clarke, Arthur C., with Wilson, Mike. *The Treasure of the Great Reef.* New York: Harper & Row Publishers, 1964.

Davis, Sir Robert H. *Deep Diving and Submarine Operations.* London: St. Catherine Press, 1951.

Dugan, James. *Man Under the Sea.* New York: Harper and Brothers, 1956.

Dumas, Frederic. *Deep Water Archaeology.* Translated by Honor Frost. London: Routledge and Kegan Paul, 1962.

Latil, Pierre de, and Rivoire, Jean. *Man and the Underwater World.* London: Jarrolds Publishers, Ltd., 1956.

Link, Marian C. *Sea Diver.* New York: Rinehart and Company, 1958.

Marx, Robert. *Shipwrecks of the Western Hemisphere: 1492–1825,* World Publishing Co., 1971.

Ohrelius, Bengt. *Vasa, the King's Ship.* Translated by Maurice Michael. Philadelphia: Chilton Company, 1963.

Peterson, Mendel. *History Under the Sea.* Washington, D.C.: Smithsonian Institution, 1969.

Stenuit, Robert. *The Deepest Days.* New York: Coward-McCann, Inc., 1966.

—— "Priceless Relics of the Spanish Armada," *National Geographic,* Vol. 135, No. 6, June, 1969, pp. 745–777.

Taylor, Joan du Plat. *Marine Archaeology.* New York: Thomas Y. Crowell Company, 1966.

Throckmorton, Peter. *The Lost Ships.* Boston: Little, Brown & Co., 1964.

Wagner, Kip, and Taylor, L. B. *Pieces of Eight.* New York: E. P. Dutton & Co., 1966.

10   The Ocean Ecosystem: Pollution and Natural Disruptions

Boughey, J. T. *Contemporary Readings in Ecology.* Belmont, California: Dickenson Publishing Company, 1969.

Ehrlich, Paul R. *The Population Bomb.* New York: Ballantine Books, 1968.

Hoult, David P., ed. *Oil on the Sea.* New York: The Plenum Press, 1969.

Kormondy, Edward J. *Concepts of Ecology.* Englewood Cliffs, New Jersey: Prentice-Hall, 1969.

Moore, Hilary B. *Marine Ecology.* New York: John Wiley & Sons, 1958.

Olson, T. A., and Burgess, F. J., eds. *Pollution and Marine Ecology.* New York: John Wiley & Sons, 1967.

Petrow, Richard. *In the Wake of Torrey Canyon.* New York: David McKay Co., 1968.

Pringle, Lawrence. *Ecology: Science of Survival.* New York: Macmillan 1970.

Raymont, John E. *Plankton and Productivity in the Oceans.* New York: Pergamon Press, 1963.

Tait, R. V. *Elements of Marine Ecology.* London: Butterworth's, 1968.

Anderson, Frank J. *Submarines, Submariners, Submarining.* Hamden, Conn.: Shoe String Press, 1963.

Barber, F. M., Lieutenant, USN. *Lecture on Submarine Boats.* Pamphlet. U.S. Navy Department of Ordnance, 1875.

Beach, Edward L. *Submarine!* New York: Holt, Rinehart & Winston, Inc., 1952.

—— *Around the World Submerged.* New York: Holt, Rinehart & Winston, Inc. 1962.

Bullard, R. O., and Emery, K. O. *Research Submersibles in Oceanography.* Marine Technology Society, Washington, D.C., 1970.

Coates, Leonidas D., and Blay, Roy A. "A History of Submarine Warfare—up to 1945." *Lockheed Horizons,* January 1969.

Craven, John P. *Sea Power and the Sea Bed,* U.S. Naval Institute Proceedings, April 1966. Annapolis: U.S. Naval Institute.

Edwards, Kenneth. *We Dive at Dawn.* Chicago: Reilly & Lee Co., 1941.

Field, Cyril. *The Story of the Submarine.* London: Sampson Low, Marston & Company, Ltd., 1908.

Herzog, Bodo. *Die Deutschen U-Boote, 1906–1945.* Munich: J. F. Lehmanns Verlag, 1959.

Hezlet, Sir Arthur. *The Submarine and Seapower.* London: Peter Davies Co., 1967.

Interagency Committee on Oceanography, Federal Council for Science and Technology. *Undersea Vehicles for Oceanography.* ICO Pamphlet No. 18. Washington, D.C.: U.S. Government Printing Office, October 1965.

Kuenne, Robert E. *The Attack Submarine—A Study in Strategy.* New Haven and London: Yale University Press, 1965.

Lockwood, Charles A. *Down to the Sea in Subs.* New York: W. W. Norton & Co., Inc., 1967.

Marx, Robert F. "The Early History of Diving." *Oceans,* September–October 1971. Volume 4, Number 5. Menlo Park, California.

Morris, Richard K. *John P. Holland.* Annapolis: U.S. Naval Institute, 1966.

Norlin, F. E. *A Short History of Undersea Craft.* Pamphlet, 1946. Privately printed by the author, then employed at the Naval Torpedo Station, Newport, R.I.

O'Neal, H. A., ed. *Project Sealab Report—Sealab I Project Group.* Office of Naval Research Report ACR–108. Washington, D.C.: U.S. Government Printing Office, June 1965.

Parsons, William Barclay. *Robert Fulton and the Submarine.* New York: Columbia University Press, 1922.

Pauli, D. C., and Clapper, G. P., ed. *Project Sealab Report.—Sealab II Project Group,* Office of Naval Research ACR–124, Washington, D.C.: U.S. Government Printing Office, March 1967.

—— and Cole, H. A., ed. *Project TEKTITE I.* Office of Naval Research Report DR 153. Washington, D.C.: U.S. Government Printing Office, January 1970.

Perry, Milton F. *Infernal Machines: The Story of Confederate Submarines and Mine Warfare.* Baton Rouge: Louisiana State University Press, 1965.

Pesce, G.-L. *La Navigation Sous-Marine.* Paris: Librairie de Sciences Générales, 1897.

Piccard, A. *Earth, Sky and Sea.* New York: Oxford Univ. Press, 1956.

Piccard, Jacques. *Sun Beneath the Sea.* New York: Charles Scribner's Sons, 1971.

President's Science Advisory Committee, Panel on Oceanography. *Effective Use of the Sea.* Pamphlet. Washington, D.C.: U.S. Government Printing Office, June 1966.

Soule, Gardner. *The Ocean Adventure.* New York: Appleton-Century-Crofts, Inc., 1966.

Stafford, Edward P. *The Far and the Deep.* New York: G. P. Putnam's Sons, 1966.

Thomas, Lowell. *Raiders of the Deep.* London: William Heinemann, Ltd., 1929.

U.S. Naval History Division. *The Submarine in the United States Navy.* Pamphlet. Washington, D.C.: U.S. Government Printing Office, 1969.

U.S. Naval Research Laboratory, Washington, D.C. *Proceedings of 4th U.S. Navy Symposium on Military Oceanography.* 2 vols. (vol. 2 classified Confidential). Alexandria, Va.: The Oceanographer of the Navy (Defense Documentation Center, Cameron Station), 1967.

U.S. Naval Submarine School. *Submarine Casualties Booklet.* Privately printed by the U.S. Navy, 1958.

Index

British Association for the Advancement of Science, 10, 69
British Honduras, 142–143
bromine, 104, 166–167
Brooke, John Mercer, 10, 24
Brookfield, Charles M., 206
Browne, Jack, 207
Brussels Maritime Conference, 70
bryophytes, 60
Buch, Kurt, 126
Buchanan, J. Y., 13, 99
Buckland, Frank, 17
Bulmer, Ian, 199
buoy-mounted instruments, 97
Burgoyne, Gen. John, 205
Bushnell, David, 72, 76, 250
butterfly fish, 114

cables, submarine, 10–12, 25, 33, 45, 71, 112, 173
Caesarea, Palestine, 223
Cairo, 226, 231
Calanus finmarchicus, 244
calcareous encrustation, 230
calcium, 104, 240
California: fish production in, 143; University of, 16
California Academy of Science, 16
Calumet and Hecla mines, 14, 47
Calypso, 217
camera, sea-floor, 172
Canadian Shield, 31
Cannon, Barry L., 270
Canton, Robert, 218
canyons, submarine, 32–34, 112
Cape Johnson Deep, 35–36
Cape of Good Hope, 4, 65, 72, 227
carbohydrates, 115
carbonate system, 107, 166
carbon dioxide, 107–108, 148, 163, 197, 234
carbon-14, 110, 128, 222, 228
carbon monoxide, 109
Cariaco Trench, 124
Caribbean Sea, 21, 124, 128, 141
Carlsberg Foundation, 17
Carnegie Institution, Washington, D.C., 16, 110
Carpenter, Cmdr. Scott, 272, 274
Carpenter, W. B., 12, 46
carpetsweeper dredge, 174
carrageenin, 60–61
Carson, Rachel, 37
Casteret, Norbert, 204
cations, 104
caves, underwater, 204
Central America, archaeology of, 226–228
Centropages, 241
Ceratium tripos, 241
Chaetomorpha, 162
Challenger, H.M.S., 2, 12–15, 17, 19, 25, 29, 34, 37, 46–48, 53, 72, 77, 82–83, 97, 99, 104, 108, 112, 170, 172
Challenger (new), 19, 35, 36
Challenger Deep, 19, 35–36, 79, 208, 258
Charles V, 250
Charon, 206
charting, age of, 64–68
charts, 63–65, 70–71, 270; *see also* maps

chemical oceanography, 98, 128–129
Chesapeake Biological Laboratory, 16
Chesher, Richard, 246
Chichén Itzá, 203, 227
Child, C. M., 54
Chilean nitrates, 170
China: fish farming in, 148; prawn culture in, 157; ships of, 248
Chironex fleckeri, 246
chlorine and chlorimetry, 101–104, 124, 166
Chow, T. J., 233
chronometer, 28, 66
Civil War, U. S., 10, 24, 81, 226, 251–252
clams, 42–43, 152, 155
Clarke, Arthur C., 228
Clermont, 251
Clipper Era, 69
cobalt, 174
Cochran, Drayton, 225
Codes Cohonianus, 64
codfish, 144
Colladon, Daniel, 94
College of Fisheries, U. of Wash., 16
Collier Chemical and Carbon Co., 172
color, of ocean, 93; density and, 7, 83
Colorado River, 34
"color layers," animals and, 17, 48
Columbian cultures, 227
Columbus, Christopher, 4–5, 63, 65, 72, 94, 249
commercial fisheries, 16–18
Commercial Fisheries, Bureau of, Laboratory, 53–54
Commission of Fish and Fisheries, U. S. 15
compass rose, 64
Concarneau laboratory, 55
Coney, R. L., 239
Confederate blockade runners, 226
Conklin, E. G., 54
Conshelf, 19
continental drift, 32, 39–41
continental margins, 31–32
continental shelf, 32, 93, 171
Controlled Underwater Rescue Vehicle (CURV), 265, 266
convection cells, 30
Cook, Capt. James, 5–7, 45, 66, 72, 75, 227
Copenhagen, Univ. of, 99, 107
copper, 166–167, 171
coral reefs, 8, 37–38, 49, 112–113, 115, 188
corals, 239, 240, 244, 246
corer, 19, 124, 127, 194
Coriolis effect, 90–92
Cornish, Vaughan, 85
Cornwallis, Gen. Charles, 206
Coscinodiscus, 241
Council for Underwater Archaeology, 223
Cousteau, Jacques-Yves, 15, 19, 55, 72, 117, 207, 217, 228, 258, 267, 275
Cox, C. S., 87
crab, 44, 114, 162, 168, 241
crab-canning ships, 140
Cranchiidae, 58
Crawford Marine Specialists, 194
Cretaceous period, 12, 38

Crete, 3, 223
Crile, George, Jr., 207
Croll, James, 71
Cromwell, Townsend, 20
Cromwell Current, 92
crown-of-thorns starfish, *see Acanthaster* species
crude oil, 238
crust, oceanic, 30
crustaceans, 131; *see also* crab; prawns; shrimp
Cuba, 48
current meter, 13, 92, 95–96
currents, ocean, 7, 20, 71, 90–92, 150
CURV (Controlled Underwater Rescue Vehicle), 266
Cuvier, Georges, 45
Cyclops, H.M.S., 13
Cyprus, 225

Dana, 18, 24, 57
Dana, James Dwight, 39
Dardanelles, 255
Dark Ages, *see* Middle Ages
Darwin, Charles, 7–8, 37, 49, 69, 81–83, 112–113
Darwin, Sir George, 69
Davis, E. W., 226
Davis, John, 5
DDT, 240
Dead Sea, 167, 176
dead water, 87
Dealey, Cmdr. Sam, 256
Dean, Anthony, 201
Decca navigation system, 84
decompression, 269
Deep Diver, 267–268
deep dives, 264–267, 275
Deep Dive systems, 275
Deep Jeep, 266
deep sea, *see also* undersea
deep-sea animals, 9, 14
"deep-sea clamm," 9, 22
deep-sea dredging, 9, 14, 17, 46
Deep Sea Drilling Project, 179–180
deep-sea expedition, first, 12
Deep Sea Rescue Vehicle (DSRV), 215, 260, 264–267
Deep Sea Research Vehicle (DRV), 37
deep-sea soundings, 7, 10, 26–29
deep-sea traps, 15
Deepstar, 80, 212, 266
Deep Submergence Systems Project, 174, 175, 266, 270
Demarchi, Francesco, 204
Denayrouze, Auguste, 202
Denayrouze-Rouquayrol tank, 202
Den Helder, Holland, 107
density layers, 7, 83
Department of Charts and Instruments, U. S., 10
Depths of the Sea (Thomson), 11, 22, 24, 46
desalination, 193, 239
detergents, 244
diamonds, 171
diatoms, 157, 242
Diaz, Bartholomeu, 4, 65
Dictyophaeria cavernosa, 244, 245
dieldrin concentration, 240
dinoflagellates, 242–243

nuclear power, 193, 250
nuclear submarine, 5, 256–257, 264–265
Nuka Hiva, 37
nutrients, from the sea, 60–61, 148, 150, 184, 243

ocean: animal collection from, 9, 14, 44, 48, 55–60; bathymetry of, 27; "blackness" of, 93–94; bottom samples of, 12; bottom temperatures of, 12; canyons in, 32–34; chemical distributions of, 124; chemical properties of, 100–104; color of, 93; color of animals in, 17; depths of, 19, 22, 35, 62; dredging of, 9, 14, 17; earliest contacts with, 2; "hot spots" in, 124; as last frontier, 41; "living fossils" in, 30; phosphorus distributions in, 127; photosynthesis in, 60, 107, 121, 126, 131, 148, 180; potential of, 145–147; pressures of, 9, 48, 58, 274–275; resources of, 184; salinity of, 83; vs. shallow seas, 31; sound in, 94–95; survival of animals in, 55–57; temperatures of, 83; trenches in, 35–37; uniqueness of, 29; water volume of, 62; see also sea; seawater
oceanauts, 208
ocean basin, 29–30
ocean-bottom corer, 124
ocean currents, 7, 71, 90–92, 150
Ocean Eagle, 236
ocean environment, 164–165
ocean floor, 29–32, 170–175
Oceanographer, 21, 169
oceanography: biological, 52; chemical, 98, 128–129; continuous measurements in, 129; in 1880's and 1890's, 15; "founding fathers" of, 8–11; future of, 20–21; geology and, 10; modern beginnings of, 81–84; origins of, 112; physical, 10; quest for causes in, 52–55; as science, 12, 68–69; space research and, 21; tools of, 95–97
ocean slopes, 32
Oceanus River, 3
Odontoceti, 51
offshore drilling, 166, 175–179, 191
oil, 175–176, 184; see also offshore drilling; petroleum
oil pollution, 236–237
oil rigs, 166
oil slick, California coast, 179
ooze, sampling of, 13
"open seas," 276–277
Oppian, 196
order, quest for in biology, 52
Oregon State Univ., 16
Origin of Species (Darwin), 45, 113
Ortell, Abraham, 65–66
Oswego Canal, 247
oxygen, 104, 108, 121, 234, 269–270
oysters, 43, 152–154

Pacific Ocean: archaeology of, 227; depths of, 22, 24, 35–36; as old ocean basin, 39–40; seamount chain in, 30, 37–40
Paisley, Col. Charles, 201
Paleolithic man, 2; see also Stone Age culture

Paleozoic era, 38, 176
Palomares (Spain) bomb incident, 93, 174, 208, 260–264
Pangaea, 41
Papal Demarcation Line, 250
"paralytic shellfish poisoning," 243
Parites compressa, 245
Parry, Sir William, 5
Passamaquoddy estuary, 183
Patuxent River, 239
Pawnee, 16
Peabody Museum, 203
Pearce, Jack, 235
pearl cultivation, 189
pearl divers, 208–209
Peary, Cmdr. Robert E., 50
Peckell, Andreas, 199
Pell, Senator Clairborne, 194
Pellamis platurus, 247
Penaeidae, 155
pendulum clock, 66
penguin eggs, 240
Penikese Is. (Mass.), laboratory, 15–16, 54
Pennsylvania, Univ. of, 116, 225
Peral, Isaac, 252
perch, 145, 178
Periere, Lothar de la, 254
Perry Submarine Company, 267
Persian Gulf, 176, 249
Peru, anchovies from, 139
Peruvian Sea Institute, 138
pesticides, 235–236, 240
Petén Itzá, 227
Petersen, C. G. J., 17, 135, 184
Petersen tag, 17, 163, 184, 188
Peterson, Mendel, 229
petroleum, 166, 175–180, 184, 191
Petromyzon marinus, 247
Pettersson, Hans, 19
pH, 104
Phallium granulatum, 44
pharmaceuticals, 60
Philadelphia, 204, 206, 231
Philippine Trench, 35, 47, 173
Phipps, William, 197, 218
Phoenicians, 3–4, 72, 98, 166, 196
phosphates, 125–127, 130, 149, 171–173, 184, 234, 244
phosphorus, 125, 127, 149, 172
photophores, 59
photosynthesis, in the ocean, 60, 107, 121, 126, 131, 148, 180, 241–242
Physical Geography of the Sea, The (Maury), 10, 24, 81
Physical History of the Sea (Marsigli), 7
physical oceanography, 128; see also oceanography
physiographic diagrams, 28
phytobenthos, 151
phytoplankton, 107, 125, 127, 151, 163, 235, 243, 244
Piccard, 267
Piccard, Auguste, 58, 258
Piccard, Jacques, 36, 258, 267
Piccard, Jean, 258
Pierson, W. F., 86
piezoelectricity, 25
piezometer, 82
Piggott, Charles, 110

pilchard, 163
Pillars of Hercules, 3
"pingers," 94
Pioneer, 251
plaice, 133, 136
plankton, 8, 15, 36, 46, 52, 112, 115, 144, 149, 151, 154, 160–163, 242
plankton nets, 13, 18
plant and animal communities, 234
plants, marine, 60, 122, 130–131, 153–154
Plato, 3, 45
Pliny the Elder, 2, 44, 63, 112
Plymouth Laboratory, see Marine Biological Laboratory
Pogonophora, 47
Poidebard, Père Antoine, 223
Polaris submarines, 19
pollution, 232–238, 240–244
polyethylene glycol, 222
Polynesians, 62, 248
Porcupine, 11–12, 46–47, 68, 72, 108
Porphyra plant, 153–154
porpoises, 51, 57–58, 277–278
Port Erin (England), laboratory, 15
portolanos, 64–65
Port Royal, Jamaica, 224
Portuguese fisheries, 132
Portuguese man-of-war, 44
Portuguese sea explorations, 249–250
Poseidonius, 3, 63
potassium, 104, 167
Potts, W. T. W., 61
power supply, 102
prawns, 151–152, 155–158, 162, 163; see also shrimp
predatory sea animals, 55–56
pressure, hydrostatic, 9, 48, 58, 274–275
Prester John, 249
Price, Derek, 203
Prien, Lt. Gunther, 255
Prince Henry the Navigator of Portugal, 3–4, 65, 72, 75, 248
Princesse Alice, 15
Princeton Univ., 35
Principia Mathematica (Newton), 67
Project Mohole, 178–180
proportions, relative constancy of, 111–121
protein, 130, 160
Ptolemy, 4–5, 163
puffer-fish, 43
Punt, Land of, 3
purse seines, 145–146
Pythagoras, 63
Pytheas of Massilia, 63

Q-ships, 254
quartz, 25, 171

rabbit scourge, 247
radioactive elements, 109–111, 238
radio direction finder, 83
radioisotopes, 109
radiolarian, 13
radio transmission, undersea, 93
Rahn, Hermann, 61
Rakestraw, Norris, 126
Ramage, Cmdr. Lawson P., 256
Ramapo, U.S.S., 27
Rance River, 183–184, 192

EXPLORING THE OCEAN WORLD   8462
IDYLL C.P.   551.46 IDY SJ

DATE DUE

SEP 28 '76	MR 13 '82		
OCT 21 '78			
MAY 12 '77			
OL 03 '78			
OC 18 '79			
MAR 20 '79			
APR 18 '79			
4/6			
MR 17 '80			
MR 23 '8			
SE 15 '83			

'76   3662   551.46
              Idy

PARKWAY SCHOOL DISTRICT
CHESTERFIELD, MISSOURI 63017

DEMCO